AF358811

Friction and Wear of Cutting Tools and Cutting Tool Materials

Friction and Wear of Cutting Tools and Cutting Tool Materials

Editors

Guoliang Liu
Chuanzhen Huang
Xiangyu Wang
Bin Zhao
Min Ji

Basel • Beijing • Wuhan • Barcelona • Belgrade • Novi Sad • Cluj • Manchester

Editors

Guoliang Liu
Qingdao University of
Technology
Qingdao
China

Chuanzhen Huang
Yanshan University
Qinhuangdao
China

Xiangyu Wang
University of Jinan
Jinan
China

Bin Zhao
Beijing Institute of
Technology
Beijing
China

Min Ji
Qingdao University of
Technology
Qingdao
China

Editorial Office
MDPI AG
Grosspeteranlage 5
4052 Basel, Switzerland

This is a reprint of articles from the Special Issue published online in the open access journal *Lubricants* (ISSN 2075-4442) (available at: https://www.mdpi.com/journal/lubricants/special_issues/0B429T090N).

For citation purposes, cite each article independently as indicated on the article page online and as indicated below:

Lastname, A.A.; Lastname, B.B. Article Title. *Journal Name* **Year**, *Volume Number*, Page Range.

ISBN 978-3-7258-1571-5 (Hbk)
ISBN 978-3-7258-1572-2 (PDF)
doi.org/10.3390/books978-3-7258-1572-2

Contents

lubricants

Editorial

Friction and Wear of Cutting Tools and Cutting Tool Materials

Guoliang Liu [1,*], Chuanzhen Huang [2], Xiangyu Wang [3], Bin Zhao [4] and Min Ji [1]

1   School of Mechanical & Automotive Engineering, Qingdao University of Technology, Qingdao 266520, China;
     jimin@qut.edu.cn
2   School of Mechanical Engineering, Yanshan University, Qinhuangdao 066004, China;
     huangchuanzhen@ysu.edu.cn
3   School of Mechanical Engineering, University of Jinan, Jinan 250022, China; me_wangxy@ujn.edu.cn
4   School of Mechanical Engineering, Beijing Institute of Technology, Beijing 100081, China; bin.zhao@bit.edu.cn
*   Correspondence: liuguoliang@qut.edu.cn

The friction between cutting tools and the workpiece/chip can significantly affect the tool wear, cutting force, cutting temperature, machined surface integrity, and machined parts' service performance. The friction and wear of cutting tools have long been a focus of researchers. In recent years, the development of new cutting methods and cutting tools, such as ultrasonic-vibration-assisted cutting, cryogenic cutting, MQL cutting, and micro-textured tools, has changed the friction and wear rules of cutting tools. This Special Issue presents the latest advances in the fields of the friction and wear of cutting tools and cutting tool materials.

This Special Issue contains 14 papers covering the design of anti-wear micro-textures/coatings, the development of new lubricating methods, cutting process modeling, and the analysis of friction mechanisms.

Xu et al. (Contribution 13) proposed a bionic microstructure tool based on shells, and found that composite bionic micro-textured tools have a significantly reduced cutting force compared with non-micro-textured tools. Composite bionic micro-textured tools lead to a reduction in surface roughness of 10–25%, and composite bionic micro-textured tools are more prone to enhancing the curling and breaking of chips. In addition, with the increase in the microstructure area occupancy, the cutting performance of the tool was significantly improved. Hu et al. (Contribution 8) designed micro-textured tools with five different morphologies and studied the influence of the different micro-texture morphologies on the tool wear during spray cooling. The wear area of the rake face was measured based on the infinitesimal method, and the optimal morphology with the best anti-wear ability was obtained. Warcholinski et al. (Contribution 5) presented an evaluation of the surface quality and properties of multilayer coatings. The results indicated that the melting point of the cathode material directly affected the number and size of the macroparticles on the surface of the growing coating. The number of macroparticles increased with the lowering of the melting point of the cathode material. Feng et al. (Contribution 3) proposed a multi-objective optimization design method for the micro-texture parameters of a cutting tool. Taking a relatively low cutting force, cutting temperature, and tool wear as the objectives, a genetic algorithm multi-objective optimization model for the micro-texture parameters of the tools was established, and the model was solved using the NSGA-II algorithm to obtain a Pareto solution set and micro-texture parameters with good, comprehensive cutting performance.

In terms of modeling the cutting processes, Huang et al. (Contribution 14) proposed a tool wear prediction model by combining multi-channel 1D convolutional neural networks (1D-CNNs) with temporal convolutional networks (TCNs). Zhou et al. (Contribution 12) established a cohesive element model to perform crack propagation simulation in nanocomposite ceramic tool materials by introducing cohesive elements with fracture criteria into microstructure models. The influences of nanoparticle size, microstructure type,

Citation: Liu, G.; Huang, C.; Wang, X.; Zhao, B.; Ji, M. Friction and Wear of Cutting Tools and Cutting Tool Materials. *Lubricants* **2024**, *12*, 192. https://doi.org/10.3390/lubricants12060192

Received: 24 May 2024
Accepted: 24 May 2024
Published: 28 May 2024

nanoparticle volume content, and interface fracture energy were analyzed, respectively. Yu et al. (Contribution 9) acquired regression models of the sawing force and temperature rise based on the response surface methodology (RSM) experiments of bone sawing. The sawing parameters for minimizing force and temperature were recommended. Cui et al. (Contribution 7) proposed a thermodynamic analysis model to evaluate the cutting force and tool design in milling. The model comprehensively considers the tool angle and quickly calculates the minimum load on the milling cutter based on the optimal geometric parameters. Li et al. (Contribution 4) established a multi-objective function by processing the physical signals in the grinding process, which considered grinding parameters, i.e., surface roughness, coefficient of friction, active energy consumption, and effective grinding time. The weight vector coefficients of the sub-objective functions were optimized through a multi-objective evolutionary algorithm based on the decomposition (MOEA/D) algorithm.

Meng et al. (Contribution 11) studied the tool wear mechanism considering the tool–chip interface temperature during milling of aluminum alloy. Adhesion wear is the main wear mechanism of the high-speed milling of ADC12 aluminum alloy. The most important factor affecting adhesion wear is the tool–chip interface friction, which is directly manifested in the tool–chip interface temperature. With the increase in temperature, the tool wear rate increases; the molten adhesive layer on the tool surface is accompanied by crack propagation; and adhesion wear, oxidation wear, and abrasive wear occur on the tool surface. Xu et al. (Contribution 6) fabricated a high-wear-resistance coating via the synergy of laser cladding and ultrasonic burnishing. Li et al. (Contribution 2) studied the influence of cutting parameters on tool wear during the carbide tool milling of GH4169 and a tool wear prediction model was obtained. Ultrasonic vibration milling was compared with ordinary milling, and the improvement degree of the different coating materials on carbide tool wear was explored.

Finally, Liu et al. (Contribution 10) and Wang et al. (Contribution 1) studied the influence of new lubricating methods on the tool wear and cutting performance. Liu et al. (Contribution 10) presented a new method of combining nanofluid MQL with ultrasonic vibration assistance in a turning process. The results show that the new method with graphene nanosheets can further improve the machining performance by reducing the specific cutting energy and areal surface roughness, compared with the NMQL turning process and UVAT process. Wang et al. (Contribution 1) discussed the feasibility of machining γ-TiAl in cryogenic LN2 cooling conditions. It was found that cryogenic machining shows significant advantages in reducing cutting force, suppressing heat-affected zones, improving surface quality, and inhibiting micro-lamellar deformation.

We would like to sincerely thank all of the authors for submitting exceptional research papers to this Special Issue. We would also like to thank all of the reviewers who used their precious time to carefully examine and help improve the quality of all of the submitted manuscripts. Finally, we would like to thank Ms. Faye Yin, the Section Managing Editor, for her outstanding continuous support.

Conflicts of Interest: The authors declare no conflicts of interest.

List of Contributions

1. Wang, X.; Zhang, X.; Pan, D.; Niu, J.; Fu, X.; Qiao, Y. Tool Wear and Surface Integrity of γ-TiAl Cryogenic Coolant Machining at Various Cutting Speed Levels. *Lubricants* **2023**, *11*, 238. https://doi.org/10.3390/lubricants11060238.
2. Li, X.; Zhang, W.; Miao, L.; Pang, Z. Analysis of Tool Wear in GH4169 Material Milling Process. *Lubricants* **2023**, *11*, 245. https://doi.org/10.3390/lubricants11060245.
3. Feng, X.; Fan, X.; Hu, J.; Wei, J. Multi-Objective Optimization Design of Micro-Texture Parameters of Tool for Cutting GH4169 during Spray Cooling. *Lubricants* **2023**, *11*, 249. https://doi.org/10.3390/lubricants11060249.

4. Li, Y.; Jiao, L.; Liu, Y.; Tian, Y.; Qiu, T.; Zhou, T.; Wang, X.; Zhao, B. Study on a Novel Strategy for High-Quality Grinding Surface Based on the Coefficient of Friction. *Lubricants* **2023**, *11*, 351. https://doi.org/10.3390/lubricants11080351.

5. Warcholinski, B.; Gilewicz, A.; Tarnowska, M. The Surface Assessment and the Properties of Selected Multilayer Coatings. *Lubricants* **2023**, *11*, 371. https://doi.org/10.3390/lubricants11090371.

6. Xu, N.; Jiang, X.; Shen, X.; Peng, H. Improving the Surface Integrity and Tribological Behavior of a High-Temperature Friction Surface via the Synergy of Laser Cladding and Ultrasonic Burnishing. *Lubricants* **2023**, *11*, 379. https://doi.org/10.3390/lubricants11090379.

7. Cui, J.; Shen, X.; Xin, Z.; Lu, H.; Shi, Y.; Huang, X.; Sun, B. Thermodynamic Analysis Based on the ZL205A Alloy Milling Force Model Study. *Lubricants* **2023**, *11*, 390. https://doi.org/10.3390/lubricants11090390.

8. Hu, J.; Wei, J.; Feng, X.; Liu, Z. Research on Wear of Micro-Textured Tools in Turning GH4169 during Spray Cooling. *Lubricants* **2023**, *11*, 439. https://doi.org/10.3390/lubricants11100439.

9. Yu, D.; Zou, F.; Zhang, W.; An, Q.; Nie, P. Measurement and Prediction of Sawing Characteristics Using Dental Reciprocating Saws: A Pilot Study on Fresh Bovine Scapula. *Lubricants* **2023**, *11*, 441. https://doi.org/10.3390/lubricants11100441.

10. Liu, G.; Wang, J.; Zheng, J.; Ji, M.; Wang, X. An Experimental Study on Ultrasonic Vibration-Assisted Turning of Aluminum Alloy 6061 with Vegetable Oil-Based Nanofluid Minimum Quantity Lubrication. *Lubricants* **2023**, *11*, 470. https://doi.org/10.3390/lubricants11110470.

11. Meng, X.; Lin, Y.; Mi, S.; Zhang, P. The Study of Tool Wear Mechanism Considering the Tool–Chip Interface Temperature during Milling of Aluminum Alloy. *Lubricants* **2023**, *11*, 471. https://doi.org/10.3390/lubricants11110471.

12. Zhou, T.; Meng, L.; Yi, M.; Xu, C. Simulation of Microscopic Fracture Behavior in Nanocomposite Ceramic Tool Materials. *Lubricants* **2023**, *11*, 489. https://doi.org/10.3390/lubricants11110489.

13. Xu, T.; Ma, C.; Shi, H.; Xiao, K.; Liu, J.; Li, Q. Effect of Composite Bionic Micro-Texture on Cutting Performance of Tools. *Lubricants* **2024**, *12*, 4. https://doi.org/10.3390/lubricants12010004.

14. Huang, M.; Xie, X.; Sun, W.; Li, Y. Tool Wear Prediction Model Using Multi-Channel 1D Convolutional Neural Network and Temporal Convolutional Network. *Lubricants* **2024**, *12*, 36. https://doi.org/10.3390/lubricants12020036.

 *lubricants* 

Article

Tool Wear and Surface Integrity of γ-TiAl Cryogenic Coolant Machining at Various Cutting Speed Levels

Xiangyu Wang [1], Xiaoxia Zhang [1], Duo Pan [2], Jintao Niu [1], Xiuli Fu [1] and Yang Qiao [1,*]

[1] School of Mechanical Engineering, University of Jinan, Jinan 250022, China; me_wangxy@ujn.edu.cn (X.W.); 17663751979@163.com (X.Z.); me_niujt@ujn.edu.cn (J.N.); me_fuxl@ujn.edu.cn (X.F.)
[2] Jigang International Engineering & Technology Co., Ltd., Jinan 250098, China; me_pand@163.com
* Correspondence: me_qiaoy@ujn.edu.cn

Abstract: High-speed machining of γ-TiAl alloy is a significant challenge due to high cutting temperatures. From the perspective of environmental protection and improving tool life, appropriate cooling strategies should be adopted. Compared with dry and conventional flood cooling conditions, the feasibility of machining γ-TiAl in cryogenic LN_2 cooling conditions was discussed. The cutting force, tool wear and its mechanism, and surface roughness, as well as sub-surface morphology characteristics, were studied by combining macro and micro techniques. The results revealed that the wear morphology of the rake and flank face under the three cooling media shows different degrees. The crater wear of the rake face is expanded at high speeds and then progresses into more serve flaking and notching wear. The main wear pattern on the flank face is gradually transformed from adhesive wear to diffusion and oxidation wear at high speeds in dry machining. In the LN_2 condition, the diffusion of workpiece elements and cutting-edge oxidation were restrained. The wear pattern is still mainly adhesive wear. In addition, cryogenic machining shows significant advantages in reducing cutting force, suppressing heat-affected zone, improving surface quality, and inhibiting micro-lamellar deformation.

Keywords: cryogenic cooling; γ-TiAl alloy; high-speed machining; tool wear; surface integrity

Citation: Wang, X.; Zhang, X.; Pan, D.; Niu, J.; Fu, X.; Qiao, Y. Tool Wear and Surface Integrity of γ-TiAl Cryogenic Coolant Machining at Various Cutting Speed Levels. *Lubricants* **2023**, *11*, 238. https://doi.org/10.3390/lubricants11060238

Received: 5 May 2023
Revised: 24 May 2023
Accepted: 25 May 2023
Published: 27 May 2023

1. Introduction

Gamma titanium aluminides (γ-TiAl) alloys are titanium aluminides with an Al content of 42–49 at.%. γ-TiAl alloys exhibit low density (~4 g·cm^{-3}), high service temperature (650–850 °C), high specific strength, excellent creep, oxidation resistance, and [1] corrosion resistance, which makes it suitable for extreme conditions [2]. Due to remarkable advantages, γ-TiAl alloys have been commercially utilized in aero engines to partially replace Ni-based superalloys in the temperature range of 600–800 °C [3]. At present, γ-TiAl alloy has been used in engine parts such as compressor blades, turbine blades, and turbochargers [4].

However, In the cutting process, the machinability of γ-TiAl alloy is affected by its physical and chemical properties [2]. The disadvantages are as follows: the low ductility and low thermal conductivity at room temperature, resulting in surface cracks easily produced in cutting; the high hardness, strength, and work-hardening tendency, which increases the difficulty of subsequent machining. In addition, it has been pointed out in the literature [5,6] that γ-TiAl alloy has a strong chemical affinity with tool materials at high temperatures, which accelerates the deterioration of tools. Based on the above limitations, Beranoagirre et al. [7] studied the influence of different cutting parameters on tool wear in the turning test of γ-Ti Al alloy. The research results have shown that cutting speed is the main factor affecting tool life. Beranoagirre et al. [8] believed that cutting speed is the factor with maximum influence on tool life due to the growing wear originating from friction along all milling. Starting from a cutting speed of 60 m/min, a small increase of this speed, around 15%, decreases tool life in half. The most interesting cutting speed is 50 m/min because a double machining time implies only a reduction of 15% in productivity. The surface finishes of gamma titanium–aluminum alloys in grinding are related directly to the feed rate. The surfaces obtained at low feed rates are better [9]. Some researchers [10,11] propose that the optimal cutting speed range of this alloy in conventional dry machining is that the cutting speed does not exceed 50 m/min.

Compared with conventional cutting, the advantages of high-speed cutting are not only reflected in machining efficiency but also in machining accuracy and surface finish [12]. However, it is more challenging to apply high-speed cutting technology to γ-TiAl alloys. Aspinwall et al. [13] analyzed the cutting temperature during milling γ-TiAl alloy. The cutting speed was increased from 50 to 135 m/min, resulting in a 43–47% rise in cutting temperature. Klocke et al. [14] noted that the rise of the cutting temperature caused by high speeds could soften the material. High-quality surfaces without cracks were obtained and thus improved the machinability of the difficult-to-be-machined alloy. However, this machining strategy also aggravated the negative effects of mechanical and thermal loads on the tool life, especially in dry machining. Saketi et al. [15] proposed that the rake and flank wear were the main wear pattern in TC4 low-speed cutting. Rajashree et al. [16] proposed that tool failure accelerates due to rubbing of the flank surface caused by heat and friction, reducing tool life and surface quality. The wear mechanism at high temperatures would be changed to diffusion wear. The performance of the tool deteriorated due to diffusion wear in high-speed machining. Studies have shown that similar tool wear patterns occur in γ-TiAl alloy machining [17,18].

In the above studies, the high cutting temperature is a dominant factor affecting machinability. Appropriate cooling strategies should be adopted. A reliable condition for the possibility of increasing machinability and, in turn, improved cutting conditions is the use of cooling techniques [19]. Machining of titanium alloys and its alloys is accompanied by oxidation of the chips and the machined surface [20]. Therefore, cutting fluids used in the processing of such materials should protect the chips and the treated surfaces from oxidation [21].

Cutting fluid cooling and lubrication can effectively increase tool life. González et al. [22] used Cryo CO_2 and minimal quantity lubrication (MQL) for milling the sides of Ti6Al4V integral blade rotors (IBRs). The results showed that Cryo CO_2 machining reduced the tool temperature and improved the surface quality and that the use of low-temperature lubrication reduced the amount of lubricant used and reduced production costs. Khanna et al. [23] conducted a life cycle assessment (LCA) to evaluate the environmental impact of different cutting fluid strategies throughout the entire product life cycle. The authors suggest that minimum quantity lubrication (MQL) may be a more sustainable option that still provides adequate tool life and dimensional accuracy while reducing overall environmental impact. However, the existing literature [24,25] shows that conventional flood cooling does not have obvious advantages in high-speed machining of difficult machining materials. For one thing, it is hard for the coolant to permeate through the cutting zone owing to the influence of the chip and high spindle speed. For another, the high cutting heat generated in the cutting process will make the cutting fluid evaporate rapidly, resulting in an insignificant cooling effect. In addition, the traditional cutting fluid will harm human health and the environment. Cutting fluids' airborne particles negatively affect a number of chronic human health conditions, including asthma, allergic reactions, skin rashes, and dermatological issues [16].

Cryogenic cooling is an environmentally-friendly cooling strategy which is that the conventional cutting fluid is replaced by liquid gas [26]. Cryogenic fluids are non-toxic and do not generate harmful fumes or vapors, making them an ideal choice for machining operations where people's safety is a concern [27]. According to a report from the National Institute of Standards and Technology, a cryogenic temperature is one that is lower than $-180\ ^\circ C$ [28]. Shah et al. [29] confirmed that cryogenic CO_2 and liquid nitrogen (LN_2) is suitable coolants for difficult-to-be-machined materials from energy consumption, tool life, and surface integrity. The low-temperature LN$_2$ commonly used in machining applications is −196 °C, and the LCO$_2$ is −56.6 °C [16]. Khanna et al. [23] compared LCO$_2$, MQL, and cryogenic machining based on machinability metrics such as cutting forces, tool wear, and surface roughness. The results showed that a higher cutting force was observed for MQL machining, followed by flood and cryogenic machining using LCO$_2$ in decreasing order, respectively. LCO$_2$ has excellent heat absorption capability due to its low boiling point (−78.5 °C), a feature that allows it to effectively control tool wear. The lower tool life under MQL machining is due to its inability to extract heat, resulting in higher tool wear.

Some researchers have summarized the effect of cryogenic cooling on the tool life and surface quality of various materials, such as Ni-based superalloys [30,31], Ti-based alloys [32–34], and Iron-based alloys [35]. Its advantage is to keep the cutting temperature down and improve the heat dissipation from the cutting zone, leading to mechanical property improving and residual stress, surface roughness, and tool wear decreasing.

However, cryogenic machining has not been widely studied in view of the poor plasticity of γ-TiAl alloy at low temperatures. Klocke et al. [21] believed that cryogenic machining was a promising strategy for cutting γ-TiAl alloy, which can improve the wear resistance of tools and reduce the microstructure changes of the surface layer. Pereira et al. [36] focused on the Inconel 718 and how the internal cryo lubrication approach could improve its milling process. The researchers conducted experiments to compare this method with traditional flood cooling and found that the cryo lubrication approach resulted in lower temperatures at the cutting edge and improved tool life. They also noted that the use of a special type of coolant called a cryogenic coolant, was necessary for achieving optimal results. Priarone et al. [37] proposed that the tool wear patterns in cryogenic machining were also rake, flank, and adhesion wear. However, the influence on the wear mechanism has not been analyzed in these papers. The above studies mostly describe the positive role of the cryogenic environment in tool life and surface integrity from a macro perspective, and most of the related wear mechanism studies are based on indirect observations.

The purpose of the work is to compare and discuss the effect of dry machining, conventional flood cooling machining, and cryogenic machining on the machinability in γ-TiAl alloy high-speed cutting. The tool wear and mechanism, the cutting force, as well as its relationship with surface roughness and microstructure morphology, are studied using macro and micro technologies through turning tests at various cutting speeds.

2. Experimental Preparation

2.1. Workpiece Materials

The material was a Ti-47.5Al-2.5V-1.0Cr at% γ-TiAl alloy. The microstructure of the material after heat treatment was the γ + α$_2$ fully lamellar structure in Figure 1. The rod workpieces were made by casting with a dimension of φ110 mm. Its mechanical properties are presented at different temperatures in Table 1.

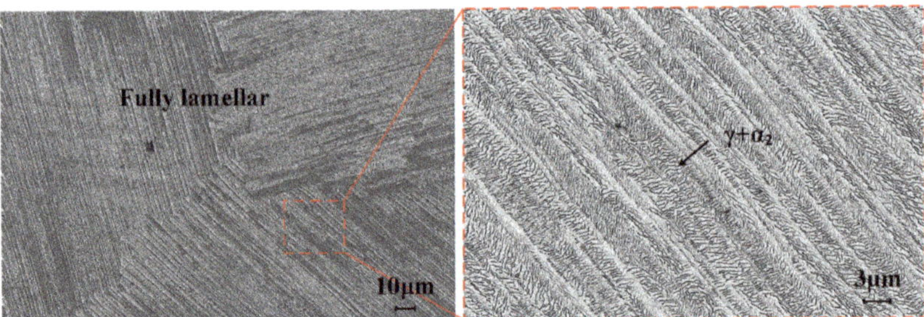

Figure 1. The SEM microstructure image of γ-TiAl alloy.

Table 1. The mechanical properties of workpiece material [18,38].

Properties	Temperature (°C)	
	25	800
Elastic modulus (GPa)	172	151
Tensile strength (GPa)	540	500
Yield strength (MPa)	440	380
Thermal conductivity (W/m·k)	18.6	23.1
Ductility (%)	1.5	6

2.2. Experimental Conditions and Processes

Face-turning tests were performed on a CNC lathe with a rated spindle speed of 4000 rpm. The uncoated carbide cutting inserts used were TPGN220408 K313 (Kennametal Inc., Pittsburgh, PA, USA). The main variables were cooling media and cutting speeds at the same cutting time. The surface roughness was measured using Surface Roughness Tester-SJ410 (MITUTOYO, Kanagawa, Japan). This device has a maximum measurement range of 800 μm (±400 μm) for the calculation of various roughness parameters. The surface force was measured using YDC-III 89A piezoelectric turning dynamometer (Dalian University of Technology, Dalian, China). The measuring force range should be $<\pm2000$ N in the three directions of X, Y, and Z. The experiment process parameters are displayed in Table 2.

Table 2. The experiment process parameters.

Item	Value
Cooling media	Dry, Emulsion, LN_2
Cutting speed (v_c)	60, 90, 120, 150, 180, 210 m/min
Feed rate (f)	0.1 mm/r
Cutting depth (a_p)	0.2 mm
Rake angle (γ)	3°
Relief angle (α)	8°

The cutting time is 1 min, and the effective machining length differs at different cutting speeds. Considering the poor machinability of the materials, the cutting speed of 210 m/min was not conducted in dry machining. The emulsion with a mixture ratio of 6% was selected as the conventional coolant. LN_2 was selected as the cryogenic coolant. The outlet pressure of the LN_2 tank was stabilized at 1.5 MPa during the operation. According to the test requirements, LN_2 and emulsion were respectively sprayed into the contact area of tool-workpieces. The schematic overview of the experimental setup is shown in Figure 2.

The influence of mechanical load on tool wear and surface integrity is studied by analyzing the variation of cutting force (F). The F was measured in three directions, that is, tangential force (Fx), radial force (Fy), and axial force (Fz).

After one minute of machining, the tool wear and failure morphology of the rake face and the flank face were observed by the ultra-depth of field microscope. The maximum flank wear (V_{max}) near the tool point was measured. The wear mechanism of the worn tools under three cooling media was analyzed, respectively, and cutting speeds at 60, 120, and 180 m/min were selected. The 9wear morphology was further observed by Scanning Electron Microscope (SEM) (Carl Zeiss AG, Oberkohen, Baden-Württemberg, Germany). To investigate the wear patterns, the mass fraction of the main elements in a certain wear area was obtained by Energy Dispersive Spectrometer (EDS) (Shimadzu Corporation, Kyoto, Japan).

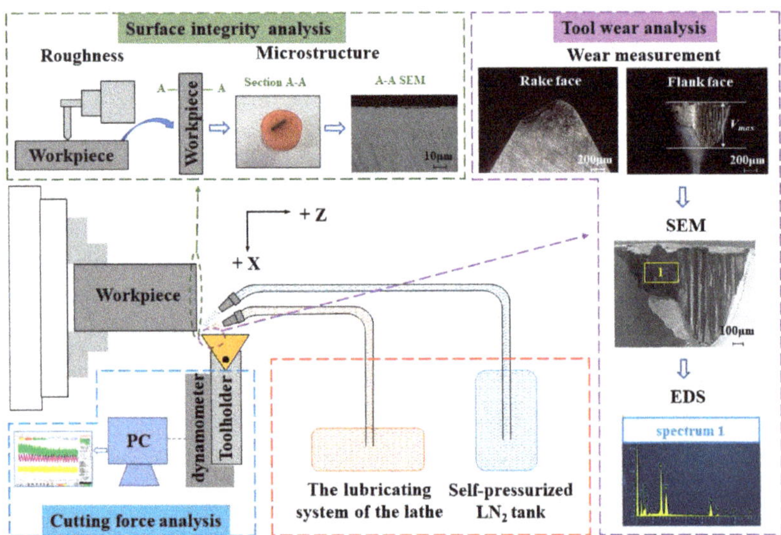

Figure 2. The experimental setup and procedure.

The related tests of surface integrity were carried out after ultrasonic cleaning for 5 min. To reduce the measurement error during the surface roughness (*Ra*), five different positions of each machined surface were measured along the feed direction, and the average values were calculated. To explore the deformation of the workpiece material, the cross-section was intercepted along the cutting speed direction by Wire Electrical Discharge Machining (WEDM) and then ground, polished, and cold-mounted. The cross-section was corroded with a Kroll reagent (ratio: 3% HF, 6% HNO_3, 91% distilled water) for 10 s, and the microstructure was observed by SEM.

3. Results and Discussion

3.1. Cutting Force

The variation of mechanical load at various cutting speeds and cooling conditions is analyzed in this part. The cutting force (*F*) is calculated by Equation (1) [39]. The variation of cutting force at various cutting speeds and cooling media is shown in Figure 3.

$$F = \sqrt{F_x^2 + F_y^2 + F_z^2} \tag{1}$$

As shown in Figure 3, the variation curve of cutting force shows an increase–decrease–increase trend along increasing cutting speeds. The cutting force under dry machining is the largest. With the increase of the cutting speed (lower than 120 m/min), the strain and strain rate of the cutting zone increase. In this case, the strain-strengthening effect is higher than the thermal softening. The degree of work hardening of the workpiece is enhanced, thereby increasing the cutting force. The value of *F* under the emulsion condition is almost the same as that under the LN_2 condition at low speeds, indicating that the cooling effect of emulsion and LN_2 is adequate to restrict the plastic deformation. In fact, the lubrication effect of emulsion in low-speed cutting is better than that of cryogenic cooling.

When the cutting speed is increased to a certain value, the cutting force is decreased. For example, the critical cutting speed is 120 m/min in dry cutting. The softening effect caused by the high temperature on the material increases gradually, resulting in a decrease in the strengthening. Given the cooling effect of coolants, the cutting force under emulsion cooling and cryogenic cooling decreases significantly at v_c = 150 m/min. Under higher speeds, the thermal softening effect tends to be stable. The work hardening trend increases continuously until it is higher than the softening again. The cutting force appears to have a certain degree of recovery. The cutting force under the LN$_2$ condition is relatively small in the higher speed range because its sufficient cooling effect can suppress plastic deformation.

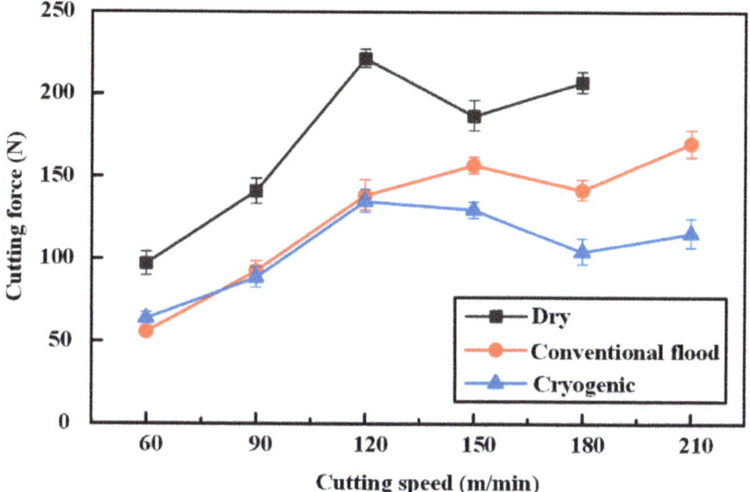

Figure 3. The cutting force under various conditions.

3.2. Tool Wear

Tool wear as a criterion to judge the machinability of materials is affected by tool material, cutting parameters, and cooling conditions. The effect of coolant on tool wear at various cutting speeds is evaluated in this part.

The maximum flank wear (V_{max}) and its morphology near the tool point under various cutting conditions are shown in Figure 4. The trends indicated that the wear degree is almost positively correlated with the cutting speeds, which is the same as the variation of the F.

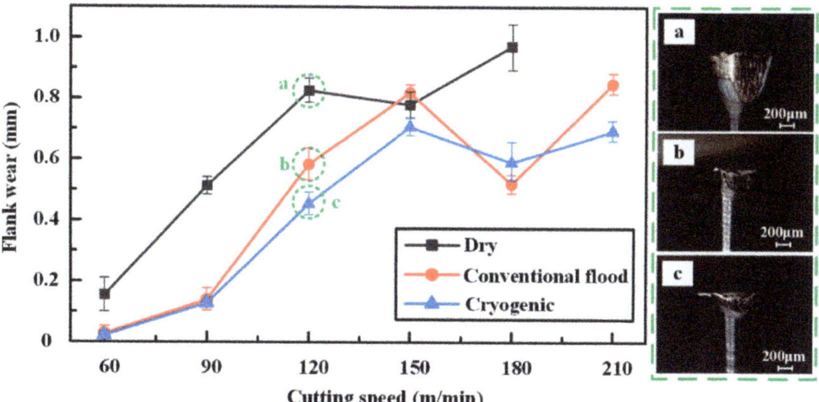

Figure 4. The maximum flank wear near the tool point and its morphology. (**a**) Dry cutting at v_c = 120 m/min, (**b**) Cutting under conventional flood at v_c = 120 m/min, (**c**) Cutting under cryogenic at v_c = 120 m/min.

According to the tool wear morphology under three cooling conditions, the flank wear is the most serious in dry machining at v_c = 120 m/min. This is due to the most violent friction between the machining surface and the tool flank face caused by the largest cutting force under dry machining. Yuan et al. [38] proposed that during the cutting process, severe friction occurs between the chip and the rake face, the workpiece, and the flank face, resulting in the high contact pressure and temperature. The workpiece material is easily bonded to the tool under high temperatures and high pressure. Part of the binder and the tool surface material is taken away by the flowing chip and the rotating workpiece, resulting in tool bonding wear, which is similar to the conclusions of this study. In addition, the flank wear is also affected by the elastic modulus of the workpiece material [40]. The temperature in the cutting zone increases in dry machining, which reduces the elastic modulus and further increases the resilience of the machining surface. The contact area of the tool flank face and the machined surface is expanded, increasing the wear area. The intervention of emulsion and LN_2 alleviate the impact of high temperature on tool life and change the friction characteristics of the tools and chips, which improves the performance of the tool. Therefore, the tool wear in conventional flood cooling and the cryogenic condition is significantly lower than that under dry conditions.

The tool wear degree decreases to a certain extent in dry machining at $v_c = 150$ m/min. Similarly, the decrease occurs under the other two conditions, but to different degrees and at $v_c = 180$ m/min. It is preliminarily considered that the cutting layer metal is softened at high temperatures. The hardness of the material and the cutting force is decreased, thus reducing the tool wear. The cooling effect of emulsion and LN_2 weakens the thermal softening effect compared to dry machining. Therefore, the tool wear curve turns at higher cutting speeds.

When $v_c = 180$ m/min, the flank wear under the cryogenic condition is slightly higher than that under the emulsion condition. Combined with the rake wear morphology in Figure 5, the flaking wear is observed. This may be due to tool embrittlement in the extremely low environment. As a result, the flaking wear greatly reduces the tool's strength. However, the overall trend shows that LN_2-assisted cutting has a significant advantage over the other two conditions in reducing tool wear at both low and high speeds.

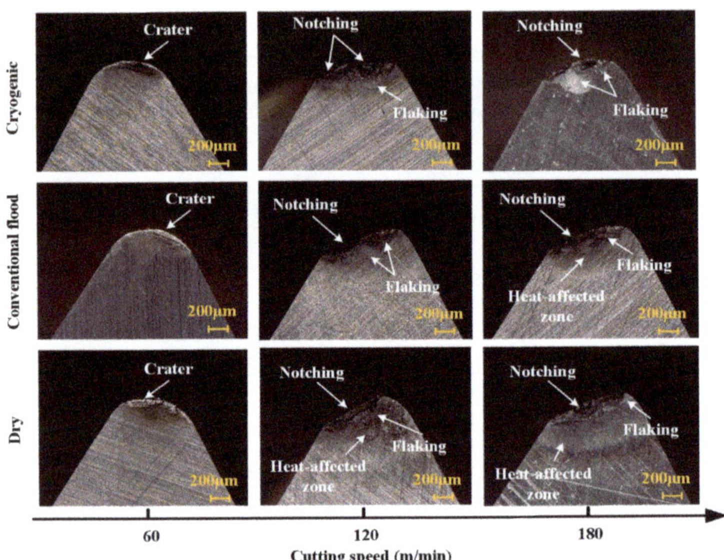

Figure 5. The rake wear morphology.

Under three cooling conditions, the crater wear is also observed in Figure 5 at $v_c = 60$ m/min. The crater is progressed into more serve flaking and notching wear at high speeds. The tool is seriously damaged at $v_c = 180$ m/min in dry machining. It can also be observed that the size of a heat-affected zone on the rake face is different in the various tools. This is the result of a large amount of cutting heat in the material removal.

The oxidation reaction occurs on the tool surface under a high-temperature environment. Comparing the morphology of the rake face under the three conditions, it can be concluded that the cryogenic condition is helpful in inhibiting the wear expansion of the crater wear and reduce the heat-affected area.

3.3. Wear Mechanism

The flank wear morphology and EDS analysis under various conditions are characterized in Figures 6–8 The obvious scratches on the tool flank face are found in the SEM morphology images. For one thing, the hard particles in the workpiece scratch the flank face in machining, resulting in abrasive wear. For another, it is clear in Figure 9 that the tool material flakes off and could stick to the flank face in a high-temperature environment. The friction between the adhered tool material and the workpiece could cause scratches. The involvement of emulsion and LN_2 achieves better friction characteristics, and their cooling capacity can help to keep the hardness of the tool, resulting in less significant scratches and wear, as shown in Figures 7 and 8.

It can be seen in Figures 6–8 that the elements of workpiece material appear on the worn tool, such as Ti and Al. However, few W elements are detected. In particular, no Co element is detected at v_c = 60 m/min. It is indicated that almost the whole wear area has been covered by the workpiece material. Moreover, this phenomenon occurs in all three cutting conditions. The strong chemical affinity between γ-TiAl alloy and cemented carbide tools and the high cutting temperature are the main reasons. Under the high temperature and pressure environment, the friction resistance of chip outflow increases, and an adhesive or chip retention layer is formed on the tool surface. Since the maximum probe depth of the EDS probe is 1 μm, the thickness of the adhesion should be more than 1 μm. As shown in Figure 10, the adhesive layer at the wear boundary falls off. The adhered materials are pulled by the flowing chips and rotating workpieces, resulting in adhesive wear.

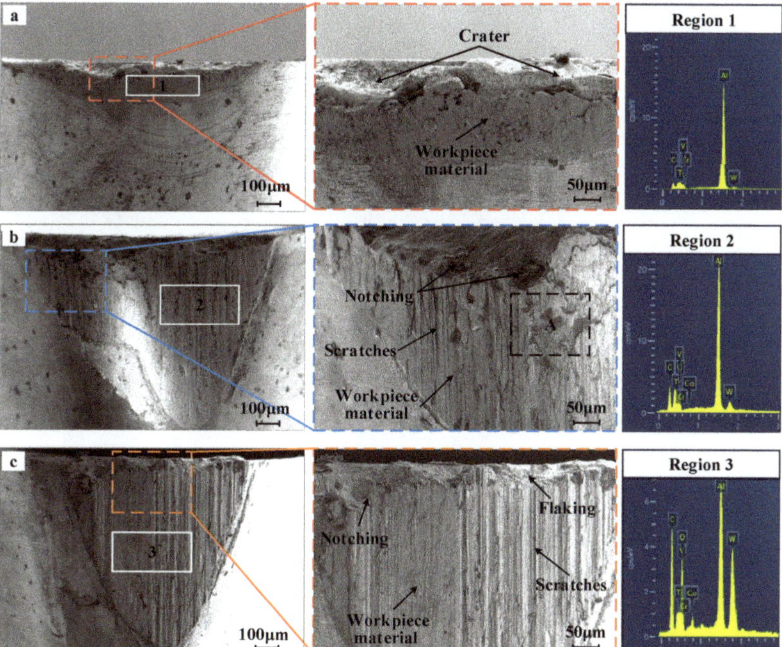

Figure 6. The flank wear morphology and EDS analysis under dry conditions. (**a**) v_c = 60 m/min; (**b**) v_c = 120 m/min; (**c**) v_c = 180 m/min.

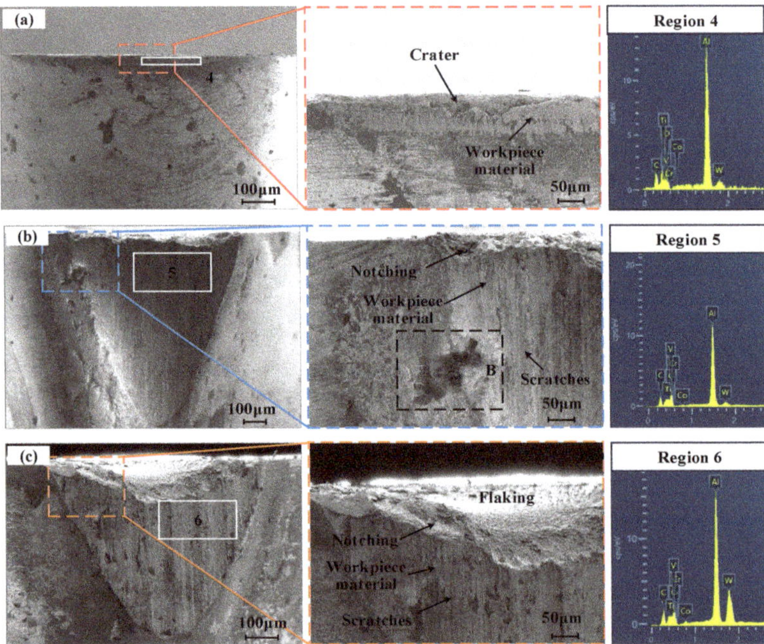

Figure 7. The flank wear morphology and EDS analysis under conventional flood cooling conditions. (**a**) $v_c = 60$ m/min; (**b**) $v_c = 120$ m/min; (**c**) $v_c = 180$ m/min.

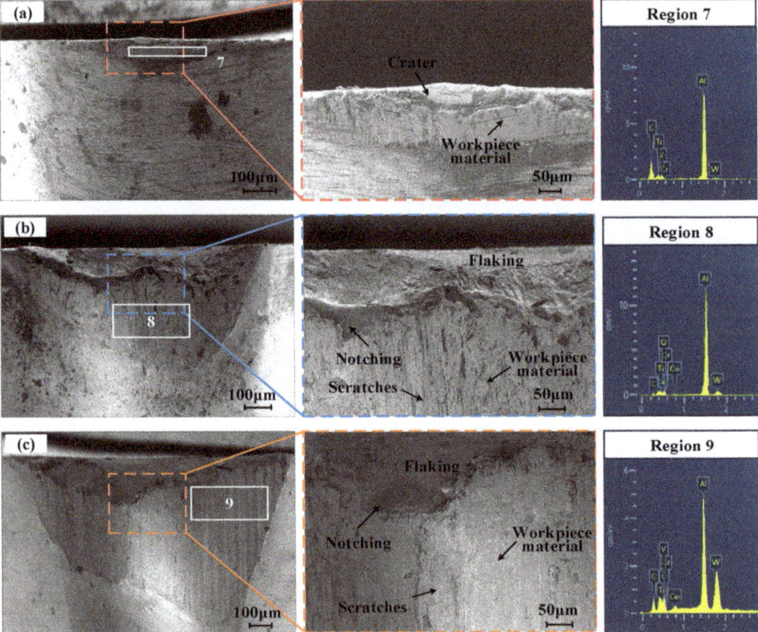

Figure 8. The flank wear morphology and EDS analysis under cryogenic cooling condition (**a**) $v_c = 60$ m/min; (**b**) $v_c = 120$ m/min; (**c**) $v_c = 180$ m/min.

By comparing the variations of element composition on the worn tool surface, as shown in Tables 3–5, it can be found that the fraction of C and O increases at high speeds overall. One reason is that the carbide in the tool is relatively more stable at high temperatures, increasing the mass fraction of C retained. The other is that the tool material has an oxidation reaction with oxygen in the air, affected by the high-temperature environment. Under the emulsion and LN_2-assisted cooling conditions, there are few O and C in the wear area. Klocke et al. [21] analyzed the effects of different cooling options (dry, overflow, high pressure, MQL, and cryogenic) on turning Gamma TiAl alloy. It has been proven that low-temperature cooling is the most effective lubrication strategy for this material. Compared with traditional lubrication, liquid nitrogen can reduce flank wear by up to 61%, which is similar to the conclusions of this study. The oxidation and diffusion can be alleviated at low temperatures. This advantage is more significant in the LN_2 conditions. Meanwhile, due to diffusion wear, elements such as Co, C, and W in the carbide tool will diffuse when the cutting temperature increases. Due to the diffusion of Co, carbide will reduce its bonding strength to the substrate due to the reduction of the binder Co, which will accelerate the wear of the tool. The metal and carbon atoms in the tool diffuse into the workpiece material bonded to the tool surface and are carried away by the chips. In addition, the mass fraction of Ti and Al decreases at high cutting speeds while the mass fraction of W increases greatly, which shows that the plasticity of the material has increased and the tool adhesion phenomenon is reduced. The degree of adhesion is negatively correlated with the cutting speed. It is also noted from Figures 6–8 that the adhesion phenomenon is the most serious in dry conditions, followed by conventional flood cooling and cryogenic conditions. The main reason is that the coolant has an inhibitory effect on the chemical activity of the workpiece material. Especially under the LN_2 conditions, the tool adhesion has been effectively suppressed with an outstanding cooling effect. However, the mass fraction of elements such as Ti, Al, and V is still higher than that of W at high speeds in LN_2 conditions, which indicates that the adhesion phenomenon is still serious. The wear pattern is still dominated by adhesive wear during high-speed cutting.

Table 3. The element composition of region 1/2/3.

Cutting Speed	Element Composition (at%)							
	Ti	Al	V	Cr	W	C	Co	O
60	30.45	40.66	1.04	4.15	0.17	23.65	-	-
120	21.88	29.89	0.27	2.36	0.10	25.49	0.47	19.53
180	17.56	28.81	0.68	1.51	0.25	29.38	0.55	21.26

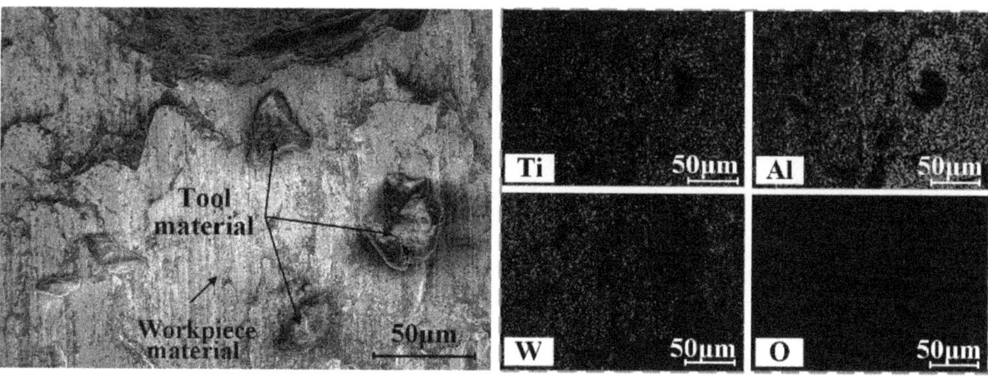

Figure 9. The morphology and EDS analysis of region A in Figure 6b.

Table 4. The Element composition of region 4/5/6.

Cutting Speed	Element Composition (at%)							
	Ti	Al	V	Cr	W	C	Co	O
60	30.23	34.31	1.67	1.44	0.04	32.31	-	-
120	17.79	30.97	0.98	1.40	0.21	32.52	0.13	15.99
180	19.51	23.63	0.80	0.57	0.31	34.87	0.50	19.81

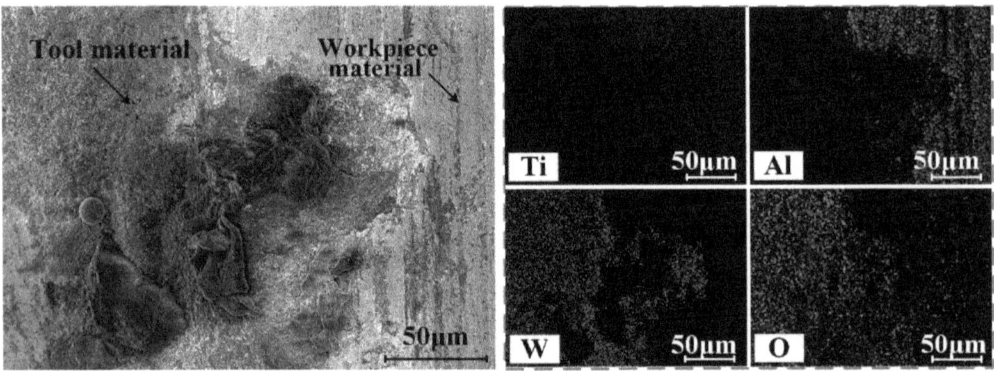

Figure 10. The morphology and EDS analysis of region B in Figure 7b.

Table 5. The Element composition of region 7/8/9.

Cutting Speed	Element Composition (at%)							
	Ti	Al	V	Cr	W	C	Co	O
60	21.78	29.92	0.83	0.21	0.03	47.23	-	-
120	14.53	24.83	0.76	0.94	0.35	46.95	0.12	11.51
180	14.32	22.35	0.73	0.78	0.51	45.43	0.80	15.08

Cutting speed has a major influence on the wear mechanism of the tool. As the cutting speed increases, the surface of the tool will be affected by factors such as higher temperature, stress, and pressure, which will easily cause thermal fatigue, structural changes, plastic deformation, and adhesive wear. Cutting forces are also critical to the wear mechanism of the tool. When cutting γ-TiAl alloy at high speed at low temperatures, the cutting force is large, which will cause plastic deformation and scratches on the surface of the tool. These deformations will gradually intensify, leading to early failure of the tool. There is currently no evidence that cutting speed and cutting force affect tool wear mechanisms significantly differently. Therefore, in the actual cutting process, it is necessary to balance the cutting speed and cutting force to ensure the life of the tool and the processing quality of the workpiece.

3.4. Surface Roughness

To further study the feasibility of cryogenic conditions in γ-TiAl alloy machining, the machined surfaces under various cutting conditions are analyzed. The average surface roughness (*Ra*) of the workpiece is obtained, as shown in Figure 11. The *Ra* measured under the other two cooling conditions is significantly lower than that under dry conditions.

When the cutting speed is 60 m/min, the *Ra* obtained by the emulsion condition is lower than that obtained by the LN_2 condition because good lubrication of emulsion makes it easier to get a smooth surface at low speeds. The material in the cutting layer is cooled rapidly in LN_2 condition, which reduces the thermal softening effect. Moreover, γ-TiAl alloy has high brittleness in a cryogenic environment. The machined surface with poor quality is obtained during low-speed cryogenic machining. When the cutting speed is increased, the cutting vibration and high cutting temperature brought by the high speeds increase the instability, aggravating the tool wear. In addition, severe work-hardening and higher cutting force increase the machining difficulty. It is not conducive to the formation of a machined surface with high quality. Therefore, when v_c = 120 m/min in dry machining, the *Ra* value reaches the peak value. The emulsion and LN_2 environment has a positive effect on reducing tool wear; the Ra reaches the peak value at v_c = 150 m/min. When the cutting speed is higher than the critical speeds, the thermal softening effect of the material is significant. The lower surface roughness is produced by the plowing effect.

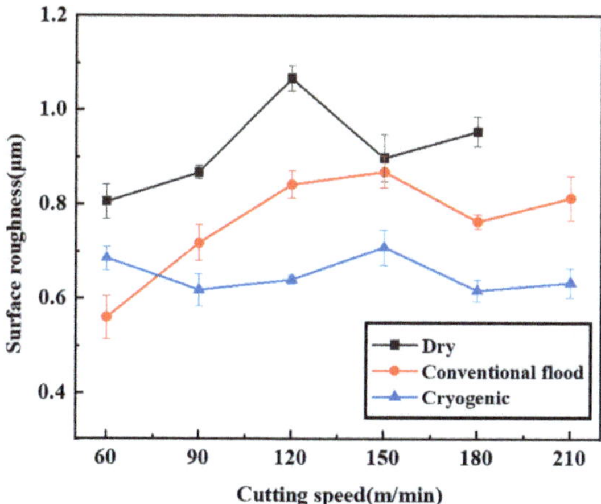

Figure 11. The surface roughness under various cutting conditions.

With the further increase of cutting speed, the tool cutting edge deteriorates under the action of mechanical and thermal load. As a result, the *Ra* value increases. Compared with the other two conditions, the variation degree of *Ra* under the LN_2 condition is the smallest in the selected cutting speed range.

Cutting in the cryogenic coolant can effectively reduce tool wear, and cutting with sharp tools can obtain a more ideal machined surface. Cryogenic cooling reduces the plasticity of the material and reduces the side flow of the material during the cutting process, which also makes the machined surface close to the ideal morphology.

3.5. Microstructure Morphology of Sub-Surface

To further explore the influence of cooling conditions on the machined surface, the microstructure of the cross-section of the workpiece is observed, as shown in Figure 12. The microstructure of the sub-surface has changed, which is mainly reflected by lamellar deformation. Obvious lamellar deformation occurs under dry and conventional flood cooling conditions. The plastic deformation will lead to residual tensile stress, which will affect the fatigue resistance of the workpiece [41]. A cryogenic cooling environment weakens the influence of the thermal effect, resulting in the reduction of microstructure deformation.

In addition, some grain boundaries protrude into adjacent grains at about 10 μm below the surface layer compared with the original obvious vertical grain boundary of the workpiece. The grain boundary is fuzzy. The reason is that some grain boundaries bend under thermal activation, and the mobility of grain boundaries is improved. The two adjacent lamellas erode each other, resulting in the blurring of the boundary between the tissues. When the cutting temperature reaches a certain degree, the $\gamma + \alpha_2$ lamellar microstructure is broken. The dynamic recrystallization of the γ and the spheroidization of the α_2 occur, which can refine the grain structure. Some studies [42–44] show that the temperature range of γ and α_2 change is the brittle–ductile transition range of γ-TiAl alloy. The transition in γ-TiAl alloy cutting from brittleness removal to plasticity removal proves the possibility of turning points in the V_{max}, F, and Ra in the above analysis. Beretta et al. [45] proposed that the fully lamellar microstructure induces higher strain localizations in a single lamellar grain, and the presence of equiaxed grains at the lamellar grain boundaries can be beneficial in limiting the high local strains, which is similar to the conclusions of this study.

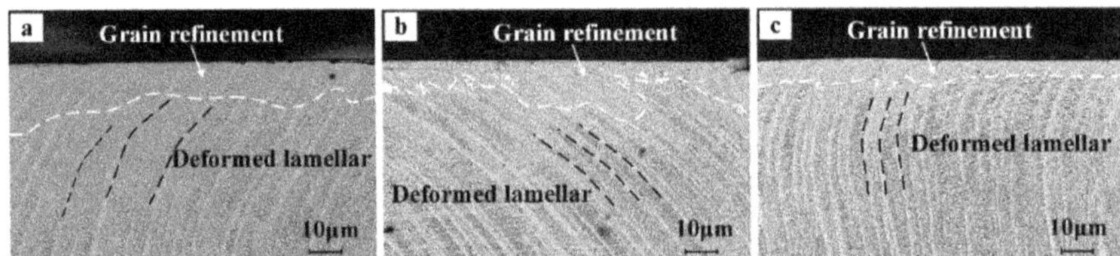

Figure 12. The microstructure morphology under various cutting conditions at $v_c = 120$ m/min (**a**) dry; (**b**) conventional flood cooling; (**c**) cryogenic cooling.

4. Conclusions

To suppress tool wear and improve surface quality, the feasibility of cryogenic cooling in γ-TiAl alloy high-speed cutting was explored. This paper studied the tool wear and mechanism, cutting force, as well as surface and sub-surface morphology characteristics under various cooling media and cutting speeds. The conclusions are as follows.

(a) The flank and rake face of the tool are worn to varying degrees at high-speed turning. The crater wear is observed on the rake face at $v_c = 60$ m/min. With the increase in cutting speed, the crater wear is expanded, progressing into more serve flaking and notching wear. LN$_2$-assisted cutting delays the occurrence of this state and greatly improves the tool life;

(b) The main wear pattern in dry machining is gradually transformed from adhesive wear to diffusion and oxidation wear at high speeds. The wear mechanism is still mainly adhesive wear in emulsion and LN$_2$ cooling conditions, accompanied by slight diffusion and oxidation wear. Cryogenic cooling has a significant effect on inhibiting adhesion, diffusion, and oxidative wear;

(c) Cryogenic cooling-assisted high-speed machining can significantly improve the surface finish and inhibit the deformation of the sub-surface microstructure. To some extent, the cooling effect of LN$_2$ inhibits thermal activation and reduces the degree of grain refinement. However, the combination of cryogenic cooling and high-speed machining technology is a field worthy of exploration from a long-term perspective;

(d) The curves of the V_{max}, F, and Ra turn at $v_c = 120$ m/min in dry conditions and at $v_c = 150$ m/min in emulsion and LN$_2$ conditions. It is preliminarily considered that the brittle–ductile transition of γ-TiAl alloy occurs within this cutting speed range. Further exploration is needed in future research work.

5. Prospect

Despite its numerous advantages, there are some potential drawbacks to the use of LN$_2$ cryogenic cooling. One of the main challenges is the cost associated with the storage and transportation of LN$_2$, and the other is that for some materials, LN$_2$ is excessively cold, and extremely low temperature is harmful to the surface integrity on the contrary.

For the first issue, the consumption of LN$_2$ should be optimized, so subsequent research on minimal-cryogenic cooling is needed. For the second issue, some other cryogenic medium may be a suitable alternative, such as LCO$_2$, which could provide a more moderate cryogenic condition. Therefore, various factors such as workpiece material, cutting conditions, and production costs should be considered when selecting cryogenic cooling process parameters in the future.

Author Contributions: Conceptualization, X.W.; data curation, D.P.; funding acquisition, X.W., X.F. and Y.Q.; investigation, D.P.; methodology, X.W. and D.P.; supervision, X.W., X.F. and Y.Q.; validation, D.P.; writing—original draft, X.W., X.Z. and D.P.; writing—review and editing, X.W., X.Z., J.N. and X.F. All authors have read and agreed to the published version of the manuscript.

Funding: This work was funded by the National Natural Science Foundation of China (Grant No. 52005215 and 52175408) and the Project of Shandong Province Higher Educational Science and Technology Program (Grant No. 2019KJB021), and the APC was funded by the National Natural Science Foundation of China (Grant No. 52005215).

Conflicts of Interest: The authors declare that they have no known competing financial interest or personal relationships that could have appeared to influence the work reported in this paper. Author Duo Pan was employed by the company Jigang International Engineering & Technology Co., Ltd. The remaining authors declare that the research was conducted in the absence of any commercial or financial relationships that could be construed as a potential conflict of interest.

References

1. Xu, R.; Li, M.; Zhao, Y. A review of microstructure control and mechanical performance optimization of γ-TiAl alloys. *J. Alloys Compd.* **2022**, *932*, 167611. [CrossRef]
2. Castellanos, S.; Cavaleiro, A.J.; Jesus, A.D.; Neto, R.; Alves, J.L. Machinability of titanium aluminides: A review. *Materials* **2019**, *233*, 426–451. [CrossRef]
3. Clemens, H.; Mayer, S. Design, processing, microstructure, properties, and applications of advanced intermetallic TiAl alloys. *Adv. Eng. Mater.* **2013**, *15*, 191–215. [CrossRef]
4. Hood, R.; Aspinwall, D.K.; Voice, W. Creep feed grinding of γ-TiAl using single layer electroplated diamond super abrasive wheels. *CIRP J. Manuf. Sci. Technol.* **2015**, *11*, 36–44. [CrossRef]
5. Yao, C.F.; Lin, J.N.; Wu, D.X.; Junxue, R.E. Surface integrity and fatigue behavior when turning γ-TiAl alloy with optimized PVD-coated carbide inserts. *Chin. J. Aeronaut.* **2018**, *145*, 214–224. [CrossRef]
6. Priarone, P.C.; Robiglio, M.; Settineri, L.; Tebaldo, V. Milling and Turning of Titanium Aluminides by Using Minimum Quantity Lubrication. *Procedia CIRP* **2014**, *24*, 62–67. [CrossRef]
7. Beranoagirre, A.; López de Lacalle, L.N. Optimizing the turning of titanium aluminide alloys. *Adv. Mater. Res.* **2012**, *498*, 189–194. [CrossRef]
8. Beranoagirre, A.; López de Lacalle, L.N. Optimising the milling of titanium aluminide alloys. *Int. J. Mechatron. Manuf. Syst.* **2010**, *3*, 425–436. [CrossRef]
9. Beranoagirre, A.; de Lacalle, L.N.L. Grinding of gamma TiAl intermetallic alloys. *Procedia Eng.* **2013**, *63*, 489–498. [CrossRef]
10. Cheng, Y.; Yuan, Q.; Zhang, B.; Wang, Z. Study on turning force of γ-TiAl alloy. *Int. J. Adv. Manuf. Technol.* **2019**, *105*, 2393–2402. [CrossRef]
11. Beranoagirre, A.; Olvera, D.; López de Lacalle, L.N. Milling of gamma titanium–aluminum alloys. *Int. J. Adv. Manuf. Technol.* **2012**, *62*, 83–88. [CrossRef]
12. Wang, B.; Liu, Z.Q.; Su, G.S.; Song, Q.; Ai, X. Investigations of critical cutting speed and ductile-to-brittle transition mechanism for workpiece material in ultra-high speed machining. *Int. J. Mech. Sci.* **2015**, *104*, 44–59. [CrossRef]
13. Aspinwall, D.K.; Mantle, A.L.; Chan, W.K.; Hood, R.; Soo, S.L. Cutting temperatures when ball nose end milling γ-TiAl intermetallic alloys. *CIRP Ann.-Manuf. Technol.* **2013**, *62*, 75–78. [CrossRef]
14. Klocke, F.; Lung, D.; Arft, M.; Priarone, P.C.; Settineri, L. On high-speed turning of a third-generation gamma titanium aluminide. *Int. J. Adv. Manuf. Technol.* **2013**, *65*, 155–163. [CrossRef]
15. Saketi, S.; Odelros, S.; Östby, J.; Olsson, M. Experimental Study of Wear Mechanisms of Cemented Carbide in the Turning of Ti6Al4V. *Materials* **2019**, *12*, 2822. [CrossRef] [PubMed]
16. Rajashree, M.; Ramanuj, K.; Amlana, P.; Kumar, S.A. Current Status of Hard Turning in Manufacturing: Aspects of Cooling Strategy and Sustainability. *Lubricants* **2023**, *11*, 108.
17. Zou, L.; Huang, Y.; Zhang, G.J.; Cui, X. Feasibility study of a flexible grinding method for precision machining of the TiAl-based alloy. *Mater. Manuf. Process.* **2019**, *34*, 1160–1168. [CrossRef]
18. Wang, Z.H.; Liu, Y.W. Study of surface integrity of milled gamma titanium aluminide. *J. Manuf. Process.* **2020**, *56*, 806–819. [CrossRef]
19. Ezugwu, E.O. High speed machining of aero-engine alloys. *J. Braz. Soc. Mech. Sci. Eng.* **2004**, *26*, 1–11. [CrossRef]
20. Liu, Z.; An, Q.; Xu, J.; Chen, M.; Han, S. Wear performance of (nc-AlTiN)/(a-Si$_3$N$_4$) coating and (nc-AlCrN)/(a-Si$_3$N$_4$) coating inhigh-speed machining of titanium alloys under dry and minimum quantity lubrication (MQL) conditions. *Wear* **2013**, *305*, 249–259. [CrossRef]
21. Klocke, F.; Settineri, L.; Lung, D.; Claudio Priarone, P.; Arft, M. High performance cutting of gamma titanium aluminides: Influence of lubricoolant strategy on tool wear and surface integrity. *Wear* **2013**, *302*, 1136–1144. [CrossRef]
22. González, H.; Pereira, O.; López de Lacalle, L.N.; Calleja, A.; Ayesta, I.; Muñoa, J. Flank-milling of integral blade rotors made in Ti$_6$Al$_4$V using Cryo CO$_2$ and minimum quantity lubrication. *J. Manuf. Sci. Eng.* **2021**, *143*, 091011. [CrossRef]

23. Khanna, N.; Shah, P.; de Lacalle, L.N.L.; Rodríguez, A.; Pereira, O. In pursuit of sustainable cutting fluid strategy for machining Ti-6Al-4V using life cycle analysis. *Sustain. Mater. Technol.* **2021**, *29*, e00301. [CrossRef]
24. M'Saoubi, R.; Axinte, D.; Soo, S.L.; Nobel, C.; Attia, H.; Kappmeyer, G.; Engin, S.; Sim, W.-M. High performance cutting of advanced aerospace alloys and composite materials. *CIRP Ann.-Manuf. Technol.* **2015**, *64*, 557–580. [CrossRef]
25. Devaraya, R.G.; Raviraj, S.; Rao, S.S.; Gaitonde, V.N. Analysis of Surface Roughness and Hardness in Titanium Alloy Machining with Polycrystalline Diamond Tool under Different Lubricating Modes. *Mater. Res.* **2014**, *17*, 1010–1022.
26. Jawahir, I.S.; Attia, H.; Biermann, D.; Duflou, J.; Klocke, F.; Meyer, D.; Newman, S.T.; Pusavec, F.; Putz, M.; Rech, J.; et al. Cryogenic manufacturing processes. *CIRP Ann.-Manuf. Technol.* **2016**, *65*, 713–736. [CrossRef]
27. Pratham, S.; Rohit, N.; Ramanuj, K.; Deepak, S.; Amlana, P.; Kumar, S.A.; Diptikanta, D. Cryogenics as a Cleaner Cooling Strategy for Machining Applications: A Concise Review. *Int. J. Energy A Clean Environ.* **2022**, *23*, 129–141.
28. Navneet, K.; Chetan, A.; Yu, P.D.; Kumar, S.A.; Rocha, M.A.; Ribeiro, S.L.R.; Kumar, G.M.; Murat, S.; Grzegorz, M.K. Review on design and development of cryogenic machining setups for heat resistant alloys and composites. *J. Manuf. Process.* **2021**, *68*, 398–422.
29. Shah, P.; Khanna, N.; Chetan. Comprehensive Machining Analysis to Establish Cryogenic LN$_2$ and LCO$_2$ as Sustainable Cooling and Lubrication Techniques. *Tribol. Int.* **2020**, *148*, 106314. [CrossRef]
30. Musfirah, A.H.; Ghani, J.A.; Haron, C. Tool wear and surface integrity of Inconel 718 in dry and cryogenic coolant at high cutting speed. *Wear* **2017**, *376–377*, 125–133. [CrossRef]
31. Yıldırım, C.V. Experimental comparison of the performance of nanofluids, cryogenic and hybrid cooling in turning of Inconel 625. *Tribol. Int.* **2019**, *137*, 366–378. [CrossRef]
32. Shokrani, A.; Dhokia, V.; Newman, S.T. Investigation of the effects of cryogenic machining on surface integrity in CNC end milling of Ti-6Al-4V titanium alloy. *J. Manuf. Process.* **2016**, *21*, 172–179. [CrossRef]
33. Sartori, S.; Ghiotti, A.; Bruschi, S. Hybrid lubricating/cooling strategies to reduce the tool wear in finishing turning of difficult-to-cut alloys. *Wear* **2017**, *376–377*, 107–114. [CrossRef]
34. Zhao, W.; Ren, F.; Iqbal, A.; Gong, L.; He, N.; Xu, Q. Effect of liquid nitrogen cooling on surface integrity in cryogenic milling of Ti-6Al-4V titanium alloy. *Int. J. Adv. Manuf. Technol.* **2020**, *106*, 1497–1508. [CrossRef]
35. Fernández, D.; Sandá, A.; Bengoetxea, I. Cryogenic Milling: Study of the Effect of CO$_2$ Cooling on Tool Wear When Machining Inconel 718, Grade EA1N Steel and Gamma TiAl. *Lubricants* **2019**, *7*, 10. [CrossRef]
36. Pereira, O.; Urbikain, G.; Rodríguez, A.; Fernández-Valdivielso, A.; Calleja, A.; Ayesta, I.; de Lacalle, L.L. Internal cryo lubrication approach for Inconel 718 milling. *Procedia Manuf.* **2017**, *13*, 89–93. [CrossRef]
37. Priarone, P.C.; Klocke, F.; Faga, M.G.; Lung, D.; Settineri, L. Tool life and surface integrity when turning titanium aluminides with PCD tools under conventional wet cutting and cryogenic cooling. *Int. J. Adv. Manuf. Technol.* **2016**, *85*, 807–816. [CrossRef]
38. Yuan, Q.; Cheng, Y.; Wang, Z.H.; Hu, X. Failure study of carbide tools for turning γ-TiAl alloys. *Tool Eng.* **2019**, *53*, 12–16.
39. Astakhov, V. *Tribology of Metal Cutting*; Elsevier: Amsterdam, The Netherlands, 2006.
40. Jiang, Z.H.; Wang, L.L.; Shi, L. Study on Tool Wear Mechanism and Characteristics of Carbide Tools in Cutting Ti6Al4V. *J. Mech. Eng.* **2014**, *50*, 178–184. [CrossRef]
41. Tian, S.; Qi, W.; Yu, H.; Sun, H.; Li, Q. Microstructure and creep behaviors of a high Nb-TiAl intermetallic compound based alloy. *Mater. Sci. Eng. A* **2014**, *614*, 338–346. [CrossRef]
42. Wang, Q.; Chen, R.; Gong, X.; Guo, J.; Su, Y.; Ding, H.; Fu, H. Microstructure, Mechanical Properties, and Crack Propagation Behavior in High-Nb TiAl Alloys by Directional Solidification. *Metall. Mater. Trans. A* **2018**, *49*, 4555–4564. [CrossRef]
43. Wang, Q.; Chen, R.; Chen, D.; Su, Y.; Ding, H.; Guo, J.; Fu, H. The characteristics and mechanisms of creep brittle-ductile transition in TiAl alloys. *Mater. Sci. Eng.* **2019**, *767*, 138393. [CrossRef]
44. Wei, W.; Zeng, W.; Chen, X.; Liang, X.; Zhang, J. Microstructural evolution, creep, and tensile behavior of a Ti-22Al-25Nb (at%) orthorhombic alloy. *Mater. Sci. Eng. A* **2014**, *603*, 176–184.
45. Patriarca, L.; Içöz, C.; Filippini, M.; Beretta, S. Microscopic Analysis of Fatigue Damage Accumulation in TiAl Intermetallics. *Key Eng. Mater.* **2014**, *592–593*, 30–35.

 *lubricants*

MDPI

Article

Analysis of Tool Wear in GH4169 Material Milling Process

Xueguang Li *, Wang Zhang, Liqin Miao and Zhaohuan Pang

College of Mechanical and Electrical Engineering, Changchun University of Science and Technology, Changchun 130022, China; zhangwang5434@sina.com (W.Z.); miaoliqin@163.com (L.M.)
* Correspondence: lixueguang@cust.edu.cn

Abstract: Nickel-based superalloy GH4169 is a material with strong mechanical properties and is difficult to process. In order to reduce tool wear during material processing and improve the workpiece surface processing quality, based on the finite element simulation software DEFORM, the influence of n, a_p, and f_z parameters on tool wear during carbide tool milling GH4169 was studied, and a simulation of an orthogonal experimental model was established. The prediction model of tool wear was obtained. The ultrasonic vibration milling was compared with ordinary milling, and the improvement degree of different coating materials on carbide tool wear was explored. The results showed that the ultrasonic vibration signal is helpful to reduce tool wear, improve the surface quality of the workpiece, and improve the stability of the milling process. TiAlN/TiN (WC)-composite-coated tools have good cutting performance, help to reduce tool temperature, reduce tool wear, and improve tool life.

Keywords: tool wear; prediction model; ultrasonic vibration; coated cutting tools; tool temperature

1. Introduction

Nickel-based high-temperature alloys have high yield strength, tensile strength, and endurance strength below 700 °C, as well as good oxidation and corrosion resistance, and forming and welding performance. Therefore, they are widely used in industries such as aerospace, petroleum, chemical, and shipbuilding. Inconel 718 is one of the important nickel-based alloys, accounting for 35% of the total nickel-based alloy output [1]. It has a wide range of applications, from rotating and stationary parts of aerospace engines to high-strength bolts and fasteners, acid gas wells and pipeline petrochemical equipment, nuclear power plant components for nuclear reactors and spacecraft, marine shafting and extrusion dies, and gas turbine engines, steam turbines, and missile parts [2,3]. While nickel-based superalloy has many advantages, it also has many disadvantages in the machining process: large plastic deformation, high cutting temperature, serious tool wear, chip bending, and serious hardening of the machined surface. [4,5] At present, the demand for GH4169 superalloy in the manufacturing market is growing, and the processing requirements are also increasing, which makes the processing technology of nickel-based superalloy need to keep pace with the times. In recent years, domestic and foreign experts and scholars have studied the processing methods, cutting parameters, and tool materials in the milling process of high-temperature nickel-based alloy. Combining simulation and experiment, they continue to optimize the material processing performance, reduce tool wear, and reduce processing costs. We have achieved fruitful results.

The accurate prediction of tool wear profile plays an important role in guiding industrial production. An accurate prediction model helps the machining process to be carried out under reasonable cutting conditions, and more accurately grasp the tool change time and strategy.

On the basis of reducing processing time and cost, the requirements for the surface quality and functionality of components need to be met. The prediction formula of tool wear mainly includes an empirical formula, a formula based on the wear mechanism, the

Citation: Li, X.; Zhang, W.; Miao, L.; Pang, Z. Analysis of Tool Wear in GH4169 Material Milling Process. *Lubricants* **2023**, *11*, 245. https://doi.org/10.3390/lubricants11060245

Received: 18 May 2023
Revised: 27 May 2023
Accepted: 30 May 2023
Published: 1 June 2023

finite element method, and the artificial intelligence method [6–8]. The earliest and most commonly used method relies on empirical formulas to predict tool life and determine tool change timing, with the most famous being the Taylor tool life formula and tool life formula [9]. Considering the complexity of the machining process, which involves numerous physical quantities, these empirical formulas often have a relatively simple form and contain fewer parameters. They mainly describe the relationship between tool life and machining parameters. For machining processes where machining conditions vary within a certain range, these empirical formulas usually perform well and are widely used for predicting tool life under long-term stable cutting conditions in industrial production. In addition to the effective prediction of tool wear, experts and scholars have changed the processing methods and tool materials to reduce tool wear.

Ultrasonic-vibration-assisted milling is a machining method that applies ultrasonic vibration to the milling cutter or workpiece, and controls its vibration frequency, amplitude, and direction to achieve periodic high-frequency separation of the tool and workpiece. Compared with traditional milling, ultrasonic-vibration-assisted milling can reduce the cutting force and cutting temperature, reduce tool wear, facilitate chip breaking, and improve surface quality [10–15]. Milling is a machining method in which multiple cutting edges alternate and participate in cutting. After applying vibration, the cutting performance improves more significantly. Therefore, ultrasonic-vibration-assisted milling is widely studied and applied [15–20].

In order to achieve high-speed and efficient cutting of nickel-based high-temperature alloys, many scholars at home and abroad have begun to study cutting tools suitable for machining nickel-based alloys, that is, to explore the machinability and tool failure mechanism of different tool materials when machining Inconel 718 alloy [21–23].

In this paper, the empirical formula of tool wear prediction is established by means of multiple linear regression, and the accuracy of the prediction model is verified by range analysis. The mechanism of tool wear in the milling process of Inconel 718 nickel-based superalloy is studied. The chip shape, milling area temperature, equivalent stress of the workpiece surface, tool temperature, and tool wear in the process of vibration milling and ordinary milling are compared and analyzed, and the advantages of ultrasonic vibration milling are obtained. The influence of coating materials, TIALN, TIN, and their composite coating materials on tool wear is explored, which provides a theoretical basis for predicting tool wear and reducing tool wear.

2. Establishment of GH4169 Simulation Model for Hard Alloy Milling

In the actual machining process of nickel-based superalloy, the machining surface precision of the workpiece is low, the tool is easy to collapse, and there is serious tool wear. In the process of machining, the tool needs to be changed many times, the machining efficiency is low, and the capital cost is large. With the rapid development of computer-aided technology, the finite element simulation technology is widely used in the field of metal cutting. Through the finite element simulation technology, not only can the test cost be reduced, but also the test cycle can be shortened, Revealing the essence of the actual cutting process and predicting potential problems that may arise during the actual cutting process, the application of finite element simulation technology has good guiding significance for actual production and processing. In this paper, the finite element software DEFORM is used to establish the simulation model of the carbide tool milling of GH4169, and to analyze the influence of three cutting elements on carbide tool wear.

2.1. Establishment of Geometric Modeling of Tool and Workpiece

In order to be closer to actual machining, this simulation will make the model of the end milling cutter as close as possible to the actual situation. The specific parameters are shown in Table 1. The workpiece size is 14 mm × 14 mm × 5 mm, and the milling model is shown in Figure 1.

Table 1. Milling cutter model parameters.

Diameter	Tooth Number	Front Angle	Back Angle	Helix Angle	Radius of the Tool
10 mm	4	8°	12°	30°	1 mm

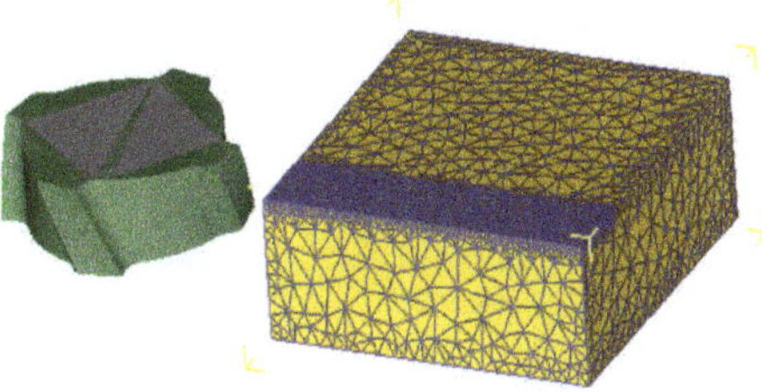

Figure 1. Milling Model.

2.2. Setting Up Workpiece and Tool Materials and Grid Division

The tool is set as a rigid body and the workpiece as a plastic body. The material of the tool selected in the experimental section of this article is WC, and nickel-based high-temperature alloy GH4169 (Inconel 718) is used as the material of the workpiece. In this model, the tool set grid is 32,000, the workpiece set grid is 30,000, the minimum grid size of the tool is 0.131 mm, and the minimum grid size of the workpiece is 0.121 mm. The parts in the contact area between the tool and the workpiece are divided into local grids, with refinement ratios of 0.1 and 0.0001, as shown in Figure 1.

2.3. Tool and Workpiece Simulation of Working Condition Settings

The milling width is set to 3 mm, the ambient temperature is set to 20 °C, the convective heat transfer coefficient is set to 0.02 N/sec/mm/C, the friction coefficient is set to 0.6, and the thermal conductivity coefficient is set to 45 N/sec/mm/C. This article uses the Usui wear model in DEFORM, which is more suitable for continuous machining processes, such as metal cutting, and its main form of wear is adhesive wear.

The following shows Usui's tool wear model:

$$w = \int a p v e^{-b/T} dt \tag{1}$$

This formula is generally used for continuous processes such as metal cutting (adhesive wear), where p—contact surface pressure; v—sliding speed; T—contact surface temperature; dt—time increment. The classic values for cutting are a = 0.00001 and b = 855, where a and b are characteristic constants (mainly determined by cutting parameters and materials).

2.4. Establishment of Tool Wear Prediction Model for Ordinary Milling

In the book *Metal Cutting Principles* [24], the process of the multiple linear regression method to solve the cutting force and tool temperature prediction model is mentioned. Based on this method, the tool wear prediction model is obtained.

First, the orthogonal experiment was established to study the influence of spindle speed, feed per tooth, and cutting depth on tool wear. Compared with multi-tooth continuous cutting, the tool wear of this intermittent cutting simulation is smaller, and the wear difference between teeth is larger.

The selection of orthogonal experimental factors and levels needs to follow certain principles. There are different orthogonal experimental tables for different factors and levels. Choosing a suitable orthogonal experimental plan can generally be divided into the following steps:

(1) Determine the number of columns, which is the number of influencing factors.

(2) Determine the number of levels, which means that each factor has several different values.

(3) Select an orthogonal table, mainly based on the number of columns and levels determined in the first two steps.

(4) After consulting the process manual related to nickel based high-temperature alloys [25] and the recommended cutting amount for hard alloy cutting tools, 16 sets of orthogonal experiments were designed with 3 factors, 4 levels, and the experimental factors were axial cutting depth ap, spindle speed n, and feed rate fz per tooth. From the actual situation, the values of each factor level are within the operating range of the machine tool, and the experimental parameter range is as follows:

(1) Spindle speed n: 3000–4500 r/min;

(2) Axial cutting depth a_p: 0.2–0.5 mm;

(3) The feed rate per tooth f_z: 0.1–0.4 mm/z.

The orthogonal experiment and simulation data are shown in Table 2:

Table 2. Orthogonal Simulation Experimental Data.

	Spindle Speed n (r/min)	Cutting Depth a_p (mm)	Feed per Tooth f_z (mm/z)	Tool Wear Amount H (μm)
1	3000	0.2	0.1	0.557
2	3000	0.3	0.2	0.499
3	3000	0.4	0.3	0.343
4	3000	0.5	0.4	0.351
5	3500	0.2	0.2	0.387
6	3500	0.3	0.1	0.778
7	3500	0.4	0.4	0.349
8	3500	0.5	0.3	0.444
9	4000	0.2	0.3	0.208
10	4000	0.3	0.4	0.323
11	4000	0.4	0.1	1.010
12	4000	0.5	0.2	0.645
13	4500	0.2	0.4	0.309
14	4500	0.3	0.3	0.363
15	4500	0.4	0.2	0.718
16	4500	0.5	0.1	0.898

The logarithm of the milling parameters H, n, a_p, and f_z in the table are taken and converted into a multiple linear regression matrix in the form of:

$$
Y_H = \begin{bmatrix} -0.2541 \\ -0.3019 \\ -0.4647 \\ -0.4547 \\ -0.4123 \\ -0.1090 \\ -0.4572 \\ -0.3526 \\ -0.6819 \\ -0.4908 \\ 0.0043 \\ -0.1904 \\ -0.5100 \\ -0.4401 \\ -0.1439 \\ -0.0467 \end{bmatrix} \quad X = \begin{bmatrix} 1 & 3.4771 & -0.6990 & -1.0000 \\ 1 & 3.4771 & -0.5229 & -0.6990 \\ 1 & 3.4771 & -0.3979 & -0.5229 \\ 1 & 3.4771 & -0.3010 & -0.3979 \\ 1 & 3.5441 & -0.6990 & -0.6990 \\ 1 & 3.5441 & -0.5229 & -1.0000 \\ 1 & 3.5441 & -0.3979 & -0.3979 \\ 1 & 3.5441 & -0.3010 & -0.5229 \\ 1 & 3.6021 & -0.6990 & -0.5229 \\ 1 & 3.6021 & -0.5229 & -0.3979 \\ 1 & 3.6021 & -0.3979 & -1.0000 \\ 1 & 3.6021 & -0.3010 & -0.6990 \\ 1 & 3.6532 & -0.6990 & -0.3979 \\ 1 & 3.6532 & -0.5229 & -0.5229 \\ 1 & 3.6532 & -0.3979 & -0.6990 \\ 1 & 3.6532 & -0.3010 & -1.0000 \end{bmatrix}
$$

The M file is established through MATLAB software, and the tool wear value and milling parameters in the formula are input into the program for multiple linear regression [26,27] calculation and then input in MATLAB:

$y = [-0.2541, -0.3019, -0.4647, -0.4547, -0.4123, -0.1090, -0.4572, -0.3526, -0.6819,$
$-0.4908, 0.0043, -0.1904, -0.5100, -0.4401, -0.1439, -0.0467];$

$x_1 = [3.4771, 3.4771, 3.4771, 3.4771, 3.5441, 3.5441, 3.5441, 3.5441, 3.6021, 3.6021, 3.6021,$
$3.6021, 3.6532, 3.6532, 3.6532, 3.6532];$

$x_2 = [-0.6990, -0.5229, -0.3979, -0.3010, -0.6990, -0.5229, -0.3979, -0.3010, -0.6990,$
$-0.5229, -0.3979, -0.3010, -0.6990, -0.5229, -0.3979, -0.3010];$

$x_3 = [-1.0000, -0.6990, -0.5229, -0.3979, -0.6990, -1.0000, -0.3979, -0.5229, -0.5229,$
$-0.3979, -1.0000, -0.6990, -0.3979, -0.5229, -0.6990, -1.0000];$

$X = [\text{ones (length (y), 1)}, x_1', x_2'\ x_3'];$

$Y = y';$

$[b, bint, r, rint, stats] = \text{regress (Y, X)};$

$b, bint, stats$

According to the results obtained from MATLAB, $a_0 = -1.9922$, $a_1 = 0.4112$, $a_2 = 0.5345$, and $a_3 = -0.6866$. Therefore, the multiple regression model is: $y = -1.9922 + 0.4112x_1 + 0.5345x_2 - 0.6866x_3$.

The empirical formula for predicting the maximum wear of cutting tools can be obtained from the above: $H = 0.01 \cdot n^{0.4112} \cdot a_p^{0.5345} \cdot f_z^{-0.6866}$.

2.5. Linear Regression Significance Test

According to the results obtained from MATLAB, it can be concluded that:

(1)   Tool wear regression system array:

In order to test the fitting degree of tool wear regression model, the significance analysis is carried out below. Use MATLAB to obtain the regression coefficients and confidence intervals, as shown in Table 3:

Table 3. Regression Model Calculation Results.

Regression Coefficient	Estimated Value of Regression Coefficient	Confidence Interval
a_0	−1.9922	−3.9917, 0.0072
a_1	0.4112	−0.1472, 0.9695
a_2	0.5345	0.2881, 0.7808
a_3	−0.6866	−0.8487, −0.5244

Obtain the statistical variable parameter stats as: $R^2 = 0.9016$, $F = 36.6678$, and $p = 0.0000$. They indicate a significant difference in $p < \alpha = 0.05$.

Residual analysis is performed on this coefficient and rcoplot (r, rint) is input in the MATLAB window to obtain the residual plot as shown in Figure 2:

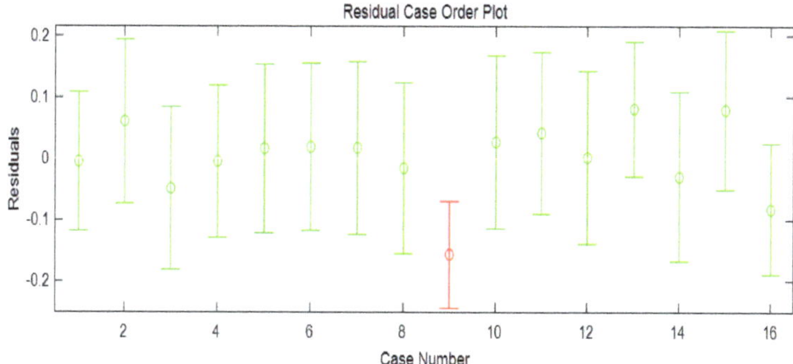

Figure 2. Residual Analysis of tool wear.

From the residual analysis chart, it can be seen that except for the ninth data point, the residuals of all other data are close to zero, and the confidence intervals of the residuals all include zero, indicating that the regression model can better match the original data. The ninth data point can be regarded as an outlier, and it is believed that the regression model is reliable [28].

From the above, it can be concluded that the regression model $y = -1.9922 + 0.4112x_1 + 0.5345x_2 - 0.6866x_3$ is valid. Therefore, the empirical formula for maximum tool wear $H = 0.01 \cdot n^{0.4112} \cdot a_p^{0.5345} \cdot f_z^{-0.6866}$ has high significance.

2.6. Analysis of Impact Patterns

The range analysis method can be used in the analysis of orthogonal experimental results to obtain the primary and secondary order of the influence of various factors on the target parameters, the optimal combination scheme, and the influence law of the influencing factors on the target parameters [29]. Therefore, it has a wide range of applications. The formula for calculating the range is as follows:

$$k_i^j = \frac{K_i^j}{j}$$
$$R_j = \max(k_i^j) - \min(k_i^j) \tag{2}$$

where k_i^j, K_i^j—range and sum of ranges at the level of factor i in column j, respectively; i, j—the level and factors of the experiment, respectively; R_j—the maximum range value of the factor.

$k_1 \sim k_4$ are the average values of the test results for factors such as spindle speed n, milling depth a_p, and feed rate f_z per tooth at levels 1–4, respectively. R_F is the maximum value of the range of different influencing factors, which reflects the degree of change in the simulation test results when the factor changes. The larger the value of R_F, the greater the impact of this influencing factor on the simulation results. According to the simulation experimental data, the range of maximum tool wear (H) is shown in Table 4:

Table 4. Analysis of Maximum Tool Wear (H) Range.

	n (Speed)	a_p (The Millin Depth)	f_z (Feed per Tooth)
K_1	1.7500	1.4610	3.2430
K_2	1.9580	1.9630	2.2490
K_3	2.1860	2.4200	1.3580
K_4	2.2880	2.3380	1.3320
k_1	0.4375	0.3653	0.8108
k_2	0.4895	0.4908	0.5623
k_3	0.5465	0.6050	0.3395
k_4	0.5720	0.5845	0.3330
R_F	0.1345	0.2398	0.4778
Primary and secondary order		$f_z > a_p > n$	

Figure 3 shows the trend of the influence of spindle speed n, axial cutting depth a_p, and feed rate f_z per tooth on the maximum wear of the tool. Based on this graph, the variation in the maximum wear of the tool when each parameter changes can be intuitively seen. By analyzing the range table and indicator factor trend chart, it can be concluded that:

(1) The range analysis method is used to determine the degree of influence of different parameters (f_z, a_p, n) on the research object (H). The R_F value of f_z is the largest through calculation, so the feed rate per tooth has the largest impact on tool wear, followed by a_p, and finally the speed n. Therefore, the order of influence on tool wear is: $f_z > a_p > n$. From the perspective of reducing tool wear, the optimal combination of milling parameters is n = 3000 r/min, $a_p = 0.2$ mm, and $f_z = 0.4$ mm/z.

(2) It can be seen from Figure 3a that the tool wear increases with the increase in the spindle speed. When the speed is between 3000 and 4000 r/min, the increment in the

maximum tool wear is relatively large. When the speed is between 4000 and 5000 r/min, the increment in tool wear decreases. As the rotational speed increases, the thickness of the workpiece material in contact with the tool per unit time also increases, and the impact force received is relatively large. The cutting force also increases, resulting in a sharp increase in tool wear. As the spindle speed increases to a certain extent, the chip is generated faster, which will take away most of the heat, the tool temperature decreases, and the tool wear speed is eased.

(3) It can be seen from Figure 3b that with the increase in axial cutting depth, the tool wear is gradually increasing. When the axial cutting depth increases from 0.2 mm to 0.4 mm, the removal of material by the tool increases, resulting in an increase in work done and an increase in cutting force and temperature in the milling area, leading to an increase in tool wear. It should be noted that when the cutting depth is small or microscopic, it can cause scraping or only cut to the hardened layer on the surface of the workpiece, resulting in significant wear of the tool and a decrease in tool life. Especially when rough-machining the workpiece, the cutting depth should be increased as much as possible within the allowable range of machine power and technology. When the axial cutting depth is 0.4–0.5 mm, the tool wear is relatively reduced. In the cutting process, more heat is taken away due to the chip falling, and the tool temperature is also reduced, which alleviates the tool wear. In general, the tool wear increases with the increase in the axial cutting depth.

(4) The quantitative analysis of Figure 3c shows that when the feed per tooth increases, the tool wear gradually decreases, and when f_z = 0.4 mm/z, the tool wear reaches the minimum of 0.34805 μm.

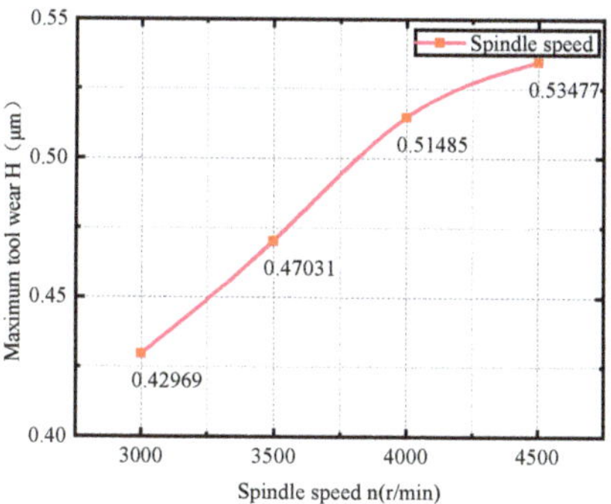

(**a**) Influence of spindle speed on tool wear

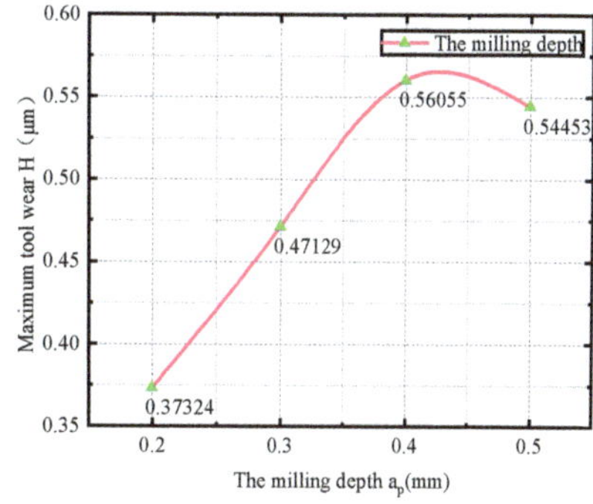

(**b**) Influence of milling depth on tool wear

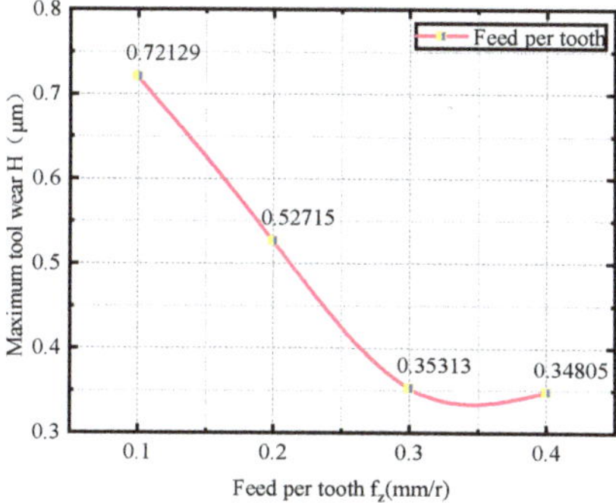

(**c**) Influence of feed per tooth on tool wear

Figure 3. Influence distribution of n, a_p, and f_z on tool wear.

3. Comparative Analysis of Ultrasonic Vibration Milling and Ordinary Milling

At present, ultrasonic-vibration-assisted machining is more and more widely used in the field of difficult-to-machine materials. Ultrasonic vibration milling is conducive to chip fracture, greatly reducing the temperature of the tool and workpiece area, reducing the vibration of sudden contact between the tool and workpiece, reducing tool wear, reducing the milling force, and improving system stability, and it is conducive to obtaining high-precision surface quality and lower surface roughness. In this paper, the GH4169 model of carbide tool vibration milling is established, and the ultrasonic vibration is compared with ordinary milling in order to reduce tool wear and improve the workpiece surface processing quality.

3.1. Establishment of GH4169 Model for Cemented Carbide Vibration Milling

According to the principle of axial ultrasonic vibration, a low-frequency and periodic vibration is added to the milling cutter along the axial direction of the tool, as shown in Figure 4. The tool movement path is planned. The motion of the tool relative to the workpiece consists of three parts, as shown in the figure: (1) the milling cutter rotates along its own axis; (2) the milling cutter moves uniformly along the feed direction of the workpiece in a straight line; and (3) the milling cutter oscillates at a low frequency along its own axis.

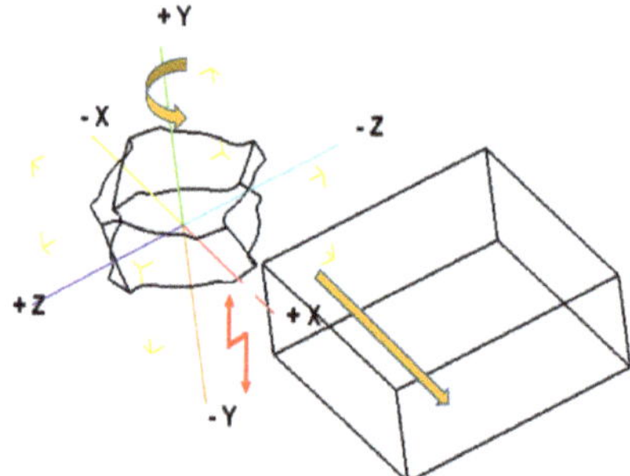

Figure 4. Schematic diagram of axial vibration milling.

The motion equation of the milling cutter center is [30]:

$$\begin{cases} X_1 = V_f \cdot t \\ Z_1 = 0 \\ y_1 = A sin(2\pi f t + \varphi_0) \end{cases} \tag{3}$$

The parameters used in this simulation are: f = 20 Hz, A = 10 μm, N = 6000 rad/min, f_z = 0.1 mm/z, a_p = 0.2 mm, and φ_0 = 0.

Then, the above formula is transformed into:

$$\begin{cases} X_1 = 40 \cdot t \\ Z_1 = 0 \\ y_1 = 0.01 sin(40\pi t) \end{cases} \tag{4}$$

MATLAB is used to generate a vibration displacement map in the y-direction, then the data are exported and imported into the DEFORM path chart. The generated tool path is shown in Figure 5.

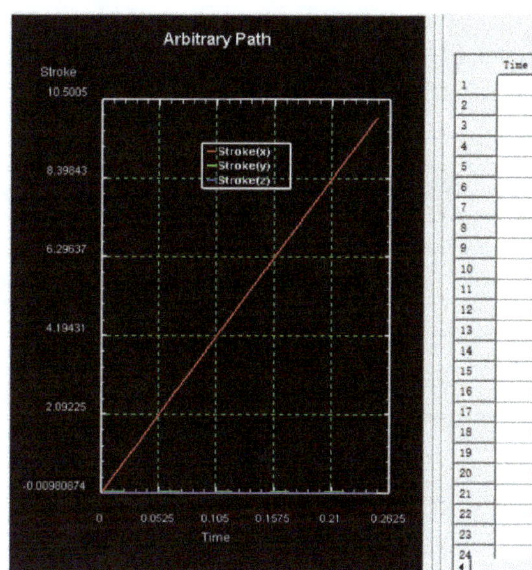

Arbitrary Path — f(Time)

	Time	Stroke(x)	Stroke(y)	Stroke(z)	An
1	0	0	0	0	
2	0. 00025	0. 01	0. 001588785	0	
3	0. 0005	0. 02	0. 001880584	0	
4	0. 00075	0. 03	0. 002172383	0	
5	0. 001	0. 04	0. 002464182	0	
6	0. 00125	0. 05	0. 002755981	0	
7	0. 0015	0. 06	0. 00304778	0	
8	0. 00175	0. 07	0. 003339579	0	
9	0. 002	0. 08	0. 003631379	0	
10	0. 00225	0. 09	0. 003923178	0	
11	0. 0025	0. 1	0. 004214977	0	
12	0. 00275	0. 11	0. 004506776	0	
13	0. 003	0. 12	0. 004798575	0	
14	0. 00325	0. 13	0. 005090374	0	
15	0. 0035	0. 14	0. 005382173	0	
16	0. 00375	0. 15	0. 005673972	0	
17	0. 004	0. 16	0. 005960689	0	
18	0. 00425	0. 17	0. 006233458	0	
19	0. 0045	0. 18	0. 006506227	0	
20	0. 00475	0. 19	0. 006778995	0	
21	0. 005	0. 2	0. 007051764	0	
22	0. 00525	0. 21	0. 007324533	0	
23	0. 0055	0. 22	0. 007597301	0	
24	0. 00575	0. 23	0. 00787007	0	

Figure 5. Axial vibration path planning.

3.2. Comparison of Results between Axial Ultrasonic Vibration Milling and Ordinary Milling

3.2.1. Comparison of Chip Shapes

When cutting GH4169 stably with hard alloy, a tool stroke of 8.5 mm is selected, as shown in Figure 6. The chips from ordinary milling appear as ribbons, while the chips from vibration milling have already broken. Under low-frequency vibration, the difference in chip shape is not significant, and the chips from vibration milling are more prone to fracture. In terms of chip breakage, ordinary milling is a physical chip, and the chips are squeezed by the tool and the workpiece. Chip fracture depends on the material's inherent performance, and uncontrollable chip length and chip breakage. Vibration milling belongs to geometric chip breaking, and low-frequency vibration will accelerate chip fracture, with controllable chip length and chip breaking. The fracture of chips not only helps to release heat but also helps to transmit stress, prevent stress concentration, and avoid chips sticking to the tool, replacing the tool for cutting, thereby affecting the quality and roughness of the machining surface. At the same time, it also reduces the adhesive wear caused by the detachment of chips bonded to the tool.

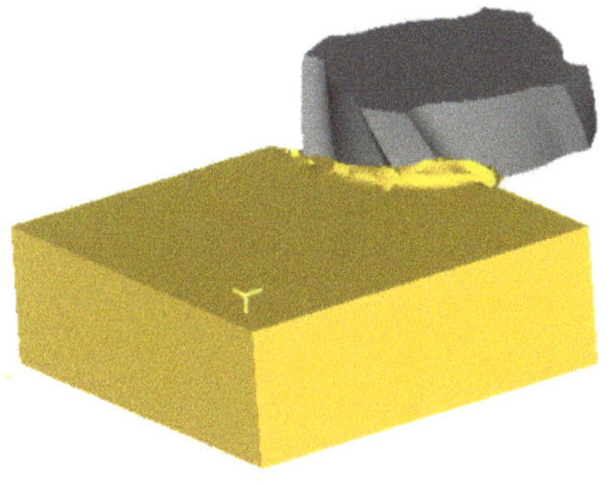

(**a**) Ordinary milling

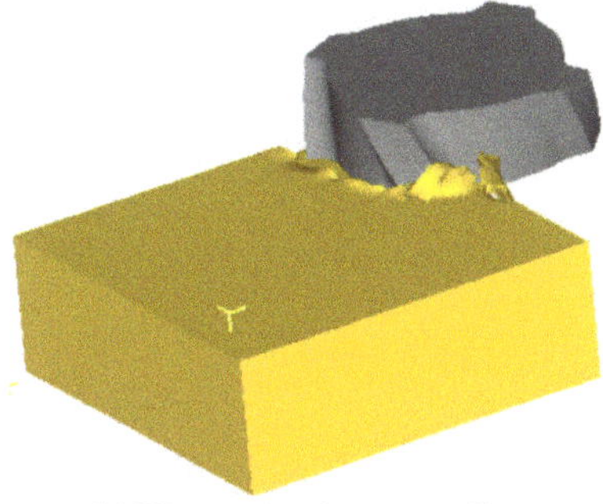

(**b**) Ultrasonic vibration milling

Figure 6. Comparison of chip cutting between ordinary milling and vibration milling.

3.2.2. Temperature Comparison in Milling Area

According to Figure 7, it can be clearly seen that when the tool first comes into contact with the workpiece, there is a significant sudden change in the temperature of ordinary milling, while the temperature of vibration milling increases relatively slowly. The range of temperature changes in the ordinary milling area is large, indicating that vibration milling alleviates the impact force of the tool on the workpiece, reduces the cutting force, and decreases the temperature of the cutting area. When the cemented carbide is stably milling GH4169, the temperature of the ordinary milling area is between 200 and 300 °C, and the temperature of vibration milling is lower than that of ordinary milling. This is because in the process of vibration milling, the chip is more likely to break, and the chip takes away a lot of heat. At the same time, the contact process between the tool and the workpiece is in a state of fluctuation, making the heat easier to transfer out.

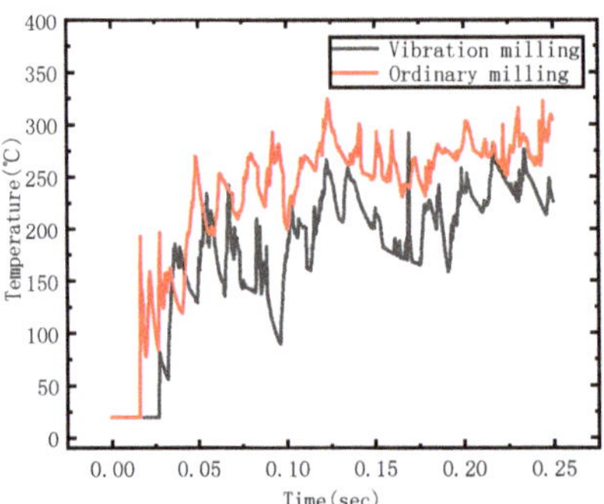

Figure 7. Temperature distribution diagram of vibration milling and ordinary milling area.

3.2.3. Tool Temperature Comparison

According to Figure 8, it can be seen that when the tool first comes into contact with the workpiece, the temperature in the ordinary milling area rises first. There is a heat conduction phenomenon between the tool and the workpiece, which causes the tool temperature to rise first. However, during vibration milling, the tool temperature rises slowly and the fluctuation range is small. The temperature of the cutting tool is closely related to the temperature of the milling area. During intermittent milling, every contact between the cutting tooth and the workpiece causes a drastic fluctuation in temperature, leading to an increase in milling force, increased work, and an increase in the temperature of the milling area. As a result, the temperature of the cutting tool increases. Vibration milling alleviates the impact force during contact, making it easier to release pressure and heat. The heat transmitted to the cutting tool is lower and the temperature of the cutting tool is relatively stable. Overall, the tool temperature fluctuates within the range of 20–60 °C.

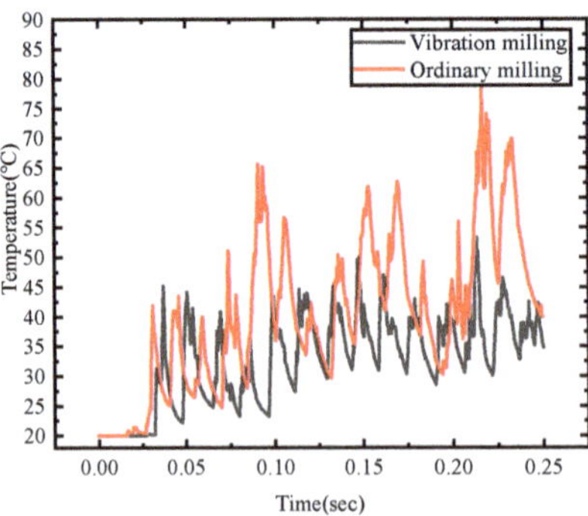

Figure 8. Temperature Distribution Diagram of Vibration Milling and Common Milling Cutters.

3.2.4. Comparison of Three-Way Milling Forces

Milling force is a type of resistance experienced by a milling cutter when cutting off the material allowance on the workpiece. It is the sum of the cutting forces on each cutting tooth that operates simultaneously. As the milling process progresses, its direction and size changes become more complex. Decomposing it into three main directions (X, Y, Z) is helpful in analyzing the impact of ultrasonic vibration on milling force.

Figure 9 shows the comparison of three-dimensional milling forces between vibration milling and ordinary milling. Overall, the milling force of ordinary milling is greater than that of vibration milling, and the milling process is intermittent. Therefore, during the ordinary milling process, the tool and workpiece suddenly come into contact multiple times, resulting in significant fluctuations in milling force. During vibration milling, chips are prone to detachment, reducing the resistance of chips to the tool. At the same time, the temperature in the milling area decreases and the tool always maintains good hardness, which is beneficial for reducing milling force. The overall milling force is: $F_X > F_Z > F_Y$. The vibration signal has a greater impact on the feed force and axial force and a smaller impact on the radial force. In this experiment, the overall milling width and depth are relatively small, so the milling force is also relatively small.

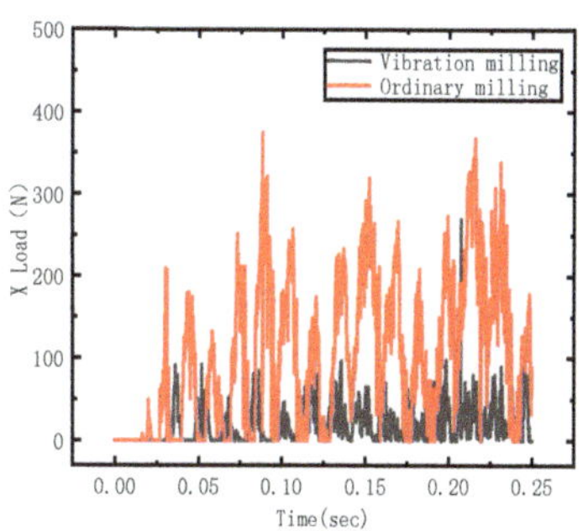

(a) Milling force in the X direction

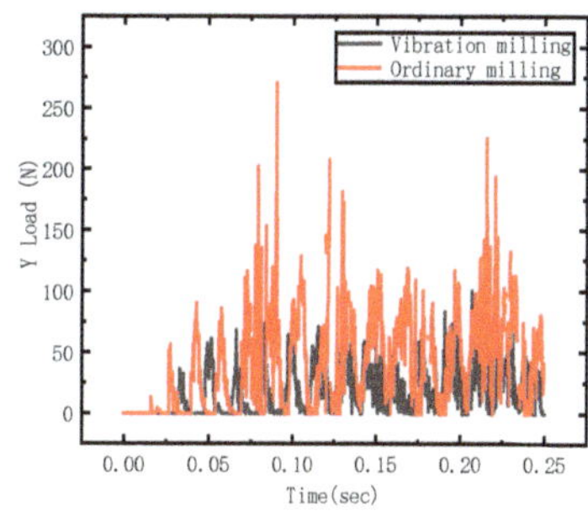

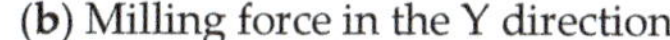

(b) Milling force in the Y direction

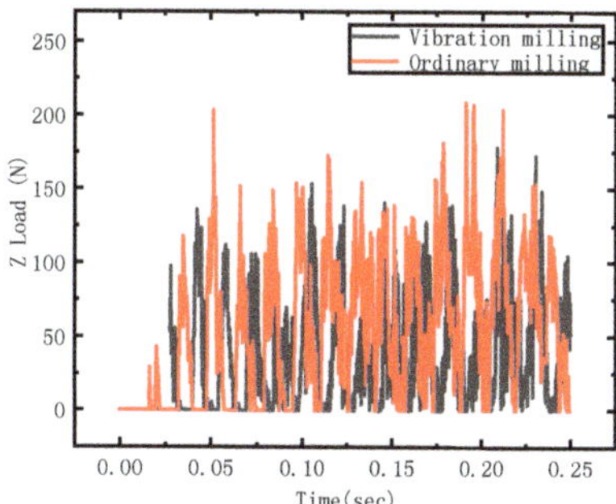

(c) Milling force in the Z direction

Figure 9. Trend of milling force between vibration milling and ordinary milling.

3.2.5. Comparison of Surface Stress of Workpieces

From Figure 10, it can be seen that the equivalent stress of the workpiece in ordinary milling is greater than that in vibration milling, and ordinary milling is more prone to severe mutations due to sudden contact and separation between the workpiece and the tool. The stress in vibration milling is relatively dense, and low-frequency vibration signals cause frequent stress fluctuations. Ultrasonic vibration milling reduces the equivalent stress on the surface of the workpiece, prevents stress concentration on the surface of the workpiece, and reduces the performance of the part, which is beneficial for enhancing the surface quality of the workpiece.

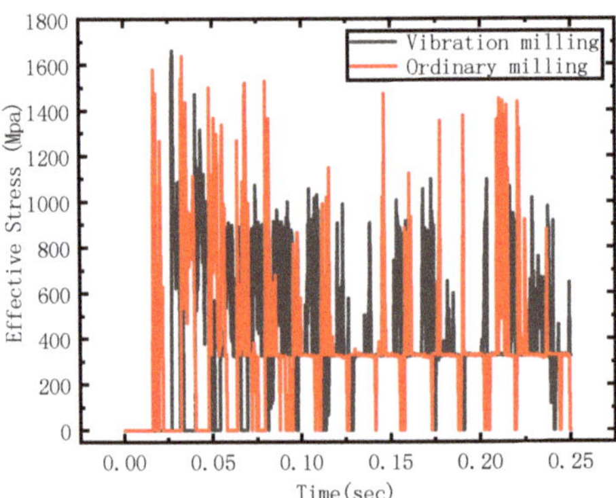

Figure 10. Equivalent Stress Trend of Vibration Milling and Ordinary Milling Workpieces.

3.2.6. Comparison of Tool Wear

As can be seen from Figure 11, when the tool travel is 10 mm, the wear amount of vibration milling is about 1/2 of that of ordinary milling. Because the chips produced by vibration milling are more likely to break, the chips are more regular, the milling process is more stable, and the tool temperature and cutting force in the cutting process are greatly reduced, which reduces the tool wear. In contrast, the temperature of the ordinary milling area is too high and the impact force of intermittent cutting is relatively large. The internal stress of the workpiece is concentrated and the deformation is more serious. The cutting force will increase and produce larger fluctuations, which will increase the tool wear.

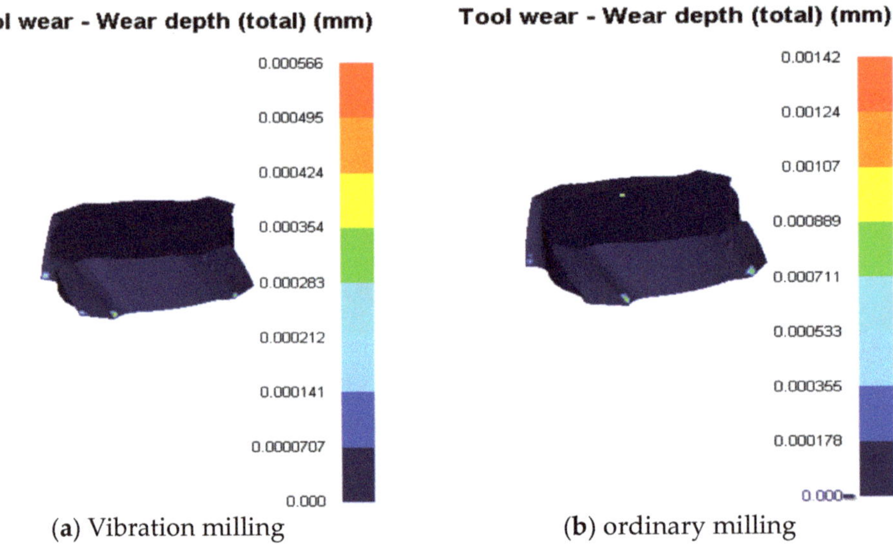

(a) Vibration milling   **(b) ordinary milling**

Figure 11. Analysis of tool wear in Vibration Milling and Ordinary Milling.

4. Effect of Coating Materials on Tool Wear in GH4169 Milling

4.1. Establishment of Simulation Model for GH4169 Milling with Coated Tools

Coated tools have high hardness, high wear resistance, high oxidation resistance, and strong film substrate bonding strength, which meet the requirements of today's high-precision and high-efficiency machining. Coated tools can improve the surface machining quality of the workpiece, reduce the milling force, reduce the surface temperature of the tool, and reduce tool wear. The physical properties of various common coatings have been encapsulated in the material library of DEFORM, and only need to be selected. Special structural coatings or newly developed coatings can be added to the material library based on their physical properties. The default bonding between coatings and between coatings and the substrate in the model is good. The coating thickness is 5 mm.

The following main study focuses on the influence of TIALN and TIN coatings on the milling process of hard alloy GH4169. This article sets the tool substrate material as WC, the axial milling depth of the tool is 2 mm, the spindle speed is 6000 r/min, and the feed rate per tooth is 0.1 mm/z. The distribution of coatings is shown in Figure 12:

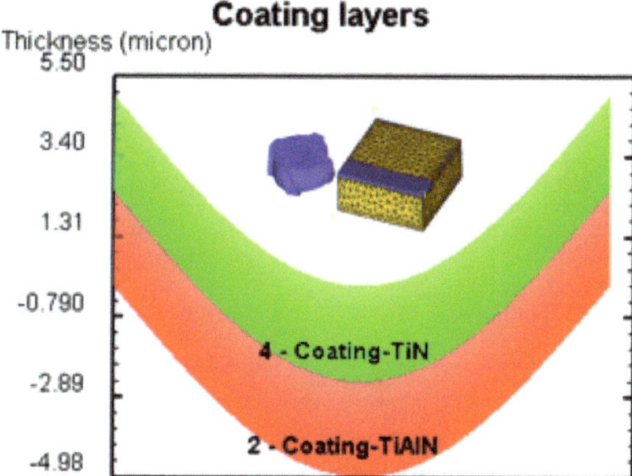

Figure 12. TIALN and TIN coating distribution diagram.

Based on existing material processing books [31], the performance parameters of TIN and TIALN coatings set in this article are shown in Table 5:

Table 5. Performance Table of TIALN and TIN Coatings.

Coating Types	TIALN	TIN
Hardness value (HV)	3300	2300
The coefficient of friction (N/s/mm/C)	0.4	0.49
Coefficient of heat conduction	10	23

4.2. Analysis of Simulation Results

4.2.1. Temperature Changes of Cutting Tools with Different Coating Types

From Figure 13, it can be seen that when the tool first comes into contact with the workpiece, the temperature difference between different types of coated and uncoated tools is not significant. At the beginning of cutting, the tool has less cutting material and less cutting force, and the temperature transmitted to the tool is relatively low. As cutting progresses, the contact area between the tool and the workpiece increases, the cutting force also increases, the temperature of the cutting area increases, and the heat transmitted to the tool also increases. The tool temperature undergoes periodic changes as chips are generated and shed. The temperature of uncoated tools is maximum. Due to the maximum thermal conductivity, a large amount of heat generated in the milling area is transmitted to the tool, causing a rapid increase in tool temperature. Secondly, the temperature of TIN-coated tools is highest, with TIALN-coated tools having the lowest temperature and TIN/TIALN-composite-coated tools having the lowest temperature. This coating has a soft and hard structure, greatly alleviating the stress on the tool. The composite coating can better prevent the temperature in the milling area from spreading inside the tool. It has a great protective effect on cutting tools and greatly reduces tool wear.

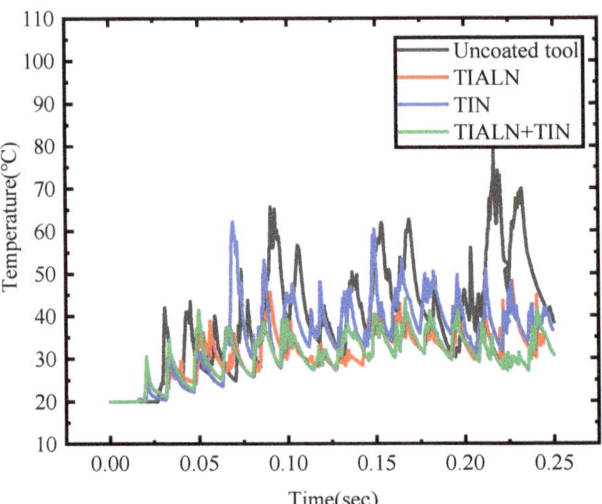

Figure 13. Temperature variation trend of cutting tools with different coating types.

4.2.2. Changes in Milling Force of Different Coating Types of Cutting Tools

From Figure 14, it can be seen that in the early stage of milling, the tool and workpiece just come into contact, the cutting distance is short, the temperature in the milling area is low, the milling force is also small, and the coating material has not yet been effective. With the progress of cutting, the temperature in the milling area increases and the thermal conductivity of the coating material is small, resulting in less heat entering the tool. The high temperature in the milling area causes the workpiece material to soften, resulting in a decrease in milling force. The material cut by the uncoated tool has a higher hardness, and the tool itself has a lower hardness than the coated material, so the milling force is relatively large. The difference in friction coefficient between TIALN and TIN coatings is not significant, and the main factor affecting milling force is the thermal conductivity coefficient. During the milling process, it can be roughly seen that $F_{TIALN+TIN} < F_{TIALN} < F_{TIN} < F_{Uncoated}$, and the thermal conductivity of the composite coating is low. The soft and hard structure can serve as a good thermal barrier, with a greater degree of material softening and less milling force. Secondly, the milling force of TIALN-coated tools is smaller than that of TIN-coated tools, with uncoated tools having the highest milling force. At the same time, it can be seen that the fluctuation range of milling force for TIN + TIALN-coated tools is relatively small, indicating that the composite coating can reduce the impact force during the milling process and increase the stability of the milling process.

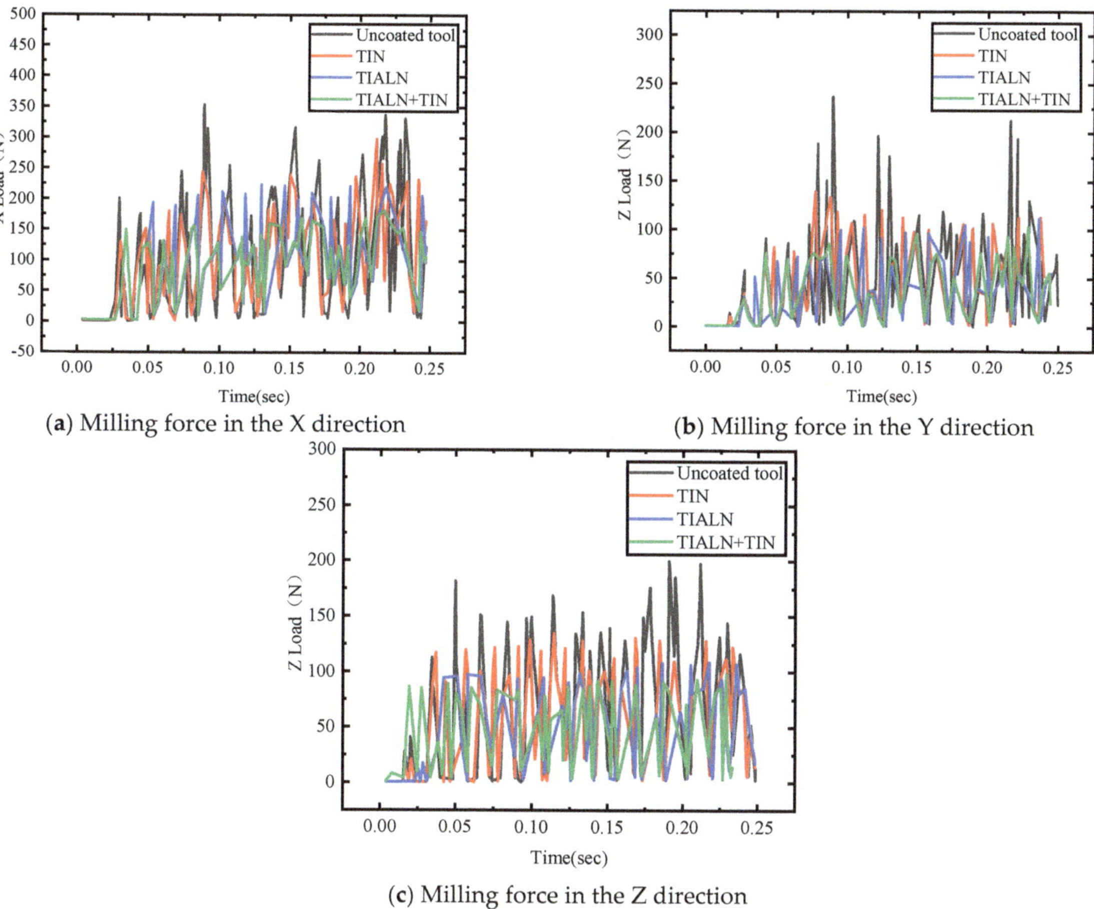

(**a**) Milling force in the X direction

(**b**) Milling force in the Y direction

(**c**) Milling force in the Z direction

Figure 14. Trend of Three Direction Milling Force Changes for Different Coating Types of Tools.

4.2.3. Comparison of Tool Wear of Different Coating Types

From Figure 15, it can be seen that when the milling stroke of the tool is 10 mm, the minimum wear of the TIN + TIALN-coated tool is 0.524 µm. Secondly, the TIALN coating tool wear is 0.576 µm. The wear amount of TIN-coated tools is 0.605 µm. The maximum uncoated tool wear is 1.42 µm. This simulation uses the Usui wear calculation formula. The tool wear is related to the temperature and pressure of the contact surface. TIN + TIALN-coated tools have good impact resistance and thermal insulation, and the milling force and tool temperature are relatively low, so the tool wear is relatively small. TIALN and TIN also reduce tool wear, which is due to the low thermal conductivity of the coating material, which maintains good hardness of the tools, and the tool temperature and cutting force are small. The impact of uncoated tools and workpiece materials is large, the workpiece surface hardens significantly, and the tool milling force and temperature are large, so the tool wear is large.

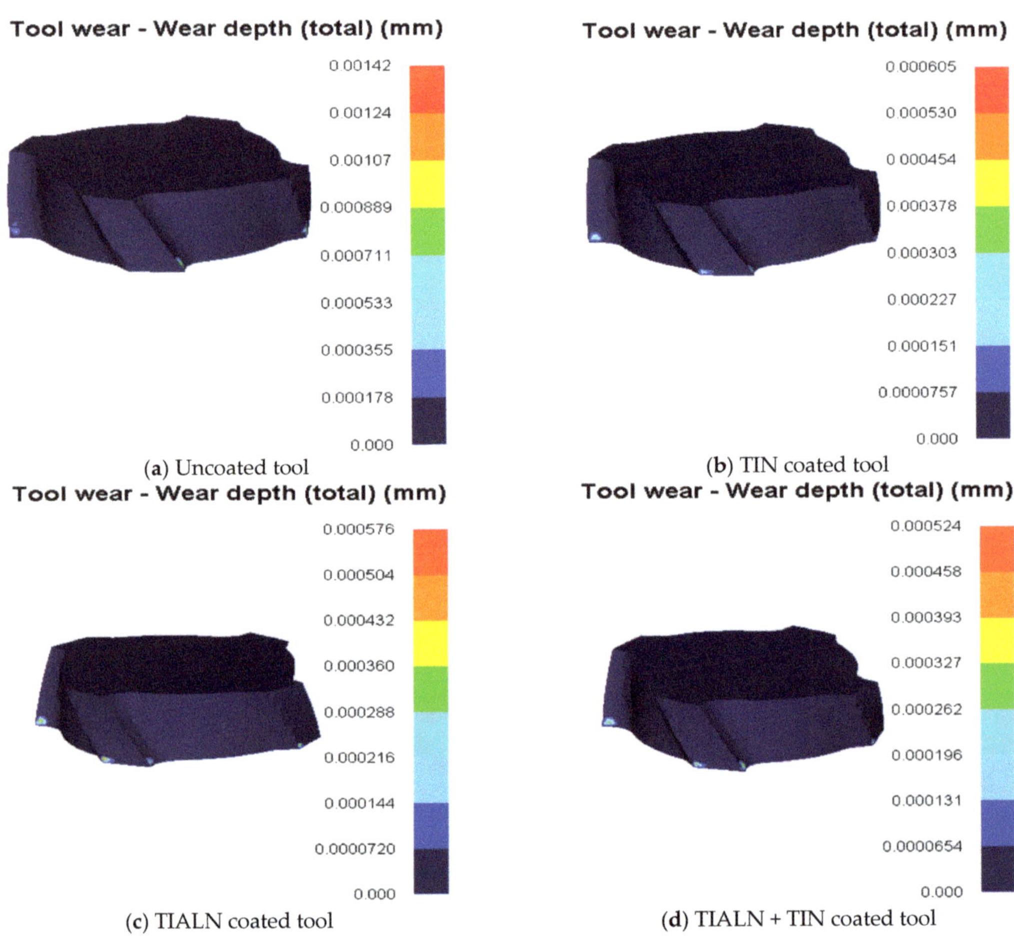

(a) Uncoated tool

(b) TIN coated tool

(c) TIALN coated tool

(d) TIALN + TIN coated tool

Figure 15. Distribution of maximum wear of different types of coated tools.

5. Conclusions

(1) Using the finite element simulation software DEFORM, the GH4169 model for carbide milling is established. The orthogonal experiments of tool wear simulation on n, a_p, and f_z parameters are carried out, and the regression prediction model of tool wear is established. The range analysis of the model is conducted, and the simulation effect is good. The empirical formula of tool wear is $H = 0.01 \cdot n^{0.4112} \cdot a_p^{0.5345} \cdot f_z^{-0.6866}$.

(2) In the process of milling GH4169, the order of the influence of n, a_p, and f_z on tool wear is: $f_z > a_p > n$. Tool wear increases with the increase in n and a_p. When f_z increases, the tool wear decreases. Taking the minimum tool wear as the target, the optimal milling parameters are obtained as $n = 3000$ r/min, $a_p = 0.2$ mm, and $f_z = 0.4$ mm/z.

(3) Based on the path planning method, a simulation model of ultrasonic vibration milling GH4169 is established. Compared with ordinary milling, it is found that ultrasonic-vibration-assisted milling is beneficial to chip breaking, reducing the temperature of the milling area, milling force, workpiece surface stress, and cutting tool temperature, which is of great significance to reduce tool wear and optimize the quality of the machined surface.

(4) The GH4169 simulation model of coated tool milling is established. By analyzing the milling conditions of cemented carbide tools, TIALN-coated tools, TIN-coated tools, and TiAlN/TiN-composite-coated tools, the influence of coating materials on tool temperature,

milling force, and tool wear is obtained. The results show that coating materials greatly reduce the tool surface temperature, three-way milling force, and tool wear. The TiAlN/TiN composite coating has the best cutting performance and is more suitable for milling difficult-to-machine material GH4169.

Author Contributions: X.L. guided the overall design and analysis of the study, W.Z. conducted simulation analysis and wrote the manuscript, L.M. and Z.P. helped collect data and simulate. All authors have read and agreed to the published version of the manuscript.

Funding: This work was financially supported by the National Natural Science Foundation of Jilin Province, China (Grant No. 20200201065JC).

Data Availability Statement: Data available in a publicly accessible repository.

Acknowledgments: This work was financially supported by the National Natural Science Foundation of Jilin Province, China (Grant No. 20200201065JC). The authors would also like to thank the anonymous reviewers for their helpful comments. All authors have agreed.

Conflicts of Interest: The authors declare that they have no conflict of interest.

References

1. Ezugwu, E.O. High speed machining of aero-engine alloys. *J. Braz. Soc. Mech. Sci. Eng.* **2004**, *26*, 1–11. [CrossRef]
2. Mei, Z.; Huang, C.; Fan, Y. Development and prospect of the aircaft digital assembly technology. *Acronautical Manuf. Technol.* **2015**, *58*, 31–37.
3. Chu, G. Research on the status quo of aircraft digi tal assembly techmology in China. *J. Xi'an Acronautical Univ.* **2016**, *1*, 40–43.
4. Ezugwu, E.Q.; Bonney, J.; Yamane, Y. An Overview of the Machinability of Aero Engine Alloys. *J. Mater. Process. Technol.* **2003**, *135*, 233–253. [CrossRef]
5. Dudzinski, D.; Devillez, A.; Moufki, A.; Larrouquère, D.; Zerrouki, V.; Vigneau, J. A review of developments towards dry and high speed machining of Inconel 718 alloy. *Int. J. Mach. Tools Manuf.* **2004**, *44*, 439–456. [CrossRef]
6. Attanasio, A.; Ceretti, E.; Fiorentino, A.; Cappellini, C.; Giardini, C. Investigation and FEM-based simulation of tool wear in turning operations with uncoated carbide tools. *Wear* **2010**, *269*, 344–350. [CrossRef]
7. Jianxin, D.; Yousheng, L.; Wenlong, S. Difusion wear in dry cutting ofTi-6Al-4V with WC/Co carbide tools. *Wear* **2008**, *265*, 1776–1783. [CrossRef]
8. List, G.; Sutter, G.; Bouthiche, A. Cutting temperature prediction in high speed machining by numerical modelling of chip formation and its dependence with crater wear. *Int. J. Mach. Tools Manuf.* **2012**, *54–55*, 1–9. [CrossRef]
9. Li, B.; Zhang, H.; Wang, H.; Chang, J.; Sun, X. Numerical simulation technology on tool wear of metal cutting. In Proceedings of the International Conference on Advanced Technology of Design & Manufacture, Beijing, China, 23–25 November 2010.
10. Chemn, G.; Chang, Y. Using two-dimensional vibration cutting for micro-milling. *Int. J. Mach. Tools Manuf.* **2006**, *46*, 659–666.
11. Hsu, C.Y.; Huang, C.K.; Wu, C.Y. Milling of MAR -M247 nickel-based superalloy with high temperature and ultrasonic aiding. *Int. J. Adv. Manuf. Technol.* **2007**, *34*, 857–866. [CrossRef]
12. Suárez, A.; Veiga, F.; de Lacalle, L.N.L.; Polvorosa, R.; Lutze, S.; Wretland, A. Effects of ultrasonics-assisted face milling on surface integrity and fatigue Life of Ni- Alloy 718. *J. Mater. Eng. Perform.* **2016**, *25*, 5076–5086. [CrossRef]
13. Biermann, D.; Kersting, P.; Surmann, T. A general approach to simulating workpiece vibrations during five-axis milling of turbine blades. *CIRP Ann. Manuf. Technol.* **2010**, *59*, 125–128. [CrossRef]
14. Shen, X.H.; Zhang, J.H.; Yin, T.J.; Dong, C.J. A study on cutting force in micro end milling with ultrasonic vibration. *Adv. Mater. Res.* **2010**, *905*, 1910–1914. [CrossRef]
15. Shen, X.H.; Wang, M.Y. Finite element analysis of temperature field in vibration milling. *Key Eng. Mater.* **2016**, *693*, 1030–1037. [CrossRef]
16. Shen, X.H.; Zhang, J.H.; Li, H.; Wang, J.J.; Wang, X.C. Ultrasonic vibration- assisted milling of aluminum alloy. *Int. J. Adv. Manuf. Technol.* **2012**, *63*, 41–49. [CrossRef]
17. Ni, C.H.; Zhu, L.; Yang, Z.H. Comparative investigation of tool wear mechanism and corresponding machined surface characterization in feed-direction ultrasonic vibration assisted milling of Ti-6Al-4V from dynamic view. *Wear* **2019**, *436*, 203006. [CrossRef]
18. Sajjady, S.A.; Abadi, H.N.H.; Amini, S.; Nosouhi, R. Analytical and experimental study of topography of surface texture in ultrasonic vibration assisted turming. *Mater. Des.* **2016**, *93*, 311–323. [CrossRef]
19. Tao, G.; Ma, C.; Bai, L.; Shen, X.; Zhang, J. Feed-direction ultrasonic vibration. assisted milling surface texture formation. *Mater. Manuf. Process* **2017**, *32*, 193–198. [CrossRef]
20. Chen, G.; Ren, C.; Zou, Y.; Qin, X.; Lu, L.; Li, S. Mechanism for material removal in ultrasonic vibration helical milling of Ti6Al4V alloy. *Int. J. Mach. Tools Manuf.* **2019**, *138*, 1–13. [CrossRef]
21. Byme, G.; Dornfeld, D.; Denkena, B. Advancing cutting technology. *CIRP Ann. Manuf. Technol.* **2003**, *52*, 483–507.

22. Pawade, R.S.; Sonawane, H.A.; Joshi, S.S. An analytical modeling to predict specific shear energy in high-speed turming of Inconel 718. *Int. J. Mach. Tools Manuf.* **2009**, *49*, 979–990. [CrossRef]
23. Fang, N.; Wu, Q. A comparative study of the cutting forces in high speed machining ofTi-6AI-4V and Inconel718 with around cutting edge tool. *J. Mater.* **2009**, *209*, 4385–4389.
24. Chen, R. *Principles of Metal Cutting*; Machine Press: Beijing, China, 2007.
25. Wu, J.; Han, R. *New Technologies in Modern Mechanical Processing*; Publishing House of Electronics Industry: Beijing, China, 2017.
26. Wang, Q. Research and Application of Functional Principal Component Analysis and Functional Linear Regression Model. Diploma Thesis, Chongqing Technology and Business University, Chongqing, China, 2020.
27. Ma, Y.; Yue, Y. Research on Surface Roughness Prediction Model of T-itanium Alloy TC25 Milling. *Manuf. Technol. Mach. Tool* **2020**, *8*, 141–145. [CrossRef]
28. Liu, A. *SPSS Basic Analysis Tutorial*; Peking University Press: Beijing, China, 2014.
29. Peng, D. Prediction of Milling Process and Analysis of Surface Quality for TC4. Diploma Thesis, Tianjin University, Tianjin, China, 2012.
30. Yang, Y. Numerical Simulation and Experimental Study on Axial Ultrasonic Vibration Assisted Milling of Titanium Alloy. Diploma Thesis, Nanjing University of Science and Technology, Nanjing, China, 2021.
31. Liu, Z.; Zhang, C.; Ren, J. *Dry Cutting Technology and Its Application*; China Machine Press: Beijing, China, 2005.

 *lubricants*

MDPI

Article

Multi-Objective Optimization Design of Micro-Texture Parameters of Tool for Cutting GH4169 during Spray Cooling

Xinmin Feng *, Xiwen Fan, Jingshu Hu and Jiaxuan Wei

Key Laboratory of Advanced Manufacturing Intelligent Technology, Ministry of Education, Harbin University of Science and Technology, Harbin 150080, China; 2120110122@stu.hrbust.edu.cn (X.F.)
* Correspondence: fxmin7301@hrbust.edu.cn; Tel.: +86-136-3364-0821

Abstract: This study explores the performance of micro-textured tools when cutting GH4169 during spray cooling. First, the morphologies of the micro-textures were selected according to the simulation and experiments. Secondly, cutting experiments were carried out during spray cooling. As appropriate for each experiment, regression models of cutting force, cutting temperature, or tool wear area were established, and variance analysis was conducted. The cutting force, cutting temperature, and tool wear area functions were obtained from the respective regression models. Based on these functions, the micro-texture parameters were optimized using the response surface method with the cutting force, cutting temperature, and rake face wear area as the objectives. Finally, a full factor experiment on the micro-texture parameters was designed using Minitab, and cutting experiments were conducted using micro-textured tools with these parameters. Taking a relatively low cutting force, cutting temperature, and tool wear as the objectives, a genetic algorithm multi-objective optimization model for the micro-texture parameters of the tools was established, and the model was solved using the NSGA-II algorithm to obtain a Pareto solution set and micro-texture parameters with a good, comprehensive cutting performance. The micro-texture morphology and parameters obtained in this study can also be used for cutting other high-temperature alloy materials with similar properties to GH4169. This research method can also be used to optimize micro-textured tools for cutting other materials.

Keywords: micro-texture parameters; GH4169; multi-objective optimization; response surface method; genetic algorithm

Citation: Feng, X.; Fan, X.; Hu, J.; Wei, J. Multi-Objective Optimization Design of Micro-Texture Parameters of Tool for Cutting GH4169 during Spray Cooling. *Lubricants* **2023**, *11*, 249. https://doi.org/10.3390/lubricants11060249

Received: 13 May 2023
Revised: 27 May 2023
Accepted: 28 May 2023
Published: 6 June 2023

1. Introduction

The nickel-based superalloy GH4169 has very stable physical and chemical properties, high creep resistance, and thermal stability [1,2]; however, due to its excellent performance, GH4169 is a material that is difficult to machine. In order to solve the problem of the difficult machining of GH4169, some scholars have started their research from the perspective of the cutting environment and applied different cooling and lubrication technologies, such as spray cooling [3], high-pressure cooling [4], liquid nitrogen cooling [5], and low-temperature cold [6], in the process of machining to obtain good machinability.

In recent years, with the continuous development of science and bionics technology, the method of machining micro-textures on the surface of tools to improve their cutting performance has been favored by most scholars. Additionally, these types of tools with micro-textures are collectively called "micro-textured tools". The existence of micro-textures can effectively reduce the contact area between the tool and the chip being cut, thereby improving the cutting performance of the tool [7–10]. Moreover, when the tool is cutting in certain lubrication conditions, the advantages of the micro-texture will be more obvious.

Scholars at home and abroad have conducted extensive research on the morphologies, size parameters, and arrangements of the micro-textures of micro-textured tools.

Some scholars have studied micro-texture morphologies that are suitable for different workpiece materials. For instance, Ali, Shafahat et al. designed micro-textured tools with rectangular and triangular microgrooves for cutting stainless steel (AISI60), simulated and analyzed the performance of the two types of micro-texture shapes, and found that the rectangular microgroove-textured tools had a better performance [11]. Tu et al. used laser technology to machine micropits and microgrooves on the rake face of Al_2O_3/TiC ceramic tools. They studied the cutting performances of two types of micro-textured self-lubricating tools on hardened steel and compared them with traditional non-textured ceramic tools. The results indicated that the reduction in the cutting force of the micropit-textured tools was greater than that of the microgroove-textured tools [12]. Rajurkar, Avadhoot et al. analyzed the performance of two types of micro-textured tools for cutting chromium nickel iron alloy 718 in different cutting conditions. The results indicated that the microchannel-textured tools showed a notable improvement in tool life up to 60% over micro dimple-textured tools at lower cutting speeds; however, almost the same tool life was found with both tools at higher cutting speeds [13]. Sagar Dhage et al. designed micro-textures in different directions on an insert and used these micro-textured inserts to cut AISI 1045 in dry cutting conditions. From the measured cutting force, it was found that the cutting force of the insert with a micro-texture parallel to the cutting edge was significantly lower than that of the insert with a micro-texture perpendicular to the cutting edge [14].

Moreover, some scholars have also explored the influence of micro-texture parameters on the cutting performance of tools. Li Kun et al. used hard alloy micro-textured tools to conduct a simulation analysis of the cutting of Al7075-T6 aluminum alloy. The results showed that the parameters of the micro-texture greatly reduced the tool wear. The optimized micro-texture parameters with the best cutting performance were as follows: width, 40 μm; pitch, 80 μm; depth, 20 μm; spacing, 70 μm [15]. Liu Yayun et al. designed carbide micro-textured tools with different texture parameters and conducted cutting experiments on Al_2O_3. From their analysis, the optimal micro-texture parameters that could achieve the optimal wear resistance and workpiece surface quality were as follows: the width of the micro-texture was 75 μm, the spacing was 100 μm, and the pitch was 75 μm [16]. V. Sharma et al. designed micro-textured tools and analyzed the influence of the micro-texture parameters on the cutting force when cutting 4340 hardened steels, finally obtaining the optimal parameters of the micro-textured tools [17]. Through an experiment, Guo et al. analyzed the influence of microgroove parameters, such as the width, depth, and spacing, on the cutting force when cutting TC4 [18].

The above analysis indicates that, at present, the research on the micro-texture parameters of inserts mainly focuses on cutting aluminum alloy and titanium alloy and that the micro-texture parameters of inserts for nickel-based superalloy have not yet been studied. This paper studied the micro-texture parameters of micro-textured tools for cutting the GH4169 superalloy.

Research on the arrangement of micro-textures has also been reported in some studies. Tong Xin et al. studied the influence of variable-density micro-textured ball-end milling cutters on cutting titanium alloy and used a fuzzy algorithm to optimize the distribution of the ball-end milling cutters, obtaining variable spacings of 200 μm, 150 μm, and 175 μm for the studied micro-textures [19]. D. Arulkirubakaran et al. simulated the cutting process of Ti-6AL-4V using WC/Co micro-textured tools while using DEFORM with SAE 40 as a semi-solid lubricant. Combined with turning experiments, it was found that the cutting temperature of the micro-textured tool decreased to varying degrees during the cutting process, and this effect was more pronounced when cutting with vertical micro-textured tools [20]. Wang et al. conducted a simulation cutting of AISI 1045 carbon steel by using micro-textured tools and concluded that compared with non-textured and micro-textured tools, the cooling effect of the lateral micro-texture was more significant, and it exhibited a good chip-breaking performance during the cutting process [21].

From the above analysis, it can be seen that the current research on the arrangement of micro-textures uses a linear arrangement parallel or perpendicular to the cutting edge, and there is no research on arc-shaped arrangements. The research on these arrangement methods also targets aluminum alloy and titanium alloy workpieces. Research on the micro-texture arrangement of tools for GH4169 has not been conducted yet.

The optimization of micro-texture parameters has also been studied. Based on Oxely's analysis model, Kishawy and Hossam A. optimized the microstructure parameters and obtained the microstructure parameters when cutting an AISI 1045 steel pipe with ceramic microgroove-textured tools, focusing on the cutting force. In addition, the examination showed no evidence of derivative cutting when using the optimized micro-textured insert, which also proved the efficacy of the proposed model [22]. Cheng Yaotian et al. simulated and experimentally analyzed the cutting force and depth of the damage under the surface of the workpiece when cutting carbon-fiber-reinforced plastic with microgroove-textured tools. The surface response method was used to optimize the groove parameters, and the microgroove parameters with the lowest cutting force and depth of damage under the surface were obtained [23]. Li Binbin et al. simulated and analyzed the cutting force of Inconel718 with sinusoidal-groove-textured tools and optimized the micro-texture parameters with the minimum cutting force by combining range analysis and effect analysis [24]. Yu Yinghua et al. studied the influence of elliptical-opening offset parabolic micro-texture feature parameters on the cutting performance of a tool for cutting Ti6Al4V based on the response surface methodology. Through neural networks and genetic algorithms, the micro-texture parameters were optimized with the main cutting force and cutting temperature as the targets, obtaining a set of optimal parameters, resulting in a reduction of 11.24% and 15.28% in the cutting force and temperature, respectively [25].

The results of different optimization methods may not necessarily be the same, and current research has mostly focused on single-objective optimization of micro-texture parameters using a single method. This paper considered the comprehensive performance of the studied tools—that is, taking the cutting force, cutting temperature, and tool wear as the targets, and considering the applicability and calculation efficiency of the optimization algorithm—and used two methods to carry out multi-objective optimization of micro-textured tools for cutting GH4169.

Based on the research of micro-textured tools mentioned above, the existing research achievements and further research on micro-textured cutting tools can be summarized as follows:

(1) Micro-textures with reasonable morphologies and size parameters on the rake/flank surfaces of the tool can reduce the contact between the tool and the chip in the cutting process. The existence of micro-textures reduces the total contact area between the tool and chip and improves the cutting performance.

(2) For the cutting performance of micro-textured tools, combinations of different morphologies and size parameters will have a positive or negative impact on the cutting performance. Generally speaking, reasonable morphologies and parameters of micro-textures can significantly improve the cutting performance of the tool; conversely, they may lead to a decrease in the cutting performance of the tool due to secondary cutting.

(3) For different workpiece materials, various micro-texture morphologies and size parameters can be used to achieve a good performance. At present, there is a lot of research on micro-textured tools for titanium, but there is relatively little research on micro-textured tools for nickel-based superalloys. The morphology that is suitable for titanium alloys is not necessarily suitable for aluminum alloys or nickel-based superalloys. For different variations of a single workpiece material, even if they have the same micro-texture morphology but different size parameters, their cutting performance will not be the same. The size parameters of a micro-texture that provide the most significant reduction in the cutting force may not necessarily have the best effect on the cutting temperature reduction or produce a good surface quality of the workpiece.

(4) Most scholars have only studied a single morphology of micro-textures or the impact of a single parameter of micro-textures on the cutting performance, rarely mentioning

the study on the effect of the interaction effect of the micro-texture parameters on the cutting performance.

Based on the above analysis, in order to improve the process ability and tool life when cutting the nickel-based superalloy GH4169, micro-textured tools were designed to cut GH4169 during spray cooling. This paper explored the cutting performance of micro-textured tools with different morphologies and parameter combinations through experimental and simulation analyses during spray cooling. The response surface method and genetic algorithm were used in the multi-objective optimization of the micro-texture parameters to obtain suitable micro-texture morphologies and size parameters for cutting GH4169, providing the tools with a good, comprehensive performance.

2. Experiment of Micro-Textured Tools for Cutting GH4169

2.1. Design and Processing of Micro-Textured Tools

Five types of micro-textured tools with different morphologies were designed and processed for the cutting experiments. Partial larger views of the five micro-textured tools with different morphologies are shown in Figure 1. T1 is a micropit-textured tool that has three rows of lattice-shaped micropits parallel to the tool nose arc on the rake face; T2, a micro-wave-groove-textured tool that has five rows of corrugated grooves parallel to the arc of the tool nose on the rake face; T3, a micro-parallel-groove-textured tool that has four rows of straight grooves that are approximately parallel to the arc of the tool nose on the rake face; T4, a micro-elliptic-line-groove-textured tool that has mixed microgrooves with a combination of circular and straight grooves on the rake face; and T5, a micro-arc-groove-textured tool that has five rows of arc grooves parallel to the tool nose arc on the rake face.

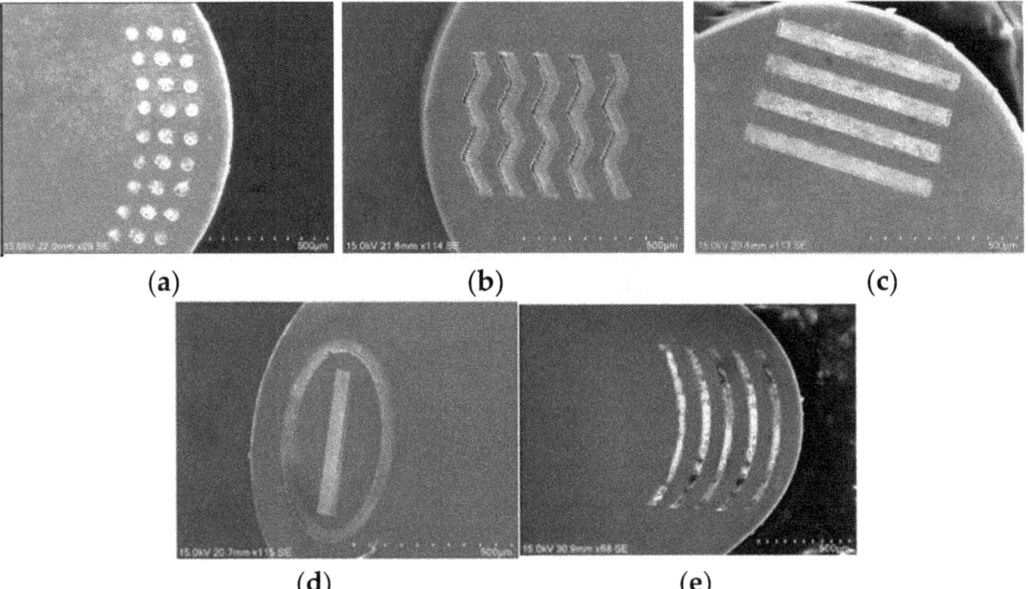

Figure 1. Partial larger views of the nose of the micro-textured tools: (**a**) micropit; (**b**) micro-wave-groove; (**c**) micro-parallel-groove; (**d**) micro-elliptic-line-groove; (**e**) micro-arc-groove.

In order to analyze the performance of the five morphologies, a simulation model of the micro-textured tools when cutting GH4169 was established using the same cutting micro-textured tools when cutting GH4169. The cutting force, cutting temperature, and wear depth of the rake face of the tools were obtained. The cutting force is shown in

Figure 2, the cutting temperature is shown in Figure 3, and the cloud map of the wear depth of the tool rake face is shown in Figure 4. The simulated average cutting force, average cutting temperature, and tool wear depth are shown in Table 1.

From Figure 2, it can be seen that compared with the non-textured tool, the micropit-textured tool had the lowest cutting force, followed by the micro-parallel-groove-textured tool. From Figure 3, it can be seen that, compared with the non-textured tool, the micro-parallel-groove-textured tool had the lowest cutting temperature, followed by the micropit-textured tool. From Figure 3, it can be seen that, compared with the non-textured tool, the micropit-textured tool had the lowest wear depth, with a maximum value of 0.182 mm, followed by the micro-parallel-groove-textured tool.

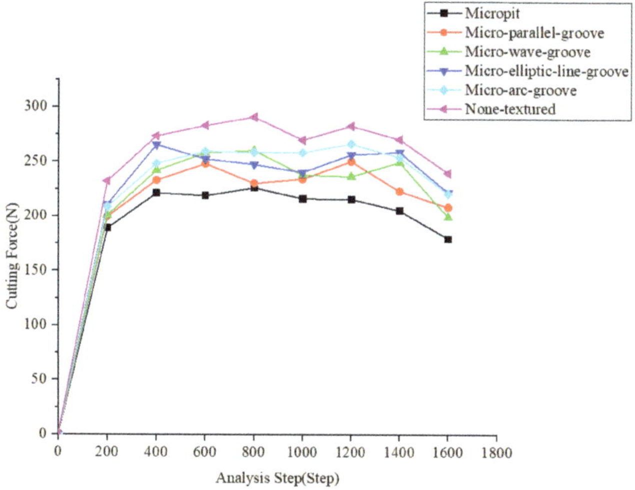

Figure 2. Cutting forces of the five micro-textured tools and a non-textured tool.

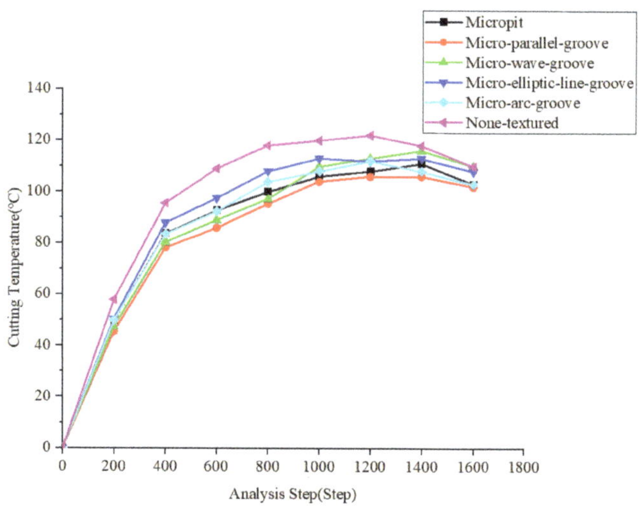

Figure 3. Cutting temperatures of the five micro-textured tools and a non-textured tool.

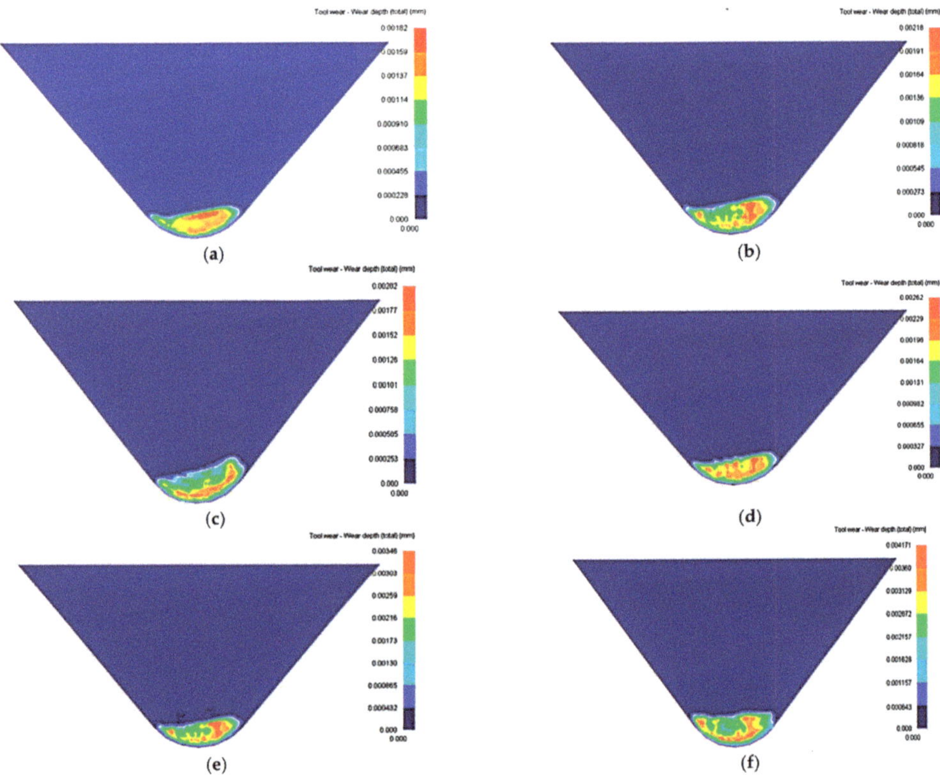

Figure 4. Wear depths of the tools: (**a**) micropit-textured tool; (**b**) micro-parallel-groove-textured tool; (**c**) micro-wave-groove-textured tool; (**d**) micro-elliptic-line-groove-textured tool; (**e**) micro-arc-groove-textured tool; (**f**) non-textured tool.

Table 1. Simulation results of the micro-textured tools with different morphologies.

Micro-Texture Morphology	Average Cutting Temperature (°C)	Average Cutting Force (N)	Wear Depth (mm)
Micropit	115	205.23	0.00182
Micro-parallel-groove	105	232.82	0.00218
Micro-wave-groove	108	247.48	0.00202
Micro-elliptic-line-groove	109	259.15	0.00262
Micro-arc-groove	115	256.41	0.00346
Non-textured	141	272.8	0.00417

From Table 1, it can be seen that among the five types of micro-textured tools, the cutting force and wear depth of the micropit-textured tool were lower than those of the other micro-textured tools. This result can be explained as follows: Among the five micro-textured tools with different morphologies, the micropits in the micropit-textured tool are distributed in a lattice. Compared with the other micro-textured tools, the contact area of the tool chips is smaller. It is believed that the friction force is proportional to the contact area of the tool chips; therefore, the friction force of the micropit-textured tool is lower, which results in a lower wear and cutting force compared with those of the other micro-textured tools with different morphologies. From Table 1, it can also be seen that the cutting temperature of the micro-parallel-groove-textured tool was lower compared with that of the other micro-textured tools.

Based on the above results, it was ultimately determined that the cutting performance of the micropit-textured tool was better than that of the other micro-textured and non-textured tools. Therefore, micropit-textured tools were used as the research object for the multi-objective optimization. Different parameters of the micropit-textured tools were processed, and cutting experiments were carried out. Through multi-objective optimization methods, the micro-texture parameters were optimized to obtain a good, comprehensive performance of the tools.

2.2. Experiment of Micro-Textured Tool for Cutting GH4169

Before the multi-objective optimization, cutting experiments were conducted on the micro-textured tools when cutting GH4169. The machine tools and equipment used in the experiment were as follows: CKA6140 CNC lathe; CNMA120408-cemented carbide inserts with a micropit texture on their rake face; DCLNR 2525 M12 tool shank; GH4169 bars with a diameter of 120 mm and a length of 300 mm; three-axis dynamic piezoelectric force measuring instrument (Kistler 9139AA) to measure the force; WPNK-191-armored thermocouple to measure the cutting temperature; KQ2200DE CNC ultrasonic cleaner (to clean the inserts); and model OoW129S composite spray cooling system (to spray atomized cutting fluid into the cutting area for cooling). The setup of the experiment is shown in Figure 5.

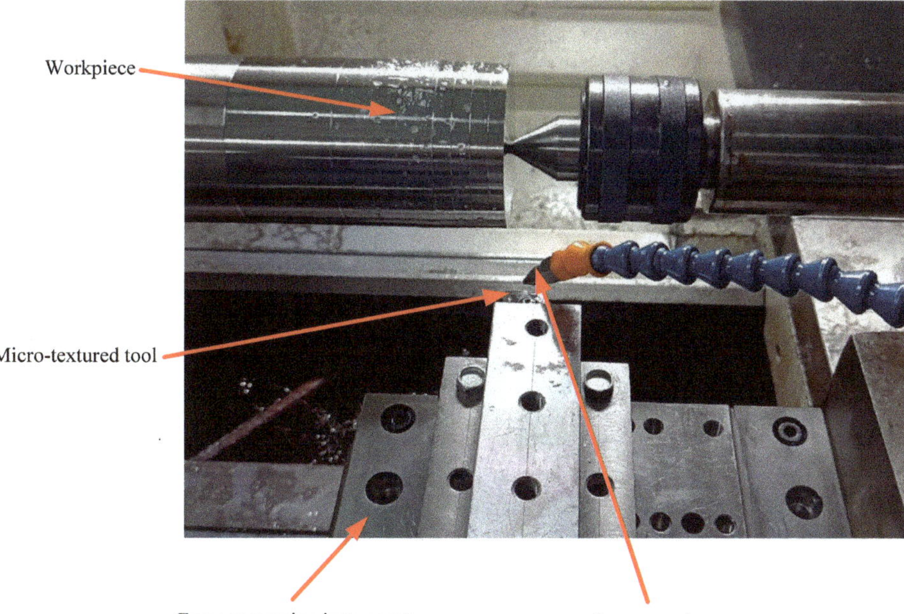

Figure 5. Machine tools and testing equipment used in the experiments.

3. Response Surface Method for Multi-Objective Optimization of Micro-Texture Parameters of Tools

3.1. Cutting Experiment and Result Analysis

Before conducting the multi-objective optimization, the first step was to determine the experimental method. In order to minimize the optimal costs and improve the optimal efficiency, this paper used the Box–Behnken design (BBD) method to design the micro-texture parameters [26].

Before conducting the experimental design, it was necessary to determine the size range of the micro-texture parameters. In this paper, the size ranges of the micropit texture parameters were as follows: the edge distance (parameter A) was 60–100 μm, the diameter (parameter B) was 40–60 μm, and the spacing (parameter C) was 100–140 μm. The division of the three parameter levels of the micro-texture is shown in Table 2. The meaning of the parameters is shown in Figure 6. The parameters of the micropit texture are shown in Table 3.

Table 2. Micro-texture parameter levels.

Level/Factor	A (μm)	B (μm)	C (μm)
−1	60	40	100
0	80	50	120
1	100	60	140

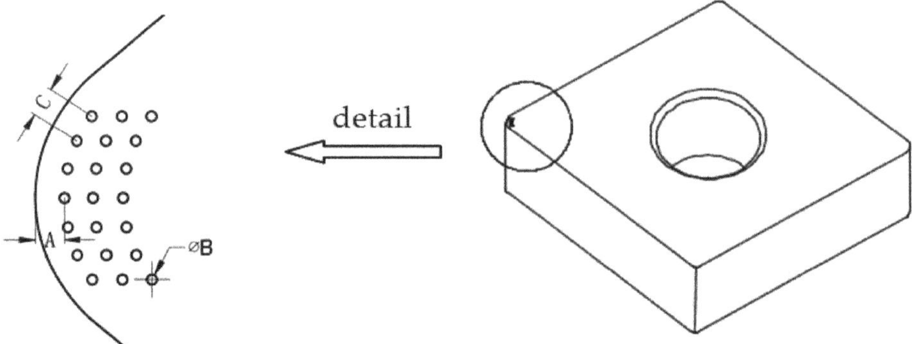

Figure 6. Schematic diagram of the micro-texture parameters.

Table 3. Combination scheme of micropit texture parameters.

NO.	A (μm)	B (μm)	C (μm)
1	80	50	120
2	60	50	100
3	80	50	120
4	100	50	100
5	80	50	120
6	80	50	120
7	60	40	120
8	80	40	140
9	80	60	140
10	100	60	120
11	60	50	140
12	100	40	120
13	60	60	120
14	80	50	120
15	80	60	100
16	100	50	140
17	80	40	100

Based on different combinations of the micropit parameters presented in the previous text, a femtosecond laser was used to process the micropit texture on the rake of the tools. Through preliminary experiments and simulations by the research group, it was found that the relatively reasonable depth of the micro-texture was between 20 and 30 μm; thus, the micro-texture depths processed in this article were all 30 μm. Images of the micropit-textured tools collected during scanning electron microscopy are shown in Figure 7.

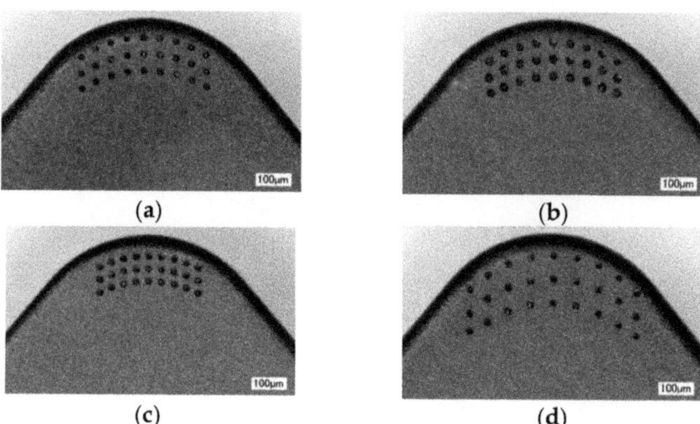

(a)   (b)   (c)   (d)

Figure 7. Partial larger views of micropit-textured tools with deferent parameters (A-B-C) processed using a femtosecond laser: (**a**) 60-50-120; (**b**) 80-60-120; (**c**) 80-50-100; (**d**) 100-40-140.

The cutting experiments were conducted at a cutting speed of 80 m/min, a feed rate of 0.2 mm/r, and a cutting depth of 0.2 mm. Moreover, the time intervals were 10–20 s, and the average cutting force and temperature of the tools under different combinations of micro-texture parameters were obtained, as shown in Table 4. In order to analyze the influence of the micro-texture parameters on the tool wear area, the micropit-textured tools were observed under a scanning electron microscope after the experiment, and it was found that the wear patterns of the tool rake faces were different and irregular. This paper used the rectangular equivalent method (micro-element method) to calculate the wear areas of the different shapes of each tool's rake face; an irregular wear area is equivalent to a regular rectangular shape. In this method, the length and width of the rectangle are measured, and the area of the rectangle equivalent to the wear area of the rake face of the tool is calculated. The calculated wear areas of the tools' rake faces are shown in Table 4, and enlarged views of the partial tool wear after the cutting experiments are shown in Figure 8.

Table 4. Cutting force, temperature, and wear area of the cutting tools under different micro-texture parameters.

No.	A (μm)	B (μm)	C (μm)	Cutting Force (N)	Cutting Temperature (°C)	Wear Area (μm²)
0	-	-	-	240.99	138	109,938
1	80	50	120	233.03	123	81,730
2	100	40	120	238.35	123	92,069
3	60	40	120	240.31	112	101,138
4	60	50	140	249.33	116	87,024
5	80	40	140	250.35	130	93,261
6	60	50	100	233.38	108	82,891
7	80	60	140	262.89	128	80,070
8	80	50	120	230.47	125	81,730
9	80	50	120	228.09	128	81,730
10	80	60	100	211.08	109	72,021
11	80	50	120	236.2	125	81,730
12	100	60	120	201.42	108	70,709
13	80	40	100	239.38	118	95,499
14	100	50	140	243.32	102	81,427
15	60	60	120	229.4	124	75,050
16	80	50	120	231.47	126	81,730
17	100	50	100	228.49	114	76,024

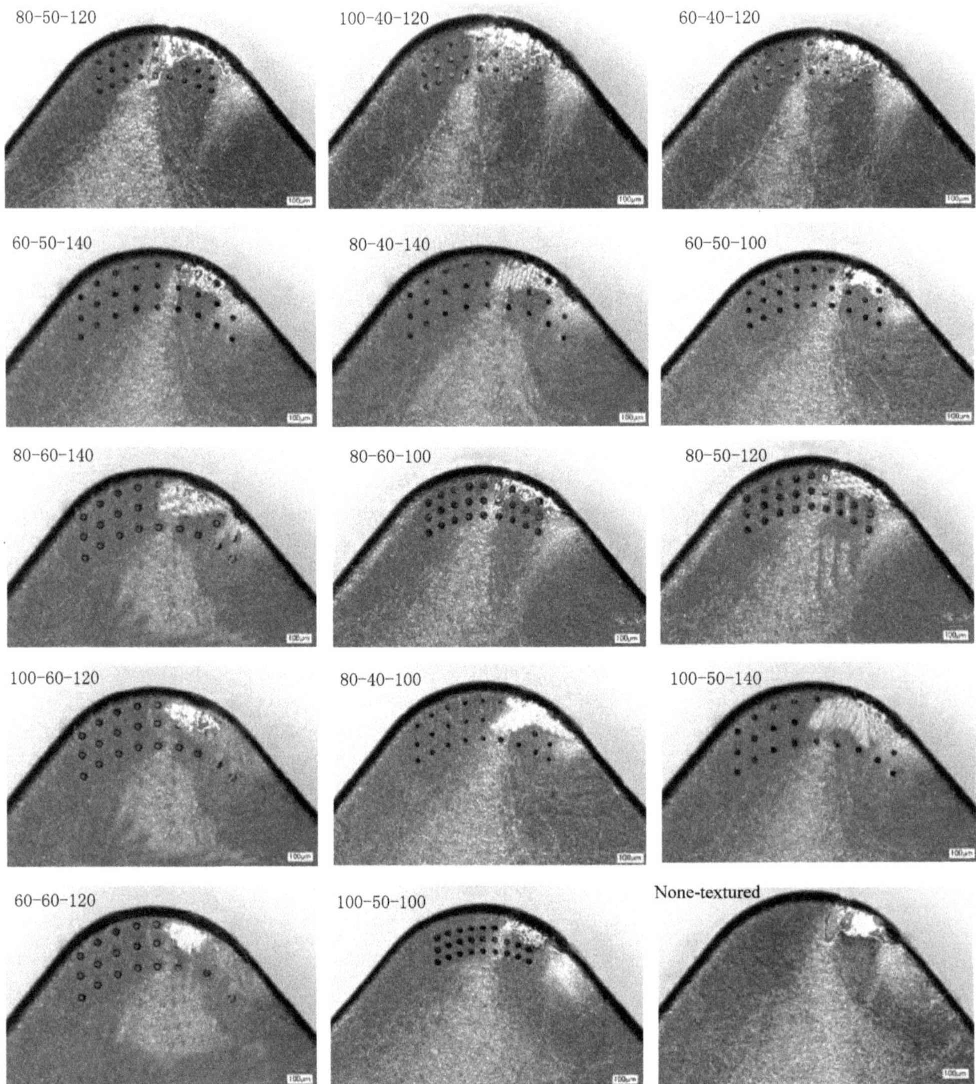

Figure 8. Wear of the rake faces of tools with different micro-texture parameters (the number in the upper right corner of the figure represents the value of parameters A, B, and C of the micro-texture).

From Table 4, it can be seen that the cutting force of tool No. 12 in the experiment was the lowest; that is, when the edge distance A was 100 μm, the diameter B was 60 μm, the spacing C was 120 μm, and the cutting force of the micropit-textured tool reached its minimum value of 201.42 N. It can also be seen that the cutting temperature of tool No. 14 in the experiment was the lowest; that is, when the edge distance A was 100 μm, the diameter B was 50 μm, the spacing C was 140 μm, and the cutting temperature of the micropit-textured tool reached its minimum value of 102 °C. The wear area of tool No. 12 was the lowest; that is, when the edge distance A was 100 μm, the diameter B was 60 μm, the spacing C was 120 μm, and the wear area of the micropit-textured tool reached its minimum value of 70,709 μm^2.

3.2. Multi-Objective Optimization of Tool Micro-Texture Parameters Using Response Surface Method

3.2.1. Regression Model Analysis

In this paper, the response surface method (RSM) was used to establish the regression equation between the three optimization variables, namely A, B, and C, and the three response indicators, namely the tool cutting force, cutting temperature, and tool wear area.

Significance analyses of the regression models of the cutting force, cutting temperature, and tool wear area of the rake faces of the tools are shown in Table 5, Table 6, and Table 7, respectively.

From Table 5, it can be seen that the interaction of the micro-texture parameter BC had a significant impact on the cutting force. From Table 6, it can be seen that the interaction of AC had a significant impact on the cutting temperature. From Table 7, it can be seen that the interaction of the micro-texture parameter BC had a very significant impact on the tool wear area.

Table 5. Variance analysis of the cutting force.

Source of Variance	Sum of Squares	Freedom	Mean Square Value	F Statistics	p
Model	2866.59	9	318.51	6.59	0.0106
A	208.49	1	208.49	4.31	0.0765
B	505.62	1	505.62	10.46	0.0144
C	1094.18	1	1094.18	22.63	0.0021
AB	169.26	1	169.26	3.50	0.1035
AC	0.31	1	0.31	0.01	0.9381
BC	416.98	1	416.98	8.62	0.0218
A2	48.34	1	48.34	1.00	0.3506
B2	5.03	1	5.03	0.10	0.7564
C2	435.19	1	435.19	9.00	0.0199
Residual	338.46	7	48.35		
Misfitting term	301.96	3	100.65	11.03	0.0210
Pure error	36.50	4	9.13	-	-
Total dispersion	3205.04	16	-	-	-

Table 6. Variance analysis of the cutting temperature.

Source of Variance	Sum of Squares	Freedom	Mean Square Value	F Statistics	p
Model	1471.69	9	163.52	7.81	0.0065
A	325.13	1	325.13	15.53	0.0056
B	2.00	1	2.00	0.10	0.7663
C	528.13	1	528.13	25.23	0.0015
AB	12.25	1	12.25	0.59	0.4693
AC	361.00	1	361.00	17.24	0.0043
BC	42.25	1	42.25	2.02	0.1984
A2	8.25	1	8.25	0.39	0.5500
B2	186.20	1	186.20	8.89	0.0204
C2	5.09	1	5.09	0.24	0.6369
Residual	146.55	7	20.94		
Misfitting term	145.75	3	48.58	242.92	0.0001
Pure error	0.8	4	0.20	-	-
Total dispersion	1618.24	16	-	-	-

The analysis results showed that the p values of the regression models of the cutting force, cutting temperature, and wear area of the rake face were 0.0106, 0.0065, and $p < 0.0001$, respectively, indicating that the regression models established were significant, and the determination coefficients R^2 of these regression models were 0.8944, 0.9094, and 0.9803, which were all close to 1; in other words, the accuracy of all experimental data explained by these models reached 89.44%, 90.94%, and 98.03%, respectively. In addition, the signal-to-noise ratio, Adeq Precision values of the models were equal to 8.9817, 5.7686, and 19.6011, all of which were greater than 4.

Table 7. Variance analysis of the tool wear area.

Source of Variance	Sum of Squares	Freedom	Mean Square Value	F Statistics	p
Model	0.4076	9	0.0453	38.82	<0.0001
A	0.1067	1	0.1067	27.07	0.0013
B	0.0794	1	0.0794	286.11	<0.0001
C	0.0003	1	0.0003	9.52	0.01766
AB	0.0272	1	0.0272	1.81	0.2207
AC	0.0025	1	0.0025	0.13	0.7286
BC	0.0756	1	0.0756	8.56	0.0222
A2	0.1601	1	0.1601	0.00	0.9935
B2	0.0007	1	0.0007	15.55	0.0056
C2	0.0421	1	0.0421	0.31	0.5940
Residual	0.0082	7	0.0012		
Misfitting term	0.0050	3	0.0017	4.14	0.2465
Pure error	0.0032	4	0.0008	-	-
Total dispersion	0.4158	16	-	-	-

Obviously, the above contents sufficiently reflect the accuracy and reliability of the established regression model of the cutting force. The regression models of the cutting force, cutting temperature, and tool wear area of the micropit-textured tools were finally established as follows:

$$R1 = 676 + 2.81 \times A - 3.23 \times B - 8.01 \times C - 0.0325 \times A \times B - 0.0007 \times A \times C + 0.0510 \times B \times C - 0.0085 \times A^2 - 0.0109 \times B^2 + 0.0510 \times C^2 \tag{1}$$

$$R2 = 122 - 3.05 \times A + 7.85 \times B - 1.34 \times C + 0.0088 \times A \times B + 0.0238 \times A \times C - 0.0163 \times B \times C - 0.0035 \times A^2 - 0.0665 \times B^2 + 0.0028 \times C^2 \tag{2}$$

$$R3 = 345{,}571 - 555 \times A - 6446 \times B - 898C^3 + 5.91 \times A \times B + 0.79 \times A \times C + 12.86 \times B \times C + 0.02 \times A^2 + 33.78 \times B^2 + 1.20 \times C^2 \tag{3}$$

Residual positive probability graphs (referred to as residual) were used to test whether the data follow certain rules and judge whether the established regression models of the micro-textured tools' cutting force, cutting temperature, and tool wear area are reasonable. As shown in Figure 9, the residual graphs of the cutting force, cutting temperature, and tool wear area show linear aggregation, and they also have the characteristics of a uniform distribution without overflow points, indicating that the data are relatively reasonable. The contour maps of the 3D response surface of the influence of the interaction between two pairs on the cutting force, cutting temperature, and tool wear area are shown in Figure 10, where the small red dots indicate the position and response values of the experimental data or optimization points in the response surface model, which is helpful for analyzing and optimizing multivariate functions.

From Table 5, it can be seen that the interaction of the micro-texture parameter BC had a significant impact on the cutting force. Figure 10a shows the contour map of the response surface of the diameter and spacing (BC). As the diameter B and spacing C decreased, the overall response surface showed a trend of an initial decrease followed by an increase. The gray area below shows an increase in the contour curvature; therefore, the interaction between the diameter and spacing (BC) had a significant impact on the cutting force of the tool. It can be seen from Table 6 that the interaction of AC had a significant impact on the cutting temperature. Figure 10b shows the contour map of the response surface of the edge distance and spacing (AC). As can be seen from the graph, with the decrease in the edge distance A and spacing C, the cutting temperature increased, and the response surface showed an overall upward trend. The curve in the gray plane below represents the contour line: the larger the curvature, the more significant the interaction between the edge distance and spacing (AC) on the cutting temperature of the tool.

It can be seen from Table 7 that the interaction of the micro-texture parameter BC had a very significant impact on the tool wear area. Figure 10c shows the contour map of the response surface of the diameter and spacing (BC). It can be seen from the figure that as the diameter B and spacing C decreased, the response surface showed an overall upward trend. The curve in the lower gray plane represents the contour. As its curvature increased, the interaction between the diameter and spacing (BC) had a more significant impact on the tool wear area.

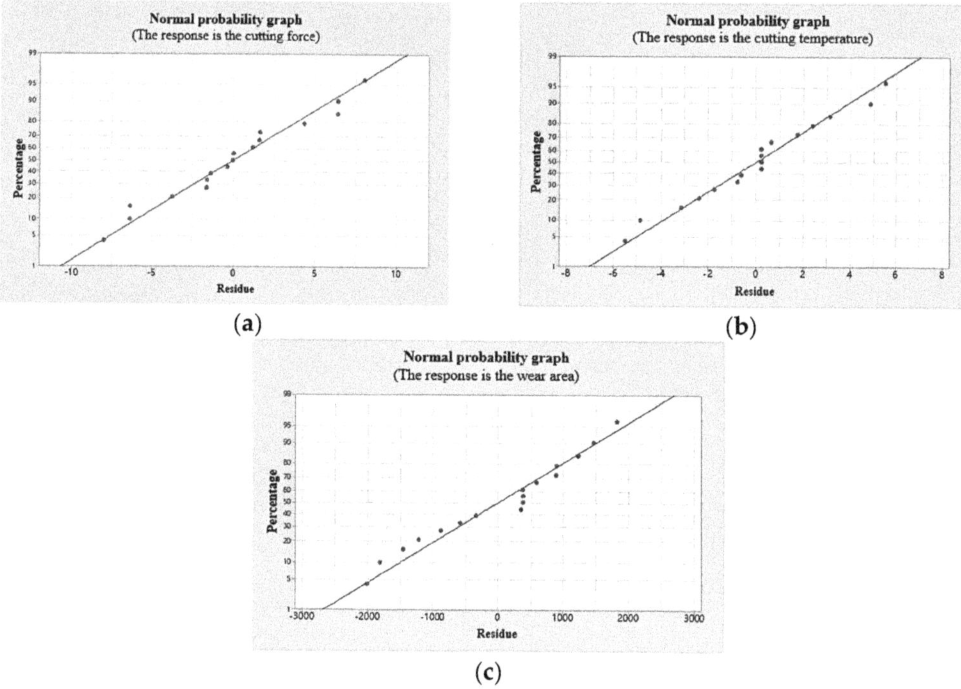

Figure 9. Residual diagrams of the cutting force, cutting temperature, and wear area: (**a**) residual diagram of the cutting force; (**b**) residual diagram of the cutting temperature; (**c**) residual diagram of the tool wear area.

3.2.2. Multi-Objective Optimization of Tools' Micro-Texture Parameters

On the basis of the above analysis, the response surface method was used to optimize the combination of micro-texture parameters. Firstly, Design-Expert was used to analyze the second-order surface model of the micro-texture parameters; optimize the extreme points of the three response variables of the micro-textured tools, namely the cutting force, cutting temperature, and tool wear area; and find a group of parameters in the process of optimization, so as to obtain the optimal cutting performance of the micro-textured tools. In the process of optimization, the three response variables of the micro-textured tools were set as follows: the cutting force ranged from 201.42 N to 280.35 N, the cutting temperature ranged from 102 °C to 130 °C, and the tool wear area ranged from 70,709 μm^2 to 101,138 μm^2.

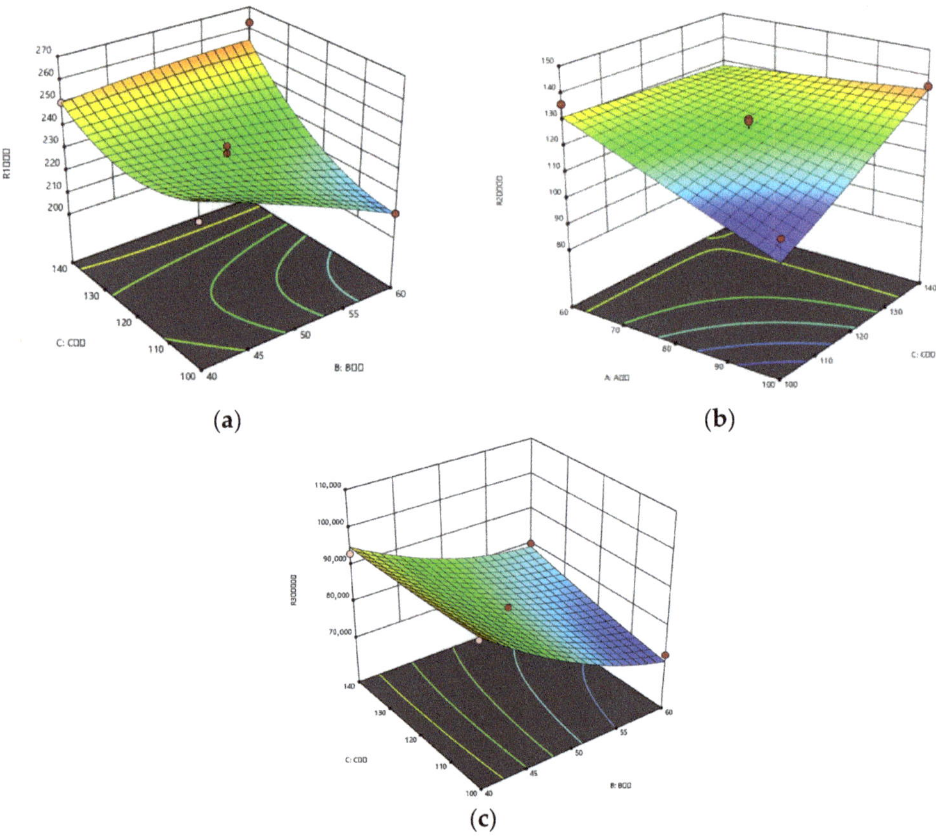

Figure 10. Three-dimensional response surface contour maps of the influence of the interaction of the micro-texture parameters on the cutting force, temperature, and wear area: (**a**) interaction of the diameter and spacing (BC) on the cutting force; (**b**) interaction of the edge distance and spacing (AC) on the cutting temperature; (**c**) interaction of the diameter and spacing (BC) on the tool wear area.

After setting the corresponding variables of the micro-texture, the optimal combination from the multi-objective optimization was obtained as follows: the edge distance A of the micro-texture was 81.3505 μm, the diameter B was 56.786 μm, and the spacing C was 102.377 μm. Correspondingly, the cutting force was 216.795 N, the cutting temperature was 117 °C, and the tool wear area was 72,697.4 μm^2.

After debugging the relevant equipment, a micropit-textured tool with the optimal parameters was processed, and a cutting experiment was carried out. The tool's cutting force was 202.5 N, the cutting temperature was 113 °C, and the wear area was 71,994 μm^2. The results show that the relative errors between the optimized results of the response surface method and the experimental results were as follows: cutting force, 6.59%; cutting temperature, 3.42%; and tool wear area, 0.94%. Under the same conditions, the cutting force of the non-textured tool was 240.98 N, the cutting temperature was 138 °C, and the wear area was 109,942 μm^2. Compared with non-textured tool, for the optimized micro-textured tool, the cutting force, cutting temperature, and tool wear area were reduced by 15.97%, 18.12%, and 34.52%, respectively.

4. Multi-Objective Optimization of Tools' Micro-Texture Parameters Using Genetic Algorithm

In order to explore the influence of the micro-texture parameters more accurately under different combinations on the cutting performance of the micro-textured tools, the NSGA-II algorithm based on approximate models was used for the multi-objective optimization of the micro-texture parameters and to intelligently solve the Pareto solution set. Firstly, 27 full-factor experiments on the micro-texture parameters were designed using Minitab, and trained and predicted using the RBF and BP approximate models in MATLAB (the analysis models; the calculation results of these models are similar to those of the analysis model from the simulation process). The two approximate models were optimized based on the predicted results. Then, the NSGA-II algorithm was used to optimize the three response target approximate models and obtain the Pareto solution set.

4.1. Experimental Design of Neural Network Optimization for Micro-Texture Parameters

When using RBF neural networks for parameter optimization, in order to achieve ideal results, it is usually necessary to use a large amount of data; therefore, this paper used Minitab to design full factor experiments on the micro-texture parameters to achieve more accurate results after training. Table 8 shows the results of the 27 full-factor experiments designed using Minitab [27,28].

Table 8. Results of 27 full-factor experiments.

NO.	A (μm)	B (μm)	C (μm)	Cutting Force (N)	Cutting Temperature (°C)	Wear Area (μm^2)
1	80	40	120	263.38	130	90,107
2	80	40	140	280.35	134	93,261
3	60	60	140	249.49	121	85,264
4	80	50	120	221.47	128	81,730
5	100	40	140	270.43	132	89,657
6	60	60	100	246.43	113	71,294
7	100	60	140	244.79	128	71,863
8	60	40	140	282.66	129	99,963
9	80	60	120	217.46	118	77,565
10	100	60	120	201.42	108	70,709
11	80	50	100	258.83	124	79,852
12	80	60	100	219.08	111	72,021
13	100	50	100	257.03	114	80,364
14	60	40	100	268.68	124	91,526
15	100	40	120	262.35	123	92,069
16	60	50	140	269.33	126	87,024
17	60	50	120	267.39	125	82,664
18	100	40	100	287.61	127	90,659
19	100	50	140	258.82	121	83,693
20	60	50	100	263.38	116	82,891
21	100	60	100	211.95	106	69,654
22	80	60	140	240.89	123	80,070
23	60	40	120	259.31	128	101,138
24	60	60	120	234.43	119	79,519
25	100	50	120	243.39	116	82,687
26	80	40	100	273.35	121	87,696
27	80	50	140	260.33	125	85,542

According to the full factor experiments shown in Table 8, the cutting experiments were conducted at a cutting speed of 80 m/min, a feed rate of 0.2 mm/r, and a cutting depth of 0.2 mm. Based on the collected experimental data (cutting force, cutting temperature, and wear area), a multi-objective approximate model for the micro-texture parameters was established using RBF and BP neural networks.

4.2. Comparison of RBF and BP Neural Network Approximate Models

By using MATLAB to program the RBF neural network, the response targets of different combinations of the micro-texture parameters were predicted using the obtained approximate model, and the predicted results of this approximate model were compared with the data obtained from the experiment. Table 9 shows the cutting force comparison between the approximate models obtained using the two neural networks and the actual simulation data. Table 10 shows the comparison of the cutting temperature, and Table 11 shows the comparison of the tool wear of the rake face. It can be seen that the error between the predicted results of the approximate model obtained using the RBF neural network and the actual simulation data were kept within a relatively small range, which indicates that the accuracy and reliability of this approximate model's prediction are relatively high. As shown in Figure 11, the line chart of the comparison between the predicted values of the last 10 groups of the RBF neural network approximate model and the simulation data intuitively show that the predicted values and the simulation data basically illustrated a consistent trend, which again proves the accuracy and reliability of the RBF neural network approximate model.

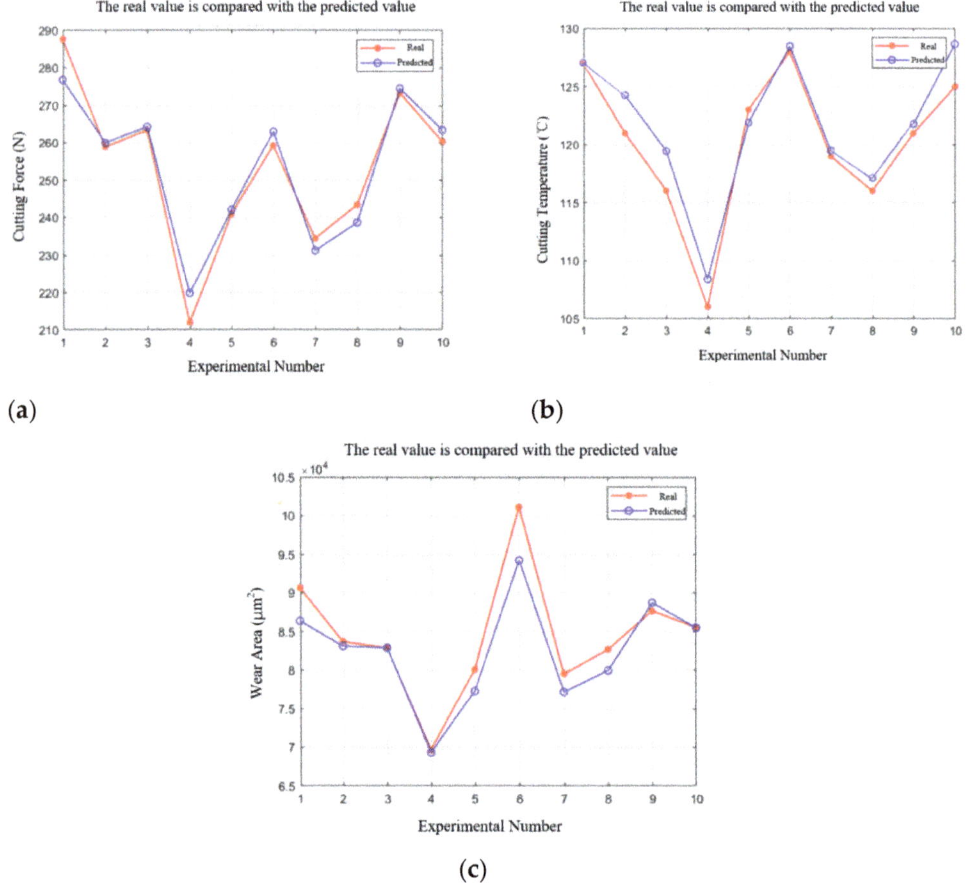

(a)

(b)

(c)

Figure 11. Real and predicted values: (**a**) cutting force; (**b**) cutting temperature; (**c**) tool wear area.

Table 9. The cutting force predicted using the RBF and BP neural networks.

NO.	Real Value (N)	RBF		BP	
		Predicted Value (N)	Relative Error	Predicted Value (N)	Relative Error
18	287.61	276.85	3.74%	284.00	1.26%
19	258.82	260.03	0.47%	256.74	0.80%
20	263.38	264.03	0.25%	258.86	1.72%
21	211.95	219.81	3.71%	225.47	6.38%
22	240.89	242.10	0.50%	248.26	3.06%
23	259.31	263.05	1.44%	261.50	0.85%
24	234.43	231.24	1.36%	233.30	0.48%
25	243.39	238.69	1.93%	236.05	1.23%
26	273.35	274.55	0.44%	279.33	2.19%
27	260.33	263.45	1.20%	266.53	2.38%

Table 10. The cutting temperature predicted using the RBF and BP neural networks.

NO.	Real Value (°C)	RBF		BP	
		Predicted Value (°C)	Relative Error	Predicted Value (°C)	Relative Error
18	127	127.45	0.35%	125.46	1.21%
19	121	124.26	2.69%	119.27	1.43%
20	116	119.43	2.96%	121.54	4.78%
21	106	108.41	2.27%	108.89	2.73%
22	123	121.89	0.90%	119.59	2.77%
23	128	128.47	0.37%	127.46	0.42%
24	119	119.48	0.40%	121.20	1.85%
25	116	117.01	0.87%	114.60	1.21%
26	121	121.79	0.65%	124.84	3.17%
27	125	128.65	2.92%	142.93	4.34%

Table 11. The wear area predicted using the RBF and BP neural networks.

NO.	Real Value (μm^2)	RBF		BP	
		Predicted Value (μm^2)	Relative Error	Predicted Value (μm^2)	Relative Error
18	90,659	86,380.91	3.72%	88,426.99	2.46%
19	83,693	83,131.80	0.67%	83,727.81	0.04%
20	82,891	82,918.22	0.03%	80,317.38	3.10%
21	69,654	69,338.83	0.45%	71,533.49	2.70%
22	80,070	77,311.25	3.44%	78,919.89	1.44%
23	101,138	94,246.54	6.81%	94,658.55	6.41%
24	79,519	77,204.46	2.91%	82,610.45	3.89%
25	82,687	79,981.91	3.27%	81,471.14	1.47%
26	87,696	88,738.77	1.19%	87,830.36	0.15%
27	85,542	85,489.67	0.61%	87,749.01	2.58%

4.3. RBF Approximate Model and NSGA-II Algorithm for Multi-Objective Optimization of Micro-Texture Parameters

By comparing the predicted and experimental values of the two neural network approximate models, it was found that the RBF neural network was more reliable and accurate than the BP neural network in predicting the cutting force, cutting temperature, and tool wear area when cutting using the micro-textured tools; therefore, the RBF neural network was selected to establish a multi-objective optimization model for the micro-texture parameters, and then combined with the NSGA-II multi-objective genetic algorithm for optimization processing.

After multiple iterations of the NSGA-II optimization algorithm, the Pareto front diagrams for the three objectives were obtained, as shown in Figure 12.

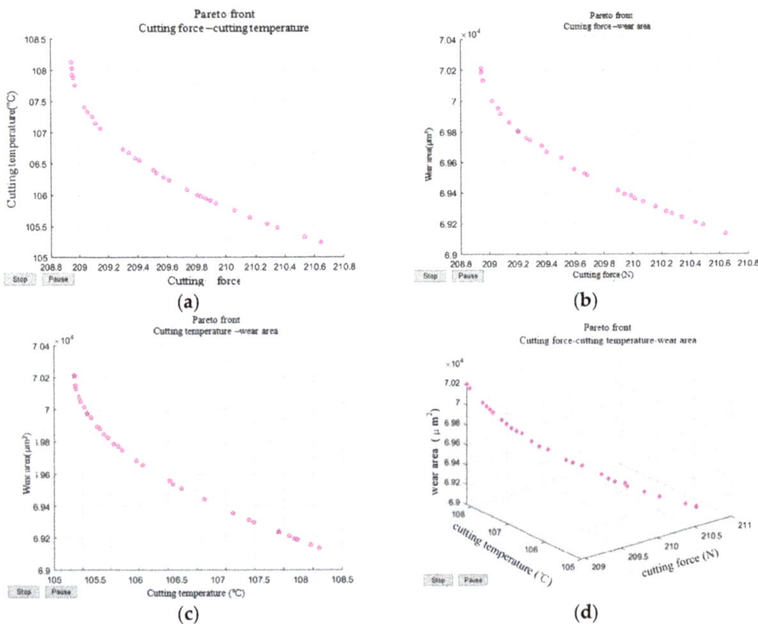

Figure 12. Pareto front diagrams: (**a**) cutting force–cutting temperature; (**b**) cutting force–wear area; (**c**) cutting temperature–wear area; (**d**) cutting temperature–wear area–cutting force.

The nonlinear relationship between each objective function can be directly seen in the Pareto diagrams. The curves in the Pareto diagrams are relatively smooth and all gather together, indicating the accuracy and reliability of this optimization algorithm. In addition, the diagrams prove that the NSGA-II optimization algorithm has good convergence. From Figure 12, it can also be seen that an increase in one variable led to a decrease in another variable, indicating the interaction and influence between various variables and verifying the convergence, reliability, and accuracy of the NSGA-II optimization algorithm.

From the Pareto solution set obtained above, several sets of optimal feasible solution sets were selected, as shown in Table 12. Through comparison, it was found that the first group of three optimal solutions was more suitable for the optimization of the micro-texture parameters in this paper compared with the other solution sets; therefore, the first group of data was used as the final result of the micropit texture optimization, where the edge distance A was 100 μm, the diameter B was 60 μm, and the spacing C was 100 μm.

Table 12. Pareto optimal solution set.

NO.	A (μm)	B (μm)	C (μm)	Cutting Force (N)	Cutting Temperature (°C)	Wear Area (μm²)
1	100	60	100	210.64	105.24	69,138.00
2	100	60	107.85	208.94	108.13	70,213.15
3	98.98	59.99	100.26	210.83	105.33	69,175.35

From Table 12, it can be seen that when the edge distance A was 100 μm, the diameter B was 60 μm, and the spacing C was 100 μm, that the corresponding cutting force was 210.64 N, the cutting temperature was 105.24 °C, and that the tool wear area was 69,138 μm². In order to verify the accuracy and reliability of the NSGA-II genetic optimization algorithm based on the RBF neural network, relevant cutting experiments were carried out to verify the optimization results, setting the same conditions as in the above experiments and cutting GH4169 with the micropit-textured tool with the optimal micro-texture parameters. The cutting force of the experiment was 204.67 N, the cutting temperature was 108 °C, and the tool wear area was 70,724 μm^2. Compared with the optimal solution set, the relative errors of the cutting force, cutting temperature, and tool wear area of the micropit-textured tool were 2.83%, 2.62%, and 0.29%, respectively. The errors were within the acceptable ranges, which verifies the accuracy and reliability of the optimization results using the RBF neural network and NSGA-II genetic optimization algorithm. Compared with the non-textured tool, of which the cutting force, cutting temperature, and tool wear area were 240.99 N, 138 °C, and 109,938 μm^2, respectively, the cutting force, cutting temperature, and tool wear area of the optimized micro-textured tool were reduced by 15.07%, 21.74%, and 35.67%, respectively.

Based on the comparison between the optimization results obtained by combining the RBF neural network with the NSGA-II genetic optimization algorithm and those obtained using the response surface method, it was found that the optimization algorithm combining the RBF neural network with the NSGA-II genetic optimization algorithm was more suitable for optimizing the micro-texture parameters. Table 13 shows the comparison results of the two optimization algorithms. In the table, the cutting performance of the tool after using the two optimization algorithms can be intuitively seen. The cutting force, cutting temperature, and tool wear area of the micro-textured tool optimized using the RBF neural network and NSGA-II genetic algorithm were lower than those of the RSM, and the relative error of RBF-NSGA-II was relatively low as well.

Table 13. Comparison of results between the two optimization algorithms.

Algorithm	Cutting Force (N)	Relative Error of Cutting Force	Cutting Temperature (°C)	Relative Error of Cutting Temperature	Wear Area (μm^2)	Relative Error of Wear Area
RSM	216.795	6.59%	117	3.42%	72,697.4	0.94%
RBF-NSGA-II	210.64	2.83%	105.24	2.62%	69,138	0.29%

5. Conclusions

This paper explored the influence of micro-texture morphologies and size parameters on the cutting performance when cutting the nickel-based superalloy GH4169 with micro-textured tools. Moreover, the optimal combination of micro-texture parameters was analyzed using a multi-objective optimization algorithm. The conclusions are as follows:

1. Simulation and cutting experiments were carried out on five types of micro-textured tools with different shapes, and the average cutting force, average cutting temperature, and tool wear depth of the tools were extracted. After comparison, it was found that the cutting temperature and tool wear depth of the micropit-textured tool were relatively low, and the cutting temperature of the micro-parallel-groove-textured tool was relatively low.

2. The traditional response surface method and Box–Behnken design method were used to optimize the micro-texture parameters of the micropits, and the optimized combination of micro-texture parameters was obtained as follows: the edge distance A was 81.3505 μm, the diameter B was 56.786 μm, and the spacing C was 102.377 μm. Compared with the non-textured tool, the cutting force, cutting temperature, and tool wear area of the optimized micro-textured tool were reduced by 15.97%, 18.12%, and 34.52%, respectively.

3. An optimization algorithm combining the RBF approximate model and NSGA-II algorithm was used to optimize the micro-texture parameters of the micropits. The optimized combination of micro-texture parameters of the micropits was as follows: the edge distance A was 100 μm, the diameter B was 60 μm, and the spacing C

was 100 µm. The experimental results show that the cutting force of the tool was 204.67 N, the cutting temperature was 108 °C, and the tool wear area was 70,724 µm^2. Compared with the non-textured tool, the cutting force of the optimized tool was reduced by 15.07%, the cutting temperature was reduced by 21.74%, and the tool wear area was reduced by 35.67%.

4. By comparing the optimization results of the response surface method and genetic algorithm with the experimental values, the results show that the RBF-NSGA-II optimization algorithm was more suitable for the multi-objective optimization of the micro-texture parameters.

Author Contributions: Methodology, X.F. (Xinmin Feng); software, X.F. (Xiwen Fan); validation, J.H.; writing—original draft preparation, X.F. (Xinmin Feng); writing—review and editing, J.W. All authors have read and agreed to the published version of the manuscript.

Funding: This study was supported by the National Natural Science Foundation of China (No. 51675144).

Data Availability Statement: Not applicable.

References

1. Hao, Z.; Yang, S.; Fan, Y.; Lu, M. Machining characteristics of cutting Inconel718 with carbide tool. *Int. J. Mater. Prod. Technol.* **2019**, *58*, 275–287. [CrossRef]
2. Huang, J.; He, Z.Z.; Yang, X.G.; Shi, D.Q.; Sun, Y.T. Improved method for creep life prediction of nickel-based directionally solidified superalloys. *J. Mech. Eng.* **2022**, *58*, 258–268.
3. Feng, X.M.; Dong, Q.S.; Hu, J.S. Simulation and experimental analysis of cutting temperature in cutting GH4169 under spray cooling. *Mech. Sci. Technol. Aerosp. Eng.* **2022**, *41*, 985–991.
4. Shi, L.M.; Zhang, C. Experimental research on turning of superalloy GH4169 under high pressure cooling condition. *Integr. Ferroelectr.* **2020**, *207*, 75–85. [CrossRef]
5. Zhao, W.; Ren, F.; Iqbal, A.; Gong, L.; He, N.; Xu, Q. Effect of liquid nitrogen cooling on surface integrity in cryogenic milling of Ti-6Al-4V titanium alloy. *Int. J. Adv. Manuf. Technol.* **2020**, *106*, 1497–1508. [CrossRef]
6. Lin, Z.; Danni, L. Simulation and experimental research of nickel base heat resisting alloy in low temperature cold air and micro lubrication cutting. *Mach. Hydraul.* **2018**, *46*, 24–26.
7. Song, W.; Wang, S.; Lu, Y.; Xia, Z. Tribological performance of microhole-textured carbide tool filled with CaF$_2$. *Materials* **2018**, *11*, 1643. [CrossRef]
8. Patel, K.; Liu, G.; Shah, S.R.; Özel, T. Effect of micro-textured tool parameters on forces, stresses, wear rate, and variable friction in titanium alloy machining. *J. Manuf. Sci. Eng.* **2020**, *142*, 021007. [CrossRef]
9. Cui, X.; Sun, N.; Guo, J.; Ma, J.; Ming, P. Performance of multi-bionic hierarchical texture in green intermittent cutting. *Int. J. Mech. Sci.* **2023**, *247*, 108203. [CrossRef]
10. Zhou, X.R.; He, L.; Yuan, S.; Zhou, T.; Tian, P.F.; Zou, Z.C. Advanced research progress of surface micro-texture in cutting process. *Surf. Technol.* **2022**, *51*, 100–127.
11. Ali, S.; Abdallah, S.; Pervaiz, S. Predicting cutting force and primary shear behavior in micro-textured tools assisted machining of AISI 630: Numerical modeling and taguchi analysis. *Micromachines* **2022**, *13*, 91. [CrossRef] [PubMed]
12. Tu, C.J.; Guo, X.H.; Guo, D.L. Comparisong of machineability of self-lubrication ceramic tools with different morphological micro-texture. *Mater. Mech. Eng.* **2018**, *42*, 47–51+57.
13. Rajurkar, A.; Chinchanikar, S. Experimental investigation on laser-processed micro-dimple and micro-channel textured tools during turning of inconel 718 alloy. *J. Mater. Eng. Perform.* **2022**, *31*, 4068–4083. [CrossRef]
14. Dhage, S.; Jayal, A.D.; Sarkar, P. Effects of surface texture parameters of cutting tools on friction conditions at tool-chip interface during dry machining of AISI 1045 steel. *Procedia Manuf.* **2019**, *33*, 794–801. [CrossRef]
15. Li, K.; Du, J.X.; Liu, L.L.; Shen, F.H.; Ma, L.J.; Pang, M.H. Effect of texture parameters on main cutting force and temperature of cutting tools. *Tool Eng.* **2019**, *53*, 42–46.
16. Liu, Y.; Deng, J.; Wang, W.; Duan, R.; Meng, R.; Ge, D.; Li, X. Effect of texture parameters on cutting performance of flank-faced textured carbide tools in dry cutting of green Al$_2$O$_3$ ceramics. *Ceram. Int.* **2018**, *44*, 13205–13217. [CrossRef]
17. Sharma, V.; Pandey, P.M. Geometrical design optimization of hybrid textured self-lubricating cutting inserts for turning 4340 hardened steel. *Int. J. Adv. Manuf. Technol.* **2017**, *89*, 1575–1589. [CrossRef]
18. Guo, S.B.; Duan, X.Y.; Yi, Z.Y. Optimization of surface micro-texture parameters of Al$_2$O$_3$ /La$_2$O$_3$/ (W, Mo) C cemented carbide tool. *Lub RIcation Eng.* **2020**, *45*, 60–71.
19. Tong, X.; Yang, S.C.; He, C.S.; Zheng, M.L. Multi-objective optimization of cutting performance of variable density micro-texture ball-end mlilling tool. *J. Mech. Eng.* **2019**, *55*, 221–232.

20.   Arulkirubakaran, D.; Senthilkumar, V.; Kumawat, V. Effect of micro-textured tools on machining of Ti–6Al–4V alloy: An experimental and numerical approach. *Int. J. Refract. Met. Hard Mat.* **2016**, *54*, 165–177. [CrossRef]
21.   Wang, B.; Zhang, J.Y.; Wu, C.L.; Deng, W.J. A numeric investigation of rectangular groove cutting with different lateral micro textured tools. *KeyEng. Mat.* **2016**, *4283*, 693. [CrossRef]
22.   Kishawy, H.A.; Salem, A.; Hegab, H.; Hosseini, A.; Elbestawi, M. An analytical model for the optimized design of micro-textured cutting tools. *CIRP Annals.* **2022**, *71*, 49–52. [CrossRef]
23.   Cheng, Y.T.; Zhang, X.; Xie, C.Y. Research on texture parameters optimization of Micro-Textured cutting tool for cutting CFRP. *Mech. Sci. Technol. Aerosp. Eng.* **2022**, *6*, 1–9.
24.   Li, B.B.; Chen, L.; Zhao, W.; Yu, M.; Zhang, K. Simulation study on texture parameter optimization of micro-textured tool. *Mod. Manuf. Eng.* **2020**, *481*, 91–96.
25.   Yu, Y.H.; Yang, S.B.; Ruan, W.X.; Xu, P.; Shen, J.X. Parameter optimization of offset parabolic micro texture. *J. Ordnance Equip. Eng.* **2023**, *44*, 225–233.
26.   Li, Y.B.; Wang, H.X.; Zhuang, K.J. Study on multi-objective optimization and high efficiency plunge milling experiment of titanium alloy for aviation. *Mech. Sci. Technol. Aerosp. Eng.* **2022**, *41*, 1921–1927.
27.   Xu, X.D.; Gong, Y.L.; Xia, T.F. Response surface experiment and multiobjective optimization of processing parameters based on deep hole. *Manuf. Autom.* **2022**, *44*, 1–4+37.
28.   Luan, X.N.; Zhang, Q.; Zhang, S.; Li, J.F. Power analysis and multi objective cutting parameter optimization in milling based on energy saving. *Tool Eng.* **2022**, *56*, 81–87.

lubricants

Article

Study on a Novel Strategy for High-Quality Grinding Surface Based on the Coefficient of Friction

Yang Li [1], Li Jiao [1], Yanhou Liu [2], Yebing Tian [2], Tianyang Qiu [1], Tianfeng Zhou [1,3], Xibin Wang [1,3] and Bin Zhao [1,3,*]

[1] School of Mechanical Engineering, Beijing Institute of Technology, Beijing 100081, China; liyang_6118@bit.edu.cn (Y.L.)
[2] School of Mechanical Engineering, Shandong University of Technology, Zibo 255049, China
[3] Chongqing Innovation Center, Beijing Institute of Technology, Chongqing 401120, China
* Correspondence: bin.zhao@bit.edu.cn; Tel.: +86-1501-0481-560

Abstract: Surface quality has a significant impact on the service life of machine parts. Grinding is often the last process to ensure surface quality and accuracy of material formation. In this study, a high-quality surface was developed by determining the coefficient of friction in grinding a quartz fiber-reinforced silica ceramic composite. By processing the physical signals in the grinding process, a multi-objective function was established by considering grinding parameters, i.e., surface roughness, coefficient of friction, active energy consumption, and effective grinding time. The weight vector coefficients of the sub-objective functions were optimized through a multi-objective evolutionary algorithm based on the decomposition (MOEA/D) algorithm. The genetic algorithm was used to optimize the process parameters of the multi-objective function, and the optimal range for the coefficient of friction was determined to be 0.197~0.216. The experimental results indicated that when the coefficient of friction tends to 0.197, the distribution distance of the microscopic data points on the surface profile is small and the distribution uniformity is good. When the coefficient of friction tends to 0.216, the surface profile shows a good periodic characteristic. The quality of a grinding surface depends on the uniformity and periodicity of the surface's topography. The coefficient of friction explained the typical physical characteristics of high-quality grinding surfaces. The multi-objective optimization function was even more important for the subsequent high-quality machining of mechanical parts to provide guidance and reference significance.

Keywords: surface quality; coefficient of friction; distribution uniformity; periodic components; signal processing

Citation: Li, Y.; Jiao, L.; Liu, Y.; Tian, Y.; Qiu, T.; Zhou, T.; Wang, X.; Zhao, B. Study on a Novel Strategy for High-Quality Grinding Surface Based on the Coefficient of Friction. *Lubricants* **2023**, *11*, 351. https://doi.org/10.3390/lubricants11080351

Received: 20 July 2023
Revised: 6 August 2023
Accepted: 15 August 2023
Published: 17 August 2023

1. Introduction

Quartz fiber-reinforced silica ceramic composite has been successfully used in building materials, the chemical industry, national defense, and other sectors [1,2] due to its low thermal conductivity, small expansion coefficient, and high-temperature resistance. However, low mechanical strength and high brittleness make it a difficult-to-cut material with poor surface quality [3]. In surface processing of this material, grinding is one of the most widely employed machining processes due to its high precision and stable surface quality. The range of force ratios taken during grinding is crucial due to the brittleness of the ceramic composite. In comparison with other machining processes, grinding consumes high amounts of energy with a low efficiency for the same level of material volume removal [4]. Some strategies have been used to improve the quality of grinding surfaces, such as surface modification or the application of coating techniques on the substrate surface [5]. However, both surface modification and surface application coating technologies pose some environmental problems to a greater or lesser extent [6]. Under the strategic target of "carbon neutrality", energy saving and carbon emission reduction have become part of a global consensus in the surface machining process. Therefore, it is particularly important to investigate the physical mechanisms of molding for high surface quality.

With the advancement of science and technology, intelligent monitoring technology has developed rapidly [7,8]. By analyzing and processing the signals collected by various sensors, the grinding process can be better monitored. Wang et al. [9] innovatively proposed a signal processing method based on fuzzy C-average clustering. This method can accurately predict the subsurface damage depth and surface quality in ultra-precision grinding of single-crystal silicon. Wang et al. [10] utilized multiple transformation forms of raw AE signals to monitor materials' surface behavior and developed a force model to explain forces under different removal modes. Ling et al. [11] used different types of new sandpaper to grind the surface of an alloy and investigated the effect of the grinding treatment on the surface properties and deformation microstructure. Zhang et al. [12] introduced a novel data-driven model using an optimized pruned extreme learning machine. Real-time quantitative monitoring of abrasive belt conditions can be achieved in robotic grinding systems by a novel method based on acoustic signals. Tian et al. [13] developed a portable power monitoring system for grinding signal acquisition, feature extraction, and data calculation to improve the surface quality achieved by grinding.

To characterize the actual wear state of micro-grinding wheels, a novel monitoring method was proposed based on the variable cutting stiffness with a fusion analysis of forces and system vibration signals [14,15]. Warren et al. [16] applied a discrete wavelet decomposition procedure to extract discriminative features from original acoustic emission (AE) signals. And the state of the grinding wheels was monitored through the wavelet-based AE signal. Based on the physical characteristics of grinding, some achievements have been made in ensuring surface quality, saving energy, and improving efficiency. Ma et al. [17] first proposed the concept of "relative extreme value error" to judge the influence sensitivity of technical factors on surface roughness. Based on the analysis of cracks and grinding kinetics, Yao et al. [18] established a relationship between surface roughness and subsurface crack depth. Kong et al. [19] introduced an effective feature extraction method through the combination of PCA and KPCA_IRBF, which can be utilized for surface roughness prediction. Meng et al. [20] presented an innovative dynamic force model for precision grinding with micro-structured grinding wheels. Li et al. [21] modeled grinding temperature with a genetic association analysis tool (PGA) and regarded it as the main constraint for the inverse problem while modeling the grinding surface roughness and grinding continuity as auxiliary constraints. The ability of high-accuracy Artificial Neural Networks (ANNs) regarding feature classification, especially on nonlinear patterns, was considered [22]. Dai et al. [23] studied the effect of grinding speed on grinding temperature and power consumption and analyzed the grinding surface performance from the perspective of undeformed chip thickness. Wang et al. [24] developed a nonparametric model based on an improved adaptive Artificial Neural Network (aANN) to predict surface quality, machining time, total power consumption, and effective power consumption.

The methodological strategies adopted in the above studies have optimized the grinding surface quality to varying degrees, but few investigations have been conducted to evaluate the superiority and inferiority of the microscopic data on the machined surface. This study intends to explore the typical physical characteristics of high-quality surfaces through the coefficient of friction, under the premise of non-pollution to the environment and controllable operation. A multi-objective numerical function for comprehensive evaluation of the grinding surface was established. The value of the coefficient of friction directly affects workpiece surface quality, which is conducive to the improvement of the service performance of processed parts. Within a suitable range of the coefficient of friction, a uniformly distributed and periodic machined surface profile with a R_a of about 0.3 μm can be obtained. This research has great significance for solving problems of high surface quality and low energy consumption in grinding.

2. Materials and Methods

2.1. Experimental Equipment

A grinding experimental platform was established for multi-feature signal acquisition; see Figure 1a. It mainly consisted of a CNC surface grinder (SMART-B818III, Falcon Machine Tools Co., Ltd., Changhua County, Taiwan), a portable power cell (PPC-3, Load Controls Inc., Sturbridge, MA, USA), and a multicomponent dynamometer (Type 9527B, Kistler Holding AG, Winterthur, Switzerland). The PPC-3 has three Hall-effect current sensors and three voltage sensors. The outputs of the PPC-3 are 0–20 mA in current and 0–10 V in voltage. The in-process current and voltage signals were measured by the PPC-3, and effective powers were calculated from the measured data. The dynamometer, the measurement error of which is one percent, can simultaneously measure force and torque in the X, Y, and Z axes of the grinder. Power signals and force signals were collected online at 1000 Hz and 2000 Hz, respectively. The surface roughness Ra and surface profile curve were measured by a roughness tester (TIME3200, TIME Group Inc., Beijing, China). After each grinding test, six positions on the machined surface were selected to measure the roughness R_a; see Figure 1b. Ignoring the maximum and minimum values, the other four measured data points were averaged for each set of experiments. The workpiece material was a quartz fiber-reinforced silica ceramic composite with dimensions of $50 \times 50 \times 25$ mm. A resin-bonded diamond flat grinding wheel was used with an outer diameter of 200 mm and a thickness of 10 mm. The basic size of abrasive grains was about 20 µm [25].

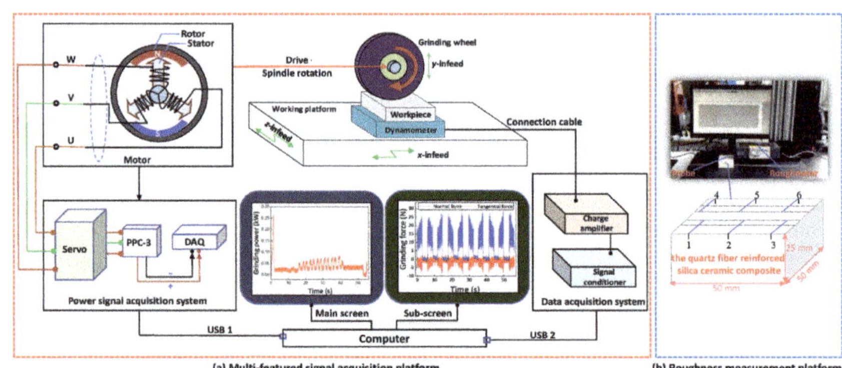

Figure 1. Grinding test configuration and data measurement.

2.2. Experimental Scheme

The grinding surface quality, energy consumption, force, and cost are influenced by the linear speed of the grinding wheel v_s (m/min), workpiece infeed speed v_w (mm/min), and grinding depth a_p (µm). In this work, a full factorial experiment with these three factors and five levels was designed, and three-fifths of the total tests were randomly selected. The linear speed of the grinding wheel was varied from 1000 m/min to 1800 m/min, the workpiece infeed speed was varied from 1000 mm/min to 5000 mm/min, and the grinding depth was varied from 4 µm to 12 µm, as listed in Table 1.

Table 1. Experimental parameters of grinding tests.

Parameters	Values
The linear speed of the grinding wheel v_s (m/min)	1000(S1), 1200(S2), 1400(S3), 1600(S4), 1800(S5)
Workpiece speed v_w (mm/min)	1000(F1), 2000(F2), 3000(F3), 4000(F4), 5000(F5)
Grinding depth a_p (µm)	4(D1), 6(D2), 8(D3), 10(D4), 12(D5)

2.3. Signal Processing

2.3.1. Grinding Force Signal Processing

The surface quality and energy consumption can be affected by grinding forces directly. It is critical to analyze and process three-component force signals. The normal force is closely related to the compressive strength of the grinding wheel and system rigidity. The tangential force has a direct impact on effective power in grinding. The axial force is the force along the axis of the grinding wheel. Insufficient contact signal removal, drift signal correction, signal denoising strategy, and effective normal force signal acquisition have been elaborated on in previous work [26]. Figure 2 shows the processing method of grinding force signals. It is observed in Figure 2a that the signals of tangential force and normal force are oppositely distributed along the time axis. The negative value of tangential force indicates the direction of the force. A discrete-time Fourier transform (DTFT) was employed to monitor the frequency domain curves of the force as follows:

$$X(\mathrm{e}^{\mathrm{j}\omega}) = \sum_{n=0}^{n-1} x[n] \cdot \mathrm{e}^{-\mathrm{j}\omega n} \tag{1}$$

where $X(\mathrm{e}^{\mathrm{j}\omega})$ is the frequency domain function of the grinding force signal, $x[n]$ is the time series of the grinding force signal, n is the index of the position of the grinding force data point, j is the imaginary unit, and ω is the grinding force signal frequency variable. The frequency distribution was analyzed to establish a low cut-off frequency value. The Chebyshev filter transfer function is as follows:

$$|G_M(j\omega)|^2 = \frac{1}{1 + \varepsilon^2 T_M^2(\omega/\omega_0)} \tag{2}$$

where ω_0 is the grinding force signal's expected cut-off frequency, ε is the grinding force signal wave coefficient which is a positive number less than 1, and T_M is a Chebyshev polynomial of order M.

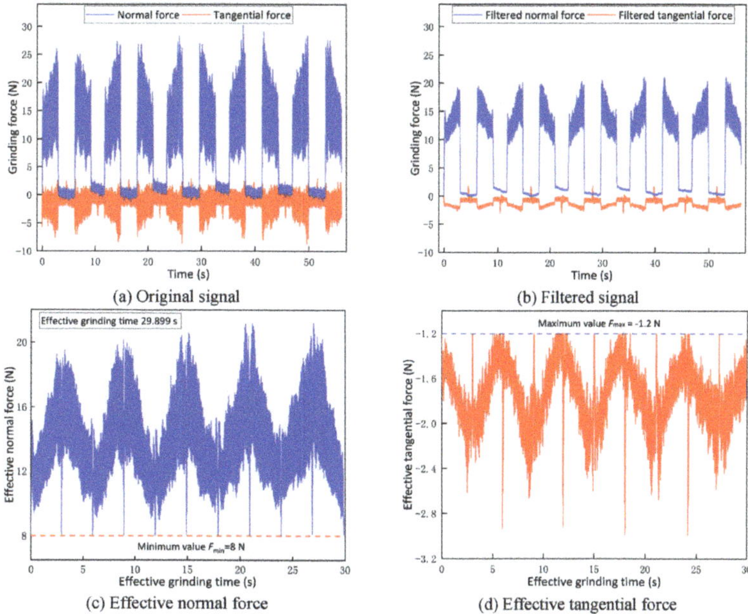

(a) Original signal

(b) Filtered signal

(c) Effective normal force

(d) Effective tangential force

Figure 2. The processing method of grinding force signals (v_s = 1000 m/min, v_w = 1000 mm/min, a_p = 10 μm).

The original force signals were filtered at an expected cut-off frequency, as shown in Figure 2b. Afterwards, a local amplification component function was used to find the minimum normal force and maximum tangential force. The signals far away from the mini/max value were extracted by the for loop. For tangential force, the absolute value of the extracted signal was taken. By removing invalid interval signals, an effective force signal was formed; see Figure 2c,d. The effective grinding time for each test can be obtained from the effective force signals. The mean coefficient of friction was calculated from the average ratio of the effective tangential force to the effective normal force.

2.3.2. Grinding Power Signal Processing

Grinding wheel abrasives with a large negative angle make the grinding process consume more energy than other machining processes under the same material removal rate [27,28]. Visualization of energy consumption by monitoring power signals online is beneficial to the energy conservation of machine tools. The active power is the actual and irreversible power of the machine tool. It reflects the amount of electric energy converted into other energy in unit time. The extraction strategy of the active power signal is crucial for the analysis of power signals detected during grinding.

In order to fully collect the power signal during grinding, PPC-3 was used to measure the power signal online in advance; see Figure 3a. The power signal was filtered through Equations (1) and (2) in the air grinding stage to achieve a denoising operation; see Figure 3b. The power signal segments during grinding were obtained by longitudinal line selection. The non-grinding power signal under the longitudinal line was removed. The minimum effective power during grinding was determined by a horizontal line; see Figure 3c. All arrays in the power signal larger than the value were extracted by the program to form an effective power array; see Figure 3d. Based on the effective power and time arrays, the grinding active energy consumption can be obtained by numerical integration:

$$E_{\text{active}} = \int_0^T P(t)dt \tag{3}$$

where E_{active} is the active power, $P(t)$ is the power signal, and t is the time.

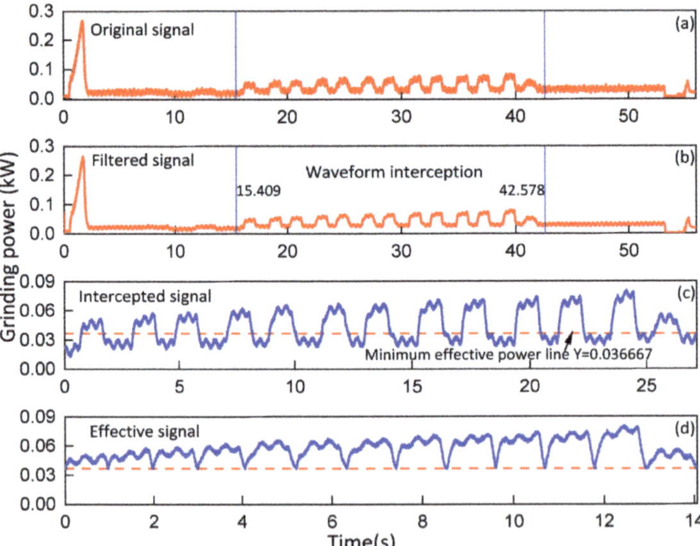

Figure 3. The processing method of grinding power signals (v_s = 1000 m/min, v_w = 3000 mm/min, a_p = 12 μm).

3. Results

3.1. Experimental Results

A full-factor experiment of v_s, v_w, and a_p was designed in five levels. A total of 125 sets of experimental parameters were obtained. Three-fifths of experimental groups, i.e., 75 groups, were selected stochastically to generate irregular samples. The processed force signal and power signal were analyzed and the experimental results are listed in Table 2. It includes surface roughness R_a (μm), coefficient of friction u (Ft/Fn), active energy consumption E_a (J), and effective grinding time T (s).

Table 2. Grinding parameters and output results.

No.	Inputs			Outputs			
	v_s (m/min)	v_w (mm/min)	a_p (μm)	R_a (μm)	u	E_a (J)	T (s)
1	1000	1000	4	0.333	0.196	520.824	29.945
2	1000	1000	10	0.400	0.227	538.088	29.690
3	1000	1000	12	0.423	0.120	626.641	29.963
4	1000	2000	4	0.426	0.159	119.739	14.982
5	1000	2000	6	0.419	0.236	147.500	14.956
6	1000	2000	8	0.424	0.181	168.877	15.019
7	1000	3000	6	0.433	0.106	72.429	10.004
8	1000	3000	8	0.465	0.132	83.263	9.987
9	1000	3000	12	0.513	0.116	116.740	9.934
10	1000	4000	4	0.427	0.130	58.609	7.510
11	1000	4000	10	0.436	0.164	95.176	7.389
12	1000	4000	12	0.529	0.156	115.372	7.481
13	1000	5000	6	0.462	0.069	72.429	5.975
14	1000	5000	8	0.466	0.177	76.690	6.342
15	1000	5000	12	0.539	0.144	96.517	5.958
16	1200	1000	4	0.305	0.140	486.283	30.353
17	1200	1000	6	0.390	0.140	511.823	30.042
18	1200	1000	8	0.401	0.134	567.206	29.926
19	1200	2000	8	0.411	0.148	167.374	11.738
20	1200	2000	10	0.453	0.251	216.624	15.019
21	1200	2000	12	0.467	0.124	317.837	15.034
22	1200	3000	4	0.440	0.053	63.669	9.980
23	1200	3000	10	0.464	0.112	71.240	9.780
24	1200	3000	12	0.479	0.111	202.890	9.967
25	1200	4000	6	0.469	0.126	69.445	7.453
26	1200	4000	8	0.501	0.127	102.964	7.454
27	1200	4000	12	0.491	0.126	108.551	7.365
28	1200	5000	4	0.445	0.112	44.026	5.310
29	1200	5000	10	0.481	0.230	46.059	5.754
30	1200	5000	12	0.533	0.126	69.332	5.961
31	1400	1000	6	0.376	0.107	518.141	29.710
32	1400	1000	8	0.400	0.167	544.445	28.751
33	1400	1000	12	0.415	0.135	568.541	28.738
34	1400	2000	4	0.313	0.071	136.096	12.854
35	1400	2000	10	0.397	0.303	146.993	13.189
36	1400	2000	12	0.437	0.260	159.001	14.559
37	1400	3000	6	0.394	0.060	162.067	9.769
38	1400	3000	8	0.432	0.149	189.511	9.805
39	1400	3000	12	0.440	0.131	144.624	9.943
40	1400	4000	4	0.349	0.180	73.428	7.117
41	1400	4000	10	0.416	0.269	91.638	7.341
42	1400	4000	12	0.475	0.149	130.246	7.446
43	1400	5000	4	0.418	0.096	43.957	6.001
44	1400	5000	6	0.430	0.218	67.381	5.984
45	1400	5000	8	0.452	0.155	84.441	5.334
46	1600	1000	4	0.237	0.205	468.124	29.853

Table 2. *Cont.*

No.	Inputs			Outputs			
	v_s (m/min)	v_w (mm/min)	a_p (μm)	R_a (μm)	u	E_a (J)	T (s)
47	1600	1000	10	0.328	0.233	548.139	30.351
48	1600	1000	12	0.403	0.175	564.990	30.643
49	1600	2000	6	0.366	0.098	138.741	15.092
50	1600	2000	8	0.388	0.343	151.043	15.099
51	1600	2000	12	0.419	0.135	178.248	15.042
52	1600	3000	4	0.311	0.064	97.888	9.906
53	1600	3000	10	0.400	0.108	111.350	9.979
54	1600	3000	12	0.440	0.081	125.653	9.958
55	1600	4000	4	0.341	0.102	27.717	7.381
56	1600	4000	6	0.434	0.112	37.726	7.533
57	1600	4000	8	0.432	0.143	89.867	7.475
58	1600	5000	8	0.450	0.141	47.901	6.010
59	1600	5000	10	0.480	0.146	63.151	5.969
60	1600	5000	12	0.471	0.141	76.120	5.944
61	1800	1000	6	0.371	0.142	492.783	30.104
62	1800	1000	8	0.393	0.154	510.063	29.023
63	1800	1000	12	0.344	0.059	688.878	30.22
64	1800	2000	4	0.287	0.139	116.907	14.702
65	1800	2000	10	0.373	0.469	144.436	14.891
66	1800	2000	12	0.416	0.174	165.680	14.857
67	1800	3000	4	0.297	0.203	71.907	9.950
68	1800	3000	6	0.393	0.355	74.378	10.046
69	1800	3000	8	0.402	0.227	75.018	10.102
70	1800	4000	6	0.426	0.106	66.879	7.464
71	1800	4000	8	0.404	0.147	69.806	7.546
72	1800	4000	12	0.419	0.137	112.556	7.619
73	1800	5000	4	0.326	0.197	45.776	5.534
74	1800	5000	10	0.413	0.228	62.568	6.008
75	1800	5000	12	0.466	0.103	78.477	5.927

3.2. Interaction among Evaluation Indicators

The evaluation indicators include surface roughness, coefficient of friction, active energy consumption, and effective grinding time, as listed in Table 2. Exploring the relationship between evaluation indicators is beneficial to promoting the grinding process toward a goal of high surface quality with high efficiency and low energy consumption. The coefficient of friction is a comprehensive reflection of tangential force and normal force, which may affect the surface quality. Another parameter that has a big influence on surface quality is the grinding depth. It is significant to investigate the relationship between the coefficient of friction and surface roughness under changes in the grinding depth for the same linear speed of the grinding wheel and the same workpiece infeed speed. A total of 21 sets of experimental data from seven groups were selected and plotted; see Figure 4. It shows that with an increasing grinding depth, the coefficient of friction increased and then decreased, and the surface roughness increased. It is clear that a high coefficient of friction leads to bad surface roughness. The smallest surface roughness is not related to the smallest coefficient of friction for each group. Therefore, it is crucial to analyze the correlation between the coefficient of friction and surface roughness.

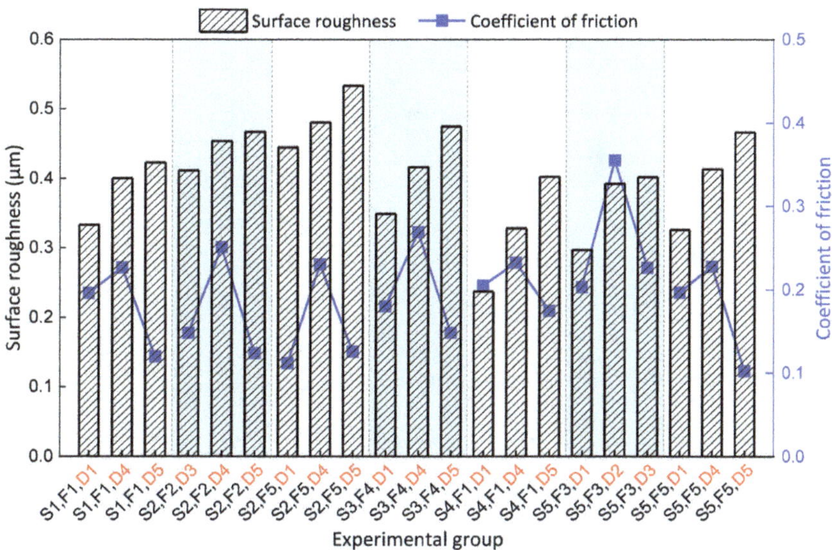

Figure 4. The relationship between coefficient of friction and surface roughness.

Compared with the linear speed of the grinding wheel and grinding depth, the workpiece infeed speed has a greater impact on the active energy consumption and effective grinding time [29]. The relationship between active energy consumption and effective grinding time was analyzed. A total of 21 experimental sets in seven control groups were selected and plotted in Figures 5 and 6. It was found that the smaller the infeed speed of the workpiece, the more active energy and grinding time were consumed. Better surface roughness was always associated with a larger active energy consumption or a longer effective grinding time.

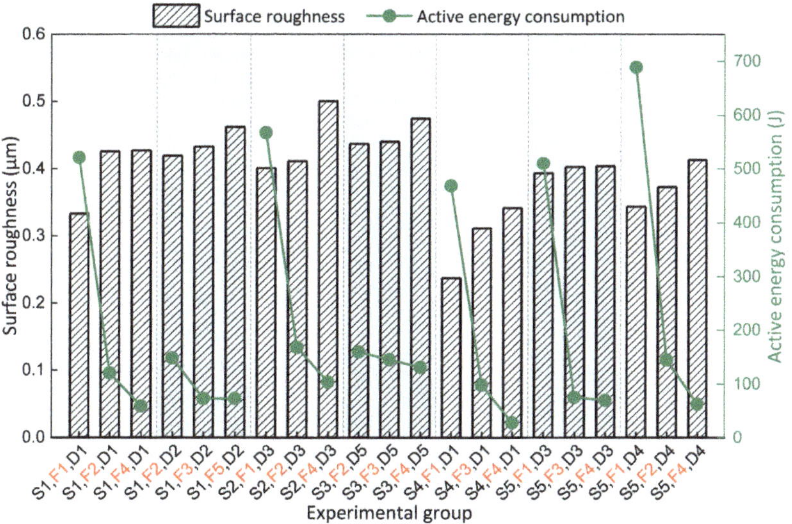

Figure 5. The relationship between active energy consumption and surface roughness.

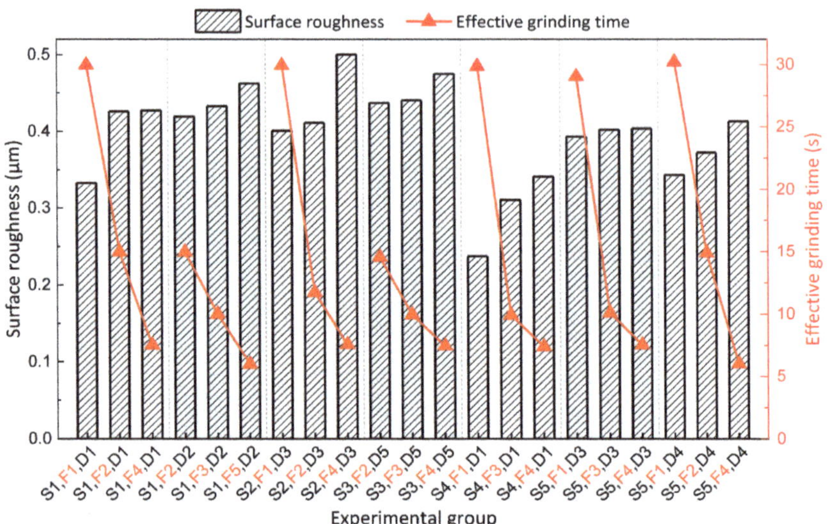

Figure 6. The relationship between effective grinding time and surface roughness.

3.3. Mathematical Modeling

In the controllable numerical control range, the different interaction patterns between the evaluation indicators show different expression forms under different processing conditions. Based on the grinding input elements, the output results of surface roughness, coefficient of friction, active energy consumption, and effective grinding time were numerically modeled. Multiple nonlinear regression refers to the unknown independent variables and unknown dependent variables presenting nonlinear characteristics upon regression. Realizing the optimal collocation of multiple independent variables to predict or evaluate dependent variables can improve the validity of the prediction results. In the grinding process, the input parameters and the evaluation results are nonlinear relationships. In this work, a novel method was proposed to judge and identify the function types of independent variables of grinding regarding the grinding evaluation indicators. In order to determine the multivariate nonlinear functional form, logarithms were taken on both sides of functions and transformed into a linear correlation function. The correlation level statistics were then calculated to determine the parameters under the corresponding mathematical mapping model. The parameter was confirmed and numerical modeling was established. The matrix calculation process is shown in Equation (4).

$$\begin{bmatrix} \lg\delta & \varepsilon_1 & \varepsilon_2 & \varepsilon_3 \\ \lg\eta & \alpha_1 & \alpha_2 & \alpha_3 \\ \lg\gamma & \beta_1 & \beta_2 & \beta_3 \\ \lg\tau & \rho_1 & \rho_2 & \rho_3 \end{bmatrix} \begin{bmatrix} 1 \\ \lg v_s \\ \lg v_w \\ \lg v_p \end{bmatrix} = \begin{bmatrix} \lg R_a \\ \lg u \\ \lg E_a \\ \lg T \end{bmatrix} \tag{4}$$

where the 4×4 matrix contains the relevant parameters of each evaluation indicator to be identified. The determination of parameters can be calculated from statistical data and bias regression coefficients by Minitab. Taking 65 groups of grinding experimental data as training samples, the multivariate nonlinear numerical functions of surface roughness R_a, coefficient of friction μ, active energy consumption E_a, and effective grinding time T were constructed. The multivariate nonlinear numerical model of the four evaluation indicators is shown below.

$$R_a = 0.8112 \cdot v_{si}^{-0.2893} \cdot v_{wi}^{0.1277} \cdot \alpha_{pi}^{0.2014} (i = 1, 2 \dots 65) \tag{5}$$

$$\mu = 0.0114 \cdot v_{si}^{0.4675} \cdot v_{wi}^{-0.0236} \cdot \alpha_{pi}^{-0.3164} \, (i = 1, 2 \ldots 65) \tag{6}$$

$$E_a = 18234879 \cdot v_{si}^{-0.0901} \cdot v_{wi}^{-1.4878} \cdot \alpha_{pi}^{0.2436} \, (i = 1, 2 \ldots 65) \tag{7}$$

$$T = 30370.3 \cdot v_{si}^{0.0016} \cdot v_{wi}^{-1.0054} \cdot \alpha_{pi}^{-0.0004} \, (i = 1, 2 \ldots 65) \tag{8}$$

Taking the other 10 experimental sets as test samples, the predicted surface roughness, coefficient of friction, active energy consumption, and effective grinding time were compared with the actual evaluation results, as shown in Figure 7.

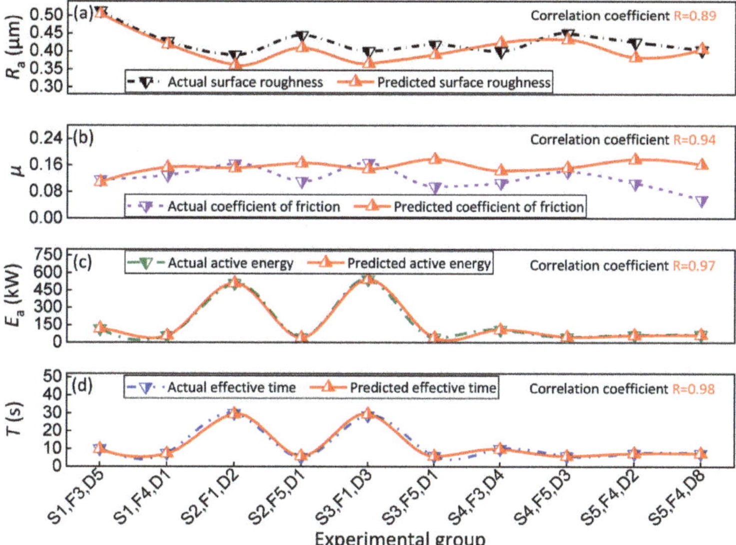

Figure 7. Validity verification of numerical model. (**a**) Validation of the surface roughness numerical model. (**b**) Validation of the coefficient of friction numerical model. (**c**) Validation of the active energy consumption numerical model. (**d**) Validation of the effective grinding time numerical model.

It can be seen that the predicted results of the established multivariate nonlinear numerical model were in good agreement with the experimental data. The correlation coefficients of surface roughness, coefficient of friction, active energy consumption, and effective grinding time were all high, being 0.89, 0.94, 0.97, and 0.98, respectively. The established numerical model has high reliability and can be used for the following modeling.

3.4. Mathematical Modeling Integration

Based on the multivariate numerical model, an optimized evaluation strategy for grinding-based machining was proposed and constructed to solve a multi-objective optimization problem in this work. In general, a single solution cannot guarantee optimal performance on multiple objectives. And the conflicting nature of the objectives makes it impossible for the algorithm to have a single optimal solution but, instead, a set of relatively good solutions, i.e., Pareto front. The MOEA/D algorithm provides an idea to decompose a multi-objective optimization problem (MOP) into multiple single-objective subproblems through a set of weight vectors. These objectives are optimized simultaneously and each subproblem is optimized by itself using the adjacent population of each subproblem. The algorithm first assigns weight vectors to all individuals randomly generated and uniformly distributed in the population to determine the neighborhood of each subproblem. The subproblem selects individuals in a determined neighborhood for crossover variation to generate new solutions. The individuals of the parent generation in the neighborhood are updated using a specific aggregation function for the next cycle. The best non-dominated individual in each generation is selected and retained as an optimal solution. All optimal solutions are filtered out through continuous iterations of solutions to produce an optimal solution.

An objective function F(x) was established and minimized by considering the following parameters: R_a, μ, E_a, and T. In this study, the number of decision variables was set to be three (v_s, v_w, and a_p), and the number of target variables was four (R_a, μ, E_a, and T). The mathematical description of minimizing MOP under unconstrained conditions can be written as:

$$\begin{cases} \text{Minimize F}(x) = [R_a(x), \mu(x), E_a(x), T(x)]^{\text{T}} \\ \text{subject to}: \begin{cases} 1000 \le x_1 \le 1800 \\ 1000 \le x_2 \le 5000 \\ 4 \le x_3 \le 12 \end{cases} \end{cases} \tag{9}$$

where $x = (x_1, x_2, x_3) \in R^3$ is a three-dimensional decision variable and $F(x) \in R^4$ is a four-dimensional objective variable. The sub-objective function preferences are in order from heavy to light. It is worth noting that the coefficient of friction is a two-level trend. Therefore, the numerical objective function for the coefficient of friction is optimized by seeking the maximum and minimum as an objective for multi-objective optimization.

3.4.1. Weight Vector Generation and Aggregation Method for MOEA/D Algorithm

The MOEA/D algorithm uses a simplex lattice point design method to generate the weight vectors of individuals in the population [30]. And it is particularly critical for the algorithm to find an optimal solution, and a relatively uniformly distributed weight vector corresponds to a higher-quality solution. The simplex lattice point design method needs to determine the parameter H (a positive integer) that affects the weight vector. Some uniformly distributed points are selected on the plane composed of $w_1 + w_2 + ... + w_s = 1$, where s is the number of objective functions. The generated weight vector requirements are given by

$$w_1^r + w_2^r + \cdots w_s^r = 1 \tag{10}$$

$$w_i^r \in \left\{ \frac{0}{H}, \frac{1}{H}, \frac{2}{H}, \cdots \frac{H}{H} \right\}, i = 1, 2, \cdots, s r = 1, 2, \ldots, N \tag{11}$$

where H is a parameter that affects the weight vector and is defined by the decision maker, s is the target number/weight vector dimension, w_i^j is the i-th component of the j-th weight vector. The number of population size/weight vectors satisfies the following equation:

$$N = C_{H+S-1}^{S-1} \tag{12}$$

The generation of weight vectors in MOEA/D algorithm requires the generation of a neighborhood by calculating the Euclidean distance of individuals in the population and the subsequent update of the solution based on the neighboring individuals in the neighborhood. The strategy used to update the solution is to calculate the value of the same aggregation function to retain both solutions on merit. The Tchebycheff approach is a nonlinear multi-objective aggregation method [31] with an aggregation function defined as follows:

$$\begin{cases} \text{Minimize } g^{\text{TA}} = (x|w, z^*) = \max_{1 \le i \le m} \left\{ w_i |F_i(x) - z_i^*| \right\} \\ \text{subject to } x \in \Omega \end{cases} \tag{13}$$

where $z^* = \min\{F(x) \mid x \in \Omega\}$, $i \in \{1, 2, \ldots, s\}$, $w = (w_1, w_2, \ldots, w_m)^{\text{T}}$ is the weight vector that satisfies $w_i \ge 0$. Each solution derived by Equation (13) maps to a Pareto optimal solution x^* in the original MOP. The decision maker can choose a different weight vector. When dealing with high-dimensional problems, the TA method can limit the convergence-receiving area and better ensure the convergence of the population.

3.4.2. MOEA/D Algorithm Framework

The MOEA/D algorithm obtains individual neighborhoods by computing the Euclidean distance between the weight vectors of individuals within the population. The relationship between the generated neighborhoods and the subproblems was used to perform simultaneous optimization of the subproblems, so that the individuals within the population keep approximating the ideal Pareto optimal surface. To measure the convergence and distribution of the algorithm, Generational Distance (GD) and Inverted Generational Distance (IGD) were used as evaluation metrics:

$$\mathrm{GD}(P, P^*) = \frac{\sqrt{\sum_{v \in P} (d(v, p))^2}}{|P|} \tag{14}$$

$$\mathrm{IGD}(P^*, P) = \frac{\sum_{V \in P*} d(v, p)}{|P|} \tag{15}$$

$$d_i = \left\{ \sum_{k=1}^{s} [F_{ki}(x) - F_{k\min}(x)]^2 \right\}^{1/2} \tag{16}$$

where P^* is a set of points uniformly distributed over the Pareto front, p is a set of optimal solution sets obtained by the algorithm for approximating the Pareto front, d_i is the i-th individual's Euclidean distance in this iteration, $F_{ki}(x)$ is the k-th objective function value of the i-th individual, and $F_{k\min}(x)$ is the minimum value of the m-th objective function of individuals.

The parent individual selection, the child individual crossover generation, and the population update of the MOEA/D algorithm are all carried out within the neighborhood. This method enables the rapid sharing of good genes with other individuals near the individual, which greatly enhances the search efficiency of the algorithm [32]. The multi-objective optimization strategy flow for the grinding process using the MOEA/D algorithm is shown in Figure 8.

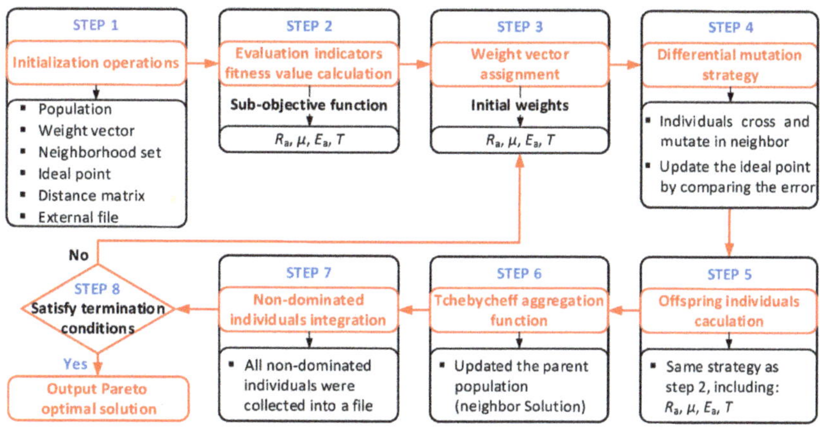

Figure 8. MOEA/D algorithm multi-objective optimization process.

The initial population P_0 satisfying the constraints of size N_p was randomly generated. N_p initial weight vectors $w_i{}^j$ were generated. The set of neighbors B_i, representing the evolved parent individuals, was selected from d according to the Euclidean distance minimization rule. Then, the ideal point z^* and the empty external file S^* were set. The fitness function values $R_a(x)$, $u(x)$, $E_a(x)$, and $T(x)$ were calculated for the parent individual. All fitness functions were selected as the reciprocal of the sub-objective function. A weight vector was assigned to each sub-objective function. Genetic recombination on individual X^i was then performed as follows: two randomly selected individuals from B_i were cross-mutated to generate new sub-individuals. The ideal point was updated by comparing the ideal point z^* with the objective function value $F_r(X^{i,\,new})$ for the offspring individual. All found non-dominated solutions were added to the external file S^* during the evolutionary process. All individuals in S^* that were dominated by new individuals were removed, leaving all solutions in S^* as non-dominated solutions. The stopping criterion was satisfied and the output S^* was used as the Pareto optimal solution set for this optimization problem.

3.5. Optimization Results for the Grinding Process

The parameters set the population size as P0 = 90, the crossover probability as 0.88, the variation probability as 0.07, the maximum number of iterations as 100, and the real-time data update period as 1 s [33]. Based on the above parameter settings, the expert weight values were obtained as (0.64, 0.26, 0.06, 0.04), (0.52, 0.34, 0.09, 0.05), (0.54, 0.15, 0.24, 0.07), and (0.53, 0.20, 0.05, 0.22), respectively. Then, the equation for the coefficient of friction of the sub-function in Equation (6) was maximized and the value of the expert vector obtained was (0.56, 0.36, 0.05, 0.03). The weight ratio of each sub-objective numerical function shows that the surface roughness sub-function has the largest percentage. This indicates surface roughness is crucial in multi-objective optimization. A good sub-objective function indicates a relatively high weight of the objective function. This is consistent with the adversarial nature between sub-objective functions in multi-objective functions.

It is impossible that all indicators satisfy the optimum at the same condition. Based on the weight vector proportion of each sub-function obtained by the MOEA/D algorithm, the genetic algorithm was used to optimize the objective function. The specific parameter settings and process of the genetic algorithm are described in our previous work [34]. The five sets of weight vectors objectively obtained by the MOEA/D algorithm were grouped. The results of the genetic algorithm-optimized grinding elements are listed in order from 1 to 5; see Table 3. According to the vector weight grouping category, the three optimized input elements of grinding were brought into Equation (5) in order. The theoretical surface roughness was 0.315 µm, 0.307 µm, 0.370 µm, 0.365 µm, and 0.311 µm. The data analysis reveals that the surface roughness predicted under the premise of minimizing the surface roughness sub-function (the first set of data) is not the minimum. The numerical functions of each sub-objective were brought in turn under the optimized parameter settings as shown in Table 3. The relevant data were obtained as shown in Figure 9. It can be seen that the weight vectors of the third and fourth groups have the best performance in terms of active energy consumption and effective grinding time, respectively. Attributed to the increased proportion of the coefficient of friction sub-function, the weight vectors of the second and fifth groups perform well in terms of surface quality. It is concluded that under the dominance of the surface roughness numerical function, the larger proportion of the numerical function for the coefficient is, the smaller the value of surface roughness from multi-objective optimization is, and the better the surface processing quality is.

Table 3. Grinding element optimization results.

Grinding Elements	Weight Vector Grouping Categories				
	1	2	3	4	5
v_s (m/min)	1560.3760	1638.3838	1753.1953	1765.5965	1724.2324
v_w (mm/min)	1094.4094	1033.6033	4708.3708	4647.1647	1286.8286
a_p (µm)	4.1712	4.0624	4.3400	4.1208	4.1016

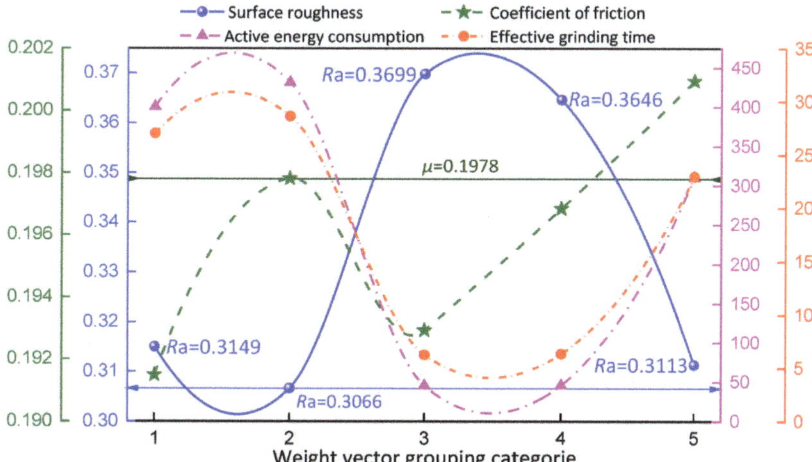

Figure 9. Sub-objective function values based on optimized parameters.

Although the weight vectors of the second and fifth groups can achieve better performance, their acquisition paths are based on the minimum and maximum coefficient of friction functions, respectively. The suitable interval for the coefficient of friction obtained by the genetic algorithm was 0.197~0.216, in which the surface roughness was small. The coefficients of friction obtained from the second and fifth sets of optimized parameters are 0.198 and 0.201, respectively. Both can be considered optimal coefficients of friction.

As shown in Figure 9, the surface quality and effective machining time show an opposite trend. This study optimized the weight ratio of surface roughness, coefficient of friction, active energy consumption, and effective grinding time using the MOEA/D algorithm, which was used as the objective function for parameter adjustment. The optimized parameters can greatly reduce the effective machining time under the premise of guaranteeing high grinding surface quality, without the need for repeated process adjustment. The advantage of this method is that the optimal surface finish quality is achieved while minimizing the active energy consumption and the machining time required for that surface quality.

4. Discussion

4.1. Comprehensive Evaluation of Optimization Parameters

4.1.1. Variation Coefficient and Surface Profile Autocorrelation Analysis

Workpiece surface quality is an important criterion for assessing the grinding process's performance [35]. Surface roughness is an important parameter of workpiece surface quality [36]. It reflects the unevenness of a machined surface with small spacing and tiny peaks and valleys [37]. Its measurement calculation is an average value determination of least squares. This may lead to a situation where even if the surface roughness is small, the peaks and troughs may differ greatly from the average roughness, indicating poor surface uniformity. So, the idea of a variation coefficient in mathematical statistics is introduced considering the holistic nature of the workpiece. The variation coefficient is used as an evaluation criterion for the degree of data dispersion with the following equation:

$$C.V = (SD/Ra) \times 100\% \tag{17}$$

where SD is the standard deviation of the measurement location.

The coefficient of friction reflects the ratio between the tangential force and the directional force. An appropriate coefficient of friction could make the workpiece surface profile contain both periodic and random components. The surface profile of a workpiece is a very complex random signal [38], and the surface roughness R_a cannot express the full information of its microscopic morphology. Therefore, an autocorrelation analysis of the workpiece surface profile is used for the microscopic analysis of workpiece surface quality [39,40]. The autocorrelation calculation and digital evaluation formula are as follows:

$$R_{xx}(\tau) = 1/D \int_0^D x(l)x(l+\tau)dl \tag{18}$$

where τ is the transverse displacement of the profile curve, D is the evaluation length, and $x(l)$ and $x(l + \tau)$ are the contour heights at corresponding coordinates. The autocorrelation function $R_{xx}(\tau)$ is a similarity measure of the workpiece surface profile and usually reaches its maximum value at $\tau = 0$. For a completely random surface profile, $R_{xx}(\tau)$ tends to zero when τ increases gradually. For a periodic profile, the $R_{xx}(\tau)$ curve is also periodic and keeps oscillating steadily as τ increases. For a random surface profile mixed with periodicity, $R_{xx}(\tau)$ gradually decays until it becomes a stable periodic oscillation.

4.1.2. Experimental Verification of Optimized Parameters

Under the same experimental conditions, the optimized machining parameters in Table 3 were used as input to the grinding process. The surface profile, roughness values, and variation coefficient of the workpiece under the above five sets of machining parameters were plotted in Figure 10.

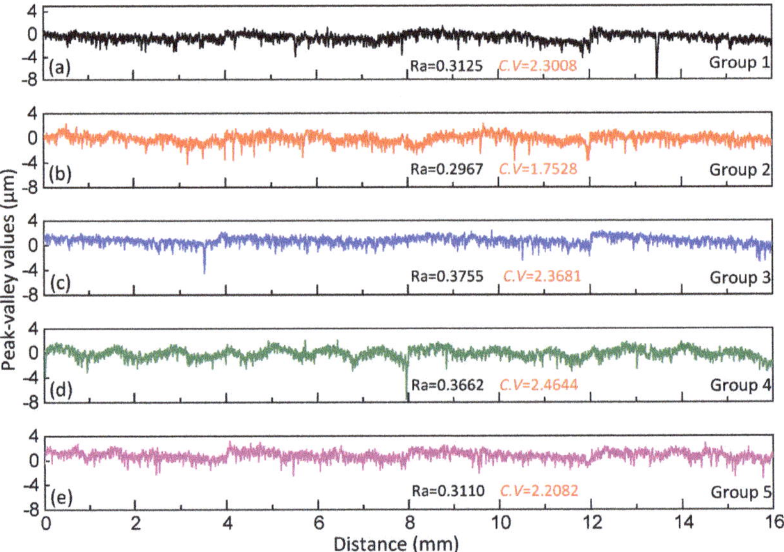

Figure 10. Surface profile curve, surface roughness, and variation coefficient. (**a**) Experimental results under optimized parameter 1 (v_s = 1560.3760 m/min, v_w = 1094.4094 mm/min, a_p = 4.1712 µm). (**b**) Experimental results under optimized parameter 2 (v_s = 1638.3838 m/min, v_w = 1033.6033 mm/min, a_p = 4.0624 µm). (**c**) Experimental results under optimized parameter 3 (v_s = 1753.1953 m/min, v_w = 4708.3708 mm/min, a_p = 4.3400 µm). (**d**) Experimental results under optimized parameter 4 (v_s = 1765.5965 m/min, v_w = 4647.1647 mm/min, a_p = 4.1208 µm). (**e**) Experimental results under optimized parameter 5 (v_s = 1724.2324 m/min, v_w = 1286.8286 mm/min, a_p = 4.1016 µm).

The variation coefficient is calculated from Equation (17). The profile curves are four surface profile curves stitched together. The experimental roughness in Figure 10 has the same change pattern as the predicted roughness values in Figure 9. Both the predicted and experimental values of roughness reach the minimum in the second and fifth groups. The established numerical model accuracy of the multi-objective function is verified. And the distribution uniformity of data points on the surface contour line is verified by the magnitude of the variation coefficient values. It was found that the highest surface quality cannot be achieved when the numerical functions of surface roughness, active energy consumption, or effective grinding time were too dominant. Both the second and the fifth groups achieved the best experimental results using the proportionally superior coefficient of friction's numerical function. It can be seen that an appropriate coefficient of friction is beneficial to profile uniformity and quality of roughness.

Figure 11 illustrates the uniformity of the data point distribution on the machined surface in terms of both the ceramic surface profile and the line texture curve. Figure 11a shows the 3D shape of the actual machined surface, while Figure 11b shows a surface texture curve. It can be seen from three-dimensional perspectives that the data points of the machined surface have a good uniformity of distribution and the surface machining quality is good. The surface profile roughness value is 0.298 μm and the line roughness value is 0.276 μm.

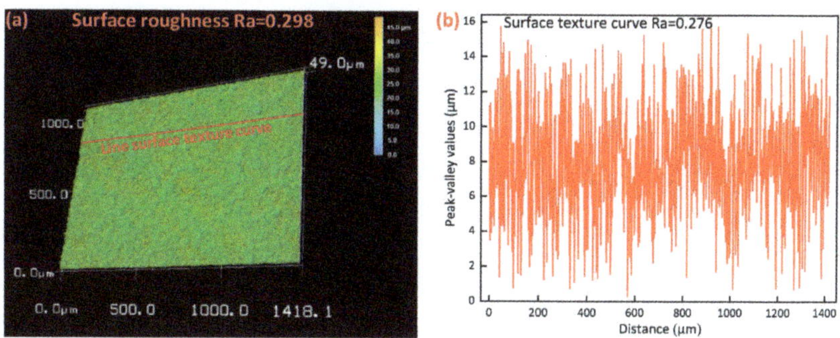

Figure 11. Ceramic surface profile and line texture curve (v_s = 1638.38 m/min, v_w = 1033.60 mm/min, a_p = 4.06 μm). (**a**) 3D topography of the machined surface. (**b**) Surface texture peak-to-valley curve.

Autocorrelation function analysis was performed on the experimental grinding surface profile of the workpiece based on the optimized parameter settings. Figure 12 shows the autocorrelation function profile curves for each optimized parameter condition, from which the following rules can be summarized:

Rule 1: The autocorrelation function profiles of the five operating conditions reached the maximum value at $\tau = 0$. The maximum values for the five working conditions were 0.702, 0.561, 0.909, 1.228, and 0.670, respectively. The maximum and minimum values of the autocorrelation coefficient were obtained for the effective machining time-dominated and friction coefficient-dominated operating conditions, respectively. It can be inferred that a smaller effective grinding time indicated a larger longitudinal parameter value of the machined surface and a worse surface quality.

Rule 2: The autocorrelation function curves obtained for all five working conditions decayed with an increase in the horizontal displacement τ. It indicated that there was a randomness signal in all five profile curves. In contrast, the autocorrelation curves of the workpiece surface in the first, third, and fourth groups were smooth and had very few wave fragments. It indicated that there were small periodic signals in the profiles of multi-objective functions dominated by surface roughness, active energy consumption, and effective grinding time. Small periodic signals of the contour curve can also be seen in the case of poor surface quality of the workpiece.

Rule 3: The autocorrelation function of the surface profile had a high-frequency oscillation (the second and fifth groups) in cases where the coefficient of friction's numerical function was proportionally dominant. The machined surface profile curve has more periodicity components, and the surface quality achieves the best results. It can be demonstrated that the coefficient of friction is important for the periodic variation of microscopic dimensions of the workpiece surface profile.

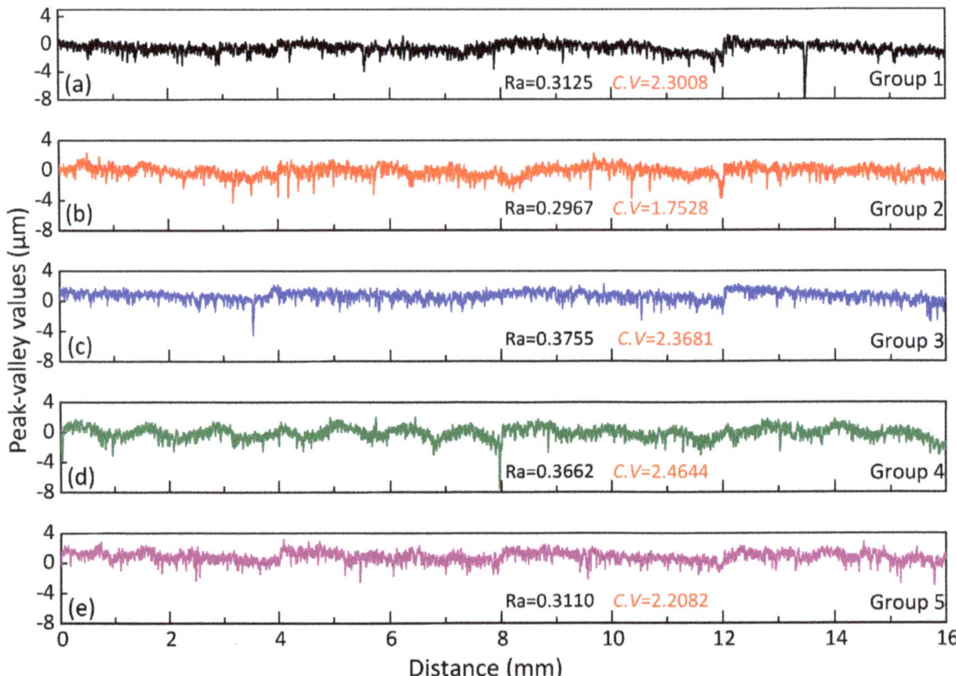

Figure 12. Surface profile autocorrelation function curves under multivariate optimization parameters. (**a**) Autocorrelation function curve under optimized parameter 1 (v_s = 1560.3760 m/min, v_w = 1094.4094 mm/min, a_p = 4.1712 μm). (**b**) Autocorrelation function curve under optimized parameter 2 (v_s = 1638.3838 m/min, v_w = 1033.6033 mm/min, a_p = 4.0624 μm). (**c**) Autocorrelation function curve under optimized parameter 3 (v_s = 1753.1953 m/min, v_w = 4708.3708 mm/min, a_p = 4.3400 μm). (**d**) Autocorrelation function curve under optimized parameter 4 (v_s = 1765.5965 m/min, v_w = 4647.1647 mm/min, a_p = 4.1208 μm). (**e**) Autocorrelation function curve under optimized parameter 5 (v_s = 1724.2324 m/min, v_w = 1286.8286 mm/min, a_p = 4.1016 μm).

Figure 13 shows the analysis of the periodic oscillation component of the signal in terms of ceramic surface profile and line texture curves. Figure 13a does not visualize the periodic components, so it is expected that the periodic components are found using the linear surface texture curve. A section of the surface texture curve (length 300 μm) in Figure 13a was selected because too many data points in the line were not conducive to the search for periodic oscillation signals. Figure 13b shows an enlarged view of the linear surface texture profile in Figure 13a. From Figure 13b, it can be seen that there are three levels of periodic oscillation signals (decreasing amplitude, light blue arrow, and blue arrow). The existence of periodic oscillation signals makes the surface obtain better surface quality (Figure 13a can be seen intuitively). The surface profile roughness value and the linear surface texture roughness value are 0.314 μm and 0.288 μm, respectively.

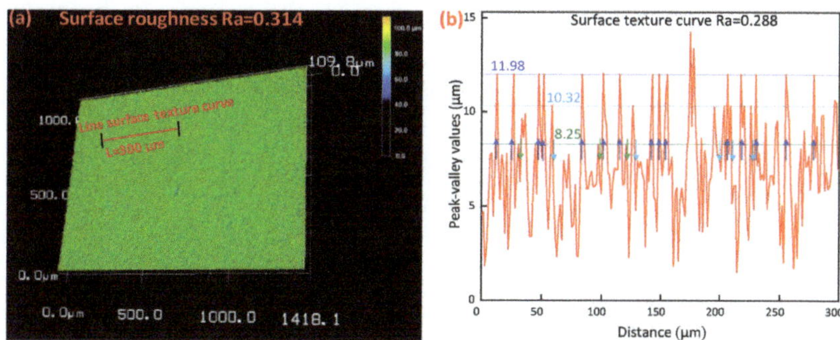

Figure 13. Ceramic surface profile and linear texture curve (v_s = 1724.23 m/min, v_w = 1286.83 mm/min, a_p = 4.10 μm). (**a**) 3D topography of the machined surface. (**b**) Surface texture peak-to-valley curve.

The analysis in this section reveals that the coefficient of friction is related to the roughness of the machined surface. By determining the appropriate coefficient of friction, it is possible to control and regulate the quality of the material production process indicators, to meet the requirements for the use of processed parts and products, which is conducive to the normal production operations of grinding enterprises. When the value for the coefficient of friction is taken between 0.197 and 0.216, the surface roughness of the workpiece is taken between 0.297 μm and 0.311 μm, the value of active energy consumption is taken between 318.769 J and 441.773 J, and effective grinding time is taken between 24.890 s and 30.207 s. The values for the coefficient of friction directly affect the roughness of the machined surface, the active energy consumption, and the effective grinding time, providing important theoretical guidance for the grinding process and providing reference significance for promoting the further development of friction mechanisms.

5. Conclusions

The performance of grinding surfaces directly affects the accuracy, service performance, and surface integrity of workpieces. A multi-objective numerical function was established by fusing and analyzing the collected multi-feature signals, taking into account the surface quality, coefficient of friction, active energy consumption, and effective grinding time. The following conclusions can be drawn:

i. The four sub-objective function models of surface roughness, coefficient of friction, active energy consumption, and effective grinding time are established with good accuracy. The correlation coefficients of them are high, with values of 0.89, 0.94, 0.97, and 0.98, respectively;

ii. The weight vectors of sub-objective functions were optimized by the MOEA/D algorithm in the multi-objective numerical function and two sets of optimal weight vectors were obtained. The weight vectors of (R_a, μ, E_a, T) are (0.52, 0.34, 0.09, 0.05) and (0.56, 0.36, 0.05, 0.03). The surface roughness R_a and coefficient of friction μ show a relatively heavy weight;

iii. Different working parameters were optimized by GA as grinding machine inputs. The optimal input parameters are experimentally verified to be (1638.38 m/min, 1033.60 mm/min, 4.06 μm) and (1724.23 m/min, 1286.83 mm/min, 4.10 μm). The surface roughness, coefficient of friction, active energy consumption, and effective grinding time obtained with the two sets of input parameters are (0.297 μm, 0.199, 441.773 J, 30.207 s) and (0.311 μm, 0.205, 318.769 J, 24.890 s), respectively;

iv. The coefficient of friction with a range of 0.197~0.216 was beneficial to the surface quality of the workpiece. Whether the friction coefficient tends to 0.197 or 0.216 will produce knowledge of chaos and bifurcation. When the coefficient of friction value tends to be 0.197, the smaller the coefficient of variation of the surface profiles, the smaller the distribution distance deviation of the microscopic data points. The distribution of data points becomes uniform. When it tends to be 0.216, the surface profile shows more periodic characteristics.

Author Contributions: Y.L. (Yang Li): conceptualization, methodology, investigation, data curation, modeling, grinding experiment, writing—original draft. L.J.: conceptualization, resources, writing—review and editing, validation, supervision. Y.L. (Yanhou Liu): methodology, writing, experiment. Y.T.: methodology, resources, financial support. T.Q.: methodology, writing—review and editing. T.Z.: methodology, writing—review and editing. X.W.: methodology, writing—review and editing. B.Z.: conceptualization, writing ideas, language correction, resources, financial support. All authors have read and agreed to the published version of the manuscript.

Funding: The authors would like to acknowledge the financial support from the National Natural Science Foundation of China (Grant No. 52205441, 51875329), Natural Science Foundation of Chongqing, China (Grant No. 2022NSCQ-MSX3775) and the Shandong Provincial Key Research & Development Project (Grant No. 2019GGX104073).

Data Availability Statement: The data presented in this study are available upon request from the first author.

Conflicts of Interest: The authors declare that they have no known competing financial interests or personal relationships that could have appeared to influence the work reported in this paper.

References

1. Lin, B.; Wang, H.; Wei, J.; Sui, T. Diamond wheel grinding characteristics of 3D orthogonal quartz fiber reinforced silica ceramic matrix composite. *Chin. J. Aeronaut.* **2021**, *34*, 404–414. [CrossRef]
2. Xia, L.; Lu, S.; Zhong, B.; Huang, L.; Yang, H.; Zhang, T.; Han, H.; Wang, P.; Xiong, L.; Wen, G. Effect of boron doping on waterproof and dielectric properties of polyborosiloxane coating on SiO_{2f}/SiO_2 composites. *Chin. J. Aeronaut.* **2019**, *32*, 2017–2027. [CrossRef]
3. Dong, W.; Ma, H.; Liu, R.; Liu, T.; Li, S.; Bao, C.; Song, S. Fabrication by stereolithography of fiber-reinforced fused silica composites with reduced crack and improved mechanical properties. *Ceram. Int.* **2021**, *47*, 24121–24129. [CrossRef]
4. Zheng, Z.; Huang, K.; Lin, C.; Zhang, J.; Wang, K.; Sun, P.; Xu, J. An analytical force and energy model for ductile-brittle transition in ultra-precision grinding of brittle materials. *Int. J. Mech. Sci.* **2022**, *220*, 107107. [CrossRef]
5. Fernández-Hernán, J.P.; López, A.J.; Torres, B.; Rams, J. Influence of roughness and grinding direction on the thickness and adhesion of sol-gel coatings deposited by dip-coating on AZ31 magnesium substrates. A Landau–Levich equation revision. *Surf. Coat. Technol.* **2021**, *408*, 126798. [CrossRef]
6. Zhu, S.Y.; Liu, Y.H.; Bai, T.B.; Shi, X.T.; Li, D.M.; Feng, L.B. High-efficient and robust fog collection through topography modulation. *Surf. Coat. Technol.* **2023**, *468*, 129747. [CrossRef]
7. Qin, R.; Zhang, Z.; Hu, Z.; Du, Z.; Xiang, X.; Wen, G.; He, W. On-line evaluation and monitoring technology for material surface integrity in laser shock peening-A review. *J. Mater. Process. Technol.* **2023**, *313*, 117851. [CrossRef]
8. Wen, J.R.; Fei, C.W.; Ahn, S.Y.; Han, L.; Huang, B.; Liu, Y.; Kim, H.S. Accelerated damage mechanisms of aluminized superalloy turbine blades regarding combined high-and-low cycle fatigue. *Surf. Coat. Technol.* **2022**, *451*, 129048. [CrossRef]
9. Wang, S.; Zhao, Q.; Wu, T. An investigation of monitoring the damage mechanism in ultra-precision grinding of monocrystalline silicon based on AE signals processing. *J. Manuf. Process.* **2022**, *81*, 945–961. [CrossRef]
10. Wang, S.; Sun, G.; Zhao, Q.; Yang, X. Monitoring of ductile-brittle transition mechanisms in sapphire ultra-precision grinding used small grit size grinding wheel through force and acoustic emission signals. *Measurement* **2023**, *210*, 112557. [CrossRef]
11. Ling, L.G.; Luo, L.; Liu, F.S. Effects of grinding treatment on surface properties and deformation microstructure in alloy 304L. *Surf. Coat. Technol.* **2021**, *408*, 126850. [CrossRef]
12. Zhang, X.; Chen, H.; Xu, J.; Song, X.; Wang, J.; Chen, X. A novel sound-based belt condition monitoring method for robotic grinding using optimally pruned extreme learning machine. *J. Mater. Process. Technol.* **2018**, *260*, 9–19. [CrossRef]
13. Tian, Y.B.; Liu, F.; Wang, Y.; Wu, H. Development of portable power monitoring system and grinding analytical tool. *J. Manuf. Process.* **2017**, *27*, 188–197. [CrossRef]
14. Feng, J.; Kim, B.S.; Shih, A.; Ni, J. Tool wear monitoring for micro-end grinding of ceramic materials. *J. Mater. Process. Technol.* **2009**, *209*, 5110–5116. [CrossRef]
15. Qin, F.; Zhang, L.X.; Chen, P.; An, T.; Dai, Y.W.; Gong, Y.P.; Yi, Z.B.; Wang, H.M. In situ wireless measurement of grinding force in silicon wafer self-rotating grinding process. *Mech. Syst. Signal Process.* **2021**, *154*, 107550. [CrossRef]
16. Warren, T.L.; Ting, C.; Qu, J.; Blau, P.J. A wavelet-based methodology for grinding wheel condition monitoring. *Int. J. Mach. Tools Manuf.* **2007**, *47*, 580–592. [CrossRef]

17. Ma, L.; Gong, Y.; Chen, X. Study on surface roughness model and surface forming mechanism of ceramics in quick point grinding. *Int. J. Mach. Tools Manuf.* **2014**, *77*, 82–92. [CrossRef]

18. Yao, Z.; Gu, W.; Li, K. Relationship between surface roughness and subsurface crack depth during grinding of optical glass BK7. *J. Mater. Process. Technol.* **2012**, *212*, 969–976. [CrossRef]

19. Kong, D.D.; Zhu, J.J.; Duan, C.Q.; Lu, L.X.; Chen, D.X. Bayesian linear regression for surface roughness prediction. *Mech. Syst. Signal Process.* **2020**, *142*, 106770. [CrossRef]

20. Meng, Q.; Guo, B.; Wu, G.; Xiang, Y.; Guo, Z.; Jia, J.; Zhao, Q.; Li, K.; Zeng, Z. Dynamic force modeling and mechanics analysis of precision grinding with microstructured wheels. *J. Mater. Process. Technol.* **2023**, *314*, 117900. [CrossRef]

21. Lei, X.; Xiang, D.; Peng, P.; Liu, G.; Li, B.; Zhao, B.; Gao, G. Establishment of dynamic grinding force model for ultrasonic-assisted single abrasive high-speed grinding. *J. Mater. Process. Technol.* **2022**, *300*, 117420. [CrossRef]

22. Liu, R.; Yang, B.; Zio, E.; Chen, X. Artificial intelligence for fault diagnosis of rotating machinery: A review. *Mech. Syst. Signal Process.* **2018**, *108*, 33–47. [CrossRef]

23. Dai, C.; Ding, W.; Zhu, Y.; Xu, J.; Yu, H. Grinding temperature and power consumption in high speed grinding of Inconel 718 nickel-based superalloy with a vitrified CBN wheel. *Precis. Eng.* **2018**, *52*, 192–200. [CrossRef]

24. Wang, J.L.; Tian, Y.B.; Hu, X.T.; Li, Y.; Zhang, K.; Liu, Y.H. Predictive modelling and Pareto optimization for energy efficient grinding based on aANN-embedded NSGA II algorithm. *J. Clean. Prod.* **2021**, *327*, 129479. [CrossRef]

25. Wang, J.L.; Tian, Y.B.; Zhang, K.; Liu, Y.H.; Cong, J.C. Online prediction of grinding wheel condition and surface roughness for the fused silica ceramic composite material based on the monitored power signal. *J. Mater. Res. Technol.* **2023**, *24*, 8053–8064. [CrossRef]

26. Li, Y.; Liu, Y.H.; Wang, J.L.; Wang, Y.; Tian, Y.B. Real-time monitoring of silica ceramic composites grinding surface roughness based on signal spectrum analysis. *Ceram. Int.* **2022**, *48*, 7204–7217. [CrossRef]

27. Wang, J.L.; Li, J.W.; Tian, Y.B.; Liu, Y.H.; Zhang, K. Methods of grinding power signal acquisition and dynamic power monitoring database establishment. *Diam. Abras. Eng.* **2022**, *42*, 356–363. [CrossRef]

28. Zhang, K.; Tian, Y.B.; Cong, J.C.; Liu, Y.H.; Yan, N.; Lu, T. Reduction grinding energy consumption by modified particle swarm optimization based on dynamic inertia weigh. *Diam. Abras. Eng.* **2020**, *41*, 71–75. [CrossRef]

29. Wang, J.L.; Tian, Y.B.; Hu, X.T.; Han, J.G.; Liu, B. Integrated assessment and optimization of dual environment and production drivers in grinding. *Energy* **2023**, *272*, 127046. [CrossRef]

30. Zhang, Q.; Li, H. MOEA/D: A Multiobjective Evolutionary Algorithm Based on Decomposition. *IEEE Trans. Evol. Comput.* **2008**, *11*, 712–731. [CrossRef]

31. Jaszkiewicz, A. On the performance of multiple-objective genetic local search on the 0/1 knapsack problem-a comparative experiment. *IEEE Trans. Evol. Comput.* **2000**, *6*, 402–412. [CrossRef]

32. Wang, W.; Li, K.; Tao, X.; Gu, F. An improved MOEA/D algorithm with an adaptive evolutionary strategy. *Inf. Sci.* **2020**, *539*, 1–15. [CrossRef]

33. Qin, S.; Sun, C.; Zhang, G.; He, X.; Tan, Y. A modified particle swarm optimization based on decomposition with different ideal points for many-objective optimization problems. *Complex Intell. Syst.* **2020**, *6*, 263–274. [CrossRef]

34. Li, Y.; Liu, Y.H.; Tian, Y.B.; Wang, Y.; Wang, J.L. Application of improved fireworks algorithm in grinding surface roughness online monitoring. *J. Manuf. Process.* **2022**, *74*, 400–412. [CrossRef]

35. Jiang, J.; Ge, P.; Hong, J. Study on micro-interacting mechanism modeling in grinding process and ground surface roughness prediction. *Int. J. Adv. Manuf. Technol.* **2013**, *67*, 1035–1052. [CrossRef]

36. Wang, Q.; Cong, W.; Pei, Z.J.; Gao, H.; Kang, R. Rotary ultrasonic machining of potassium dihydrogen phosphate (KDP) crystal: An experimental investigation on surface roughness. *J. Manuf. Process.* **2009**, *11*, 66–73. [CrossRef]

37. Zhang, Y.; Li, C.; Jia, D.; Li, B.; Wang, Y.; Yang, M.; Hou, Y.; Zhang, X. Experimental study on the effect of nanoparticle concentration on the lubricating property of nanofluids for MQL grinding of Ni-based alloy. *J. Mater. Process. Technol.* **2016**, *232*, 100–115. [CrossRef]

38. Jiang, J.L.; Ge, P.Q.; Bi, W.B.; Zhang, L.; Wang, D.X.; Zhang, Y. 2D/3D ground surface topography modeling considering dressing and wear effects in grinding process. *Int. J. Mach. Tools Manuf.* **2013**, *74*, 29–40. [CrossRef]

39. Zhang, X.; Li, C.; Zhang, Y.; Wang, Y.; Li, B.; Yang, M.; Guo, S.; Liu, G.; Zhang, N. Lubricating property of MQL grinding of Al$_2$O$_3$/SiC mixed nanofluid with different particle sizes and microtopography analysis by cross-correlation. *Precis. Eng.* **2017**, *47*, 532–545. [CrossRef]

40. Zhang, X.; Li, C.; Zhang, Y.; Jia, D.; Li, B.; Wang, Y.; Yang, M.; Hou, Y.; Zhang, X. Performances of Al$_2$O$_3$/SiC hybrid nanofluids in minimum-quantity lubrication grinding. *Int. J. Adv. Manuf. Technol.* **2016**, *86*, 3427–3441. [CrossRef]

 *lubricants*

Article

The Surface Assessment and the Properties of Selected Multilayer Coatings

Bogdan Warcholinski [1,*], Adam Gilewicz [1] and Maria Tarnowska [2]

[1] Faculty of Mechanical Engineering, Koszalin University of Technology, 75-453 Koszalin, Poland; adam.gilewicz@tu.koszalin.pl
[2] Centre for Biological Treatment, Independent Public Clinical Hospital No. 1, Pomeranian Medical University in Szczecin, 71-252 Szczecin, Poland; m.tarnowska@spsk1.szn.pl
*   Correspondence: bogdan.warcholinski@tu.koszalin.pl

Abstract: The paper presents an evaluation of the surface quality and properties of multilayer coatings, obtained using cathodic arc evaporation, of the same structure, in which the top layer is a CrN chromium nitride layer. The second components of a double-layer module with a thickness of 400 nm and a thickness of each layer about 200 nm are two component TiN, Mo_2N systems and three component TiAlN and CrCN systems. In studies using scanning electron microscopy and optical microscopy, the surface density of the macroparticles of the coating and their dimensions were estimated. The largest amount of macroparticles was recorded on the surface of the TiAlN/CrN coatings and the lowest on CrCN/CrN and Mo_2N/CrN coatings. Their adhesion to steel substrates using a scratch test and Rockwell test and wear were also investigated. The results indicated that the melting point of the cathode material directly affected the number and size of the macroparticles on the surface of the growing coating. The number of macroparticles increased with the lowering of the melting point of the cathode material. All the coatings showed good adhesion with the critical load Lc_2, greater than 60 N with a hardness above 20 GPa. The Mo_2N/CrN coating, despite its relatively low critical load compared to the other tested coatings, had the best wear-resistant properties, which was probably due to the $Mo_2N \rightarrow MoO_3$ transformation.

Keywords: PVD; multilayer coatings; roughness; friction; wear

Citation: Warcholinski, B.; Gilewicz, A.; Tarnowska, M. The Surface Assessment and the Properties of Selected Multilayer Coatings. *Lubricants* **2023**, *11*, 371. https://doi.org/10.3390/lubricants11090371

Received: 8 August 2023
Revised: 28 August 2023
Accepted: 1 September 2023
Published: 3 September 2023

1. Introduction

Some of the transition metals (Ti, V, Cr, Zr, Nb, Mo, Ta, and W) are widely used for forming composite coatings of nitrides and carbides. They are characterized by good mechanical properties, mainly a high hardness, good adhesion, toughness, wear by friction, and temperature resistance, but also a high reactivity with mating material.

The world's growing population has caused a significant increase in the demand for wood and wood composites for various applications. This has caused an increase in the interest in high-performance wood-processing techniques, and mainly, the tools used in them. Among the most effective are tools those with thin, hard coatings that allow for working in dry conditions. Thus, in the case of wood, a hygroscopic material, its treatment can be carried out without cooling lubricants. Currently, many works are being carried out aimed at developing new coatings enabling the more effective processing of wood raw material [1,2].

The coatings of transition metal nitrides synthesized using PVD (Physical Vapor Deposition) exhibit properties that enable multiple applications. They allow for increasing tool life and improving their resistance to corrosion. Their use can reduce the consumption of coolants or eliminates them from the treatment of certain types of materials. Titanium nitride TiN coatings are characterized by a high hardness, but start to oxidize at a temperature of 550 °C [3]. Chromium nitride coatings are more resistant to oxidation compared to other metal nitride coatings, and show a good wear resistance. Both coatings are characterized by a very good adhesion to substrates. A slightly worse adhesion is presented by CrCN [4] and Mo_2N [5] coatings.

The coatings formed using cathodic arc evaporation are characterized by a high hardness, good adhesion, high density, and homogeneity, and they show better properties than coatings prepared using magnetron sputtering. The disadvantage of this deposition method is the relatively high roughness of the coatings' surfaces, resulting from a large number of macroparticles on these surfaces. The number of macroparticles per unit area and their linear dimensions are dependent on the melting point of the cathode [6,7]. The studies of Creasy et al. [6] and Munz et al. [7] indicated that, as the melting point of a cathode increases the number of macroparticles, the maximum macroparticle size and fraction of sample area covered by macroparticles decrease. Harris et al. [8] investigated the effect of nitrogen pressure increase during the formation of TiN coatings on HSS (M2) jobber drills on the surface quality and efficiency of the drilling process. They showed that the coatings formed at a higher nitrogen pressure were characterized by a lower roughness and the number of macroparticles per unit area and maximum macroparticle diameter were significantly reduced, probably increasing the tool life.

Equally important as the selection of the method of formation of protective coatings (magnetron sputtering or cathodic arc evaporation) is the selection of the coating itself, including its type. Multilayer coatings generally have better properties than single-layer coatings. They have a greater hardness and adhesion to substrates, which often leads to a greater resistance to wear and corrosion. A modification of the structure of the coating [9,10] enables it to improve its functional properties. Multilayer coatings are characterized by an increased stability compared to monolayer coatings. This is due to their different mechanism of destruction [10,11], caused also by the interruption of the columnar growth of the coating. This may cause crack deflection, leading to a toughening of the coating in the interlayer zone, and may reduce stress concentration and crack propagation [11].

The properties of two-component coatings can be improved by changing their chemical composition—by doping metallic or non-metallic elements or modifying the structure of the coating—for example, reducing the grain size or forming a multilayer coating. The design of the multilayer coating can include the thickness of each layer, the chemical and phase composition, and its mechanical properties (hardness and Young's modulus) [8]. The coating used on tools should be characterized by a chemical inertness of the outer surface of the coating with processed material, a high hardness at the middle part of the coating, and a good adhesion to the substrate. A single-layer coating rarely meets all of these requirements. For example, an increase in hardness results in a reduction in coating ductility. A modification of the structure of a coating [9,10] enables it to improve its functional properties.

Santana et al. [12], examining the properties of TiAlN/CrN coatings, indicated the advantages of the use of a bilayer with a thickness of 400 nm and broad transition zones between the layers of the coating. These include increases in hardness and adhesion resulting from the interdiffusion of the layer components, due to a high energy ion bombardment, which is characteristic for arc coating processes. This reduces the mismatch of the lattice parameters of the crystal structures of the layers. Simultaneously, the temperature-activated diffusion process, with a larger thickness of modules, does not degrade the multilayer structure. In turn, Kot et al. [13], analyzing Cr/CrN multilayer coatings with thicknesses of (Cr + CrN) their bilayer from 62 to 1000 nm, found that, for the thickness of the bilayer from 250 to 500 nm, both the hardness, Young's modulus, and adhesion of the coating were the highest.

Due to their properties [14–16], chromium nitride coatings are widely used in many industrial applications [17–19]. The incorporation of Al into chromium nitride coatings improves their oxidation resistance at elevated temperatures (above 900 °C [20]) and their hardness, significantly increasing their durability. Such coatings are applied, for example, to woodworking tools [21–24].

TiAlN is characterized by a higher *temperature* resistance than *TiN*. TiAlN coatings oxidize at a temperature above 700 °C. The high wear of TiAlN coatings (higher than TiN coatings) operated at a low temperature results from their brittleness and high coefficient of friction. The main advantage of such a coating is its high-temperature stability. In TiAlN coatings "working" in an oxidizing environment at a temperature above 700 °C, a two-layer structure is formed. In the upper art of the aluminum-rich coating, a dense Al_2O_3 layer is formed, and in the inner part, a layer with a predominant titanium concentration. The top Al_2O_3 layer reduces diffusion wear during high-temperature machining and is renewed when the outer part of the coating is worn off. At a high temperature, chromium and aluminum oxides have lubricating properties. In addition, aluminum oxide acts as a diffusion barrier, blocking the diffusion of oxygen into the coating. The hardness of TiAlN coatings at a high temperature strongly depends on their chemical composition and is higher than that of TiN coatings [25,26].

It was found that the addition of chromium to TiAlN coatings improves their resistance to oxidation [27].

Molybdenum nitride coatings have excellent properties: a low wear rate and low coefficient of friction, combined with a high hardness and good adhesion, makes them a good candidate for tribological applications [28–30]. In an oxidizing environment and at an elevated temperature, molybdenum nitride oxidizes and MoO_3 is formed. This allows it to be used as a self-lubricating tribological coating operating at a high temperature [31].

There have been many studies on the properties of multi-layer coatings Me(X)N/CrN, where Me is a transition metal and X is an additional element, but few studies have included two-layer coatings with CrN as a topcoat. This is especially true for Mo_2N/CrN coatings, in which molybdenum has the ability to form oxides, known as solid lubricants, significantly reducing their wear. The small coefficient of friction of about 0.3–0.4, small compared to CrN (0.6) and TiAlN (0.8) coatings, and may be associated with the formation of MoO_3 in the friction process, characterized by favorable sliding properties. A similar effect, lowering the coefficient of friction, can be achieved by doping the coating with carbon. Hence, we attempt to compare the properties of Mo_2N/CrN and CrCN/CrN coatings.

The aim of the work was to check how the roughness of the multilayer coating changed:

(a) with the same CrN surface layer and a second layer formed from cathodes with different melting points: TiAl (approx. 1450 °C) and Mo (approx. 2650 °C),

(b) when both layers in the multilayer coating were formed from the same cathode (Cr) but one of the layers was doped with a non-metallic element, carbon.

It was also important to study the mechanical properties of these coatings and carry out industrial tests on tools covered with these coatings by planing dry pine in order to determine their suitability for this type of processing.

2. Materials and Methods

2.1. Coating Deposition

TiN, TiAlN, CrCN, Mo_2N, and CrN monolayer coatings, and related TiN/CrN and Mo_2N/CrN and TiAlN/CrN and CrCN/CrN multilayer coatings, were formed via cathodic arc evaporation using the TINA 900M device on silicon wafers (for stress measurements), HS6-5-2 steel substrates with a diameter of 32 mm (for mechanical tests), and planer knives made of the same steel, with dimensions of $160 \times 30 \times 3$ mm, manufactured by Leitz GmbH & Co. KG, Oberkochen, Germany (for industrial testing). In this project, Ti (99.999%) $Ti_{30}Al_{70}$ (99.9%) Mo (99.98%), and Cr (99.999%) cathodes with a diameter of 100 mm and thickness of 12 mm were used. The substrates (hardness above 62 HRC) were ground and polished to a roughness parameter Ra of 0.02 μm, and then they were chemically degreased and ultrasonically cleaned in a hot alkaline bath for 10 min. After washing in distilled water and hot-air drying, the substrates were mounted on the rotating holder (2 rev/min) in a vacuum chamber at a distance of 18 cm from the arc sources. The chamber was evacuated to a pressure of 1×10^{-3} Pa. Prior to the deposition, the substrates were sputter cleaned at a nitrogen pressure of 0.5 Pa using argon and chromium ions under a substrate bias voltage of -600 V for 10 min. Cr ion bombardment was used to remove contamination and the oxide layer from the surfaces of the substrates, enabling the good adhesion of the deposited coatings. A thin (about 0.1 μm) chromium layer was also deposited onto the substrate to improve its adhesion [32]. A deposition process was performed on substrates heated to a temperature of 300 °C, at a substrate bias voltage of -70 V, with the arc current and nitrogen pressure corresponding to each of the targets (Table 1). A CrCN coating was formed by introducing acetylene C_2H_2 (flow rate 10 sccm) to the vacuum chamber.

Table 1. The deposition parameters of the coatings.

	TiN	Mo$_2$N	CrN	TiAlN	CrCN
Nitrogen pressure [Pa]	0.5	1.8	1.8	1.0	1.8
Arc current [A]	80	140	80	60	80

Multilayer coatings were synthesized by alternating the deposition of monolayers from two sources placed on the opposite sides of the vacuum chamber. The thickness of the layer depended on the exposure time of the substrate in front of a given source, i.e., the rotational speed holder on which the substrate was placed. The multilayer coatings were composed of repeated bilayers with a constant thickness. One of the layers in each of the investigated multilayer coatings was a CrN layer. The coating thickness was controlled by the time of deposition. The basic parameters of the multilayer coatings' structures are summarized in Table 2.

Table 2. The structure of multilayer coatings.

	TiN/CrN	Mo$_2$N/CrN	TiAlN/CrN	CrCN/CrN
Coating thickness, (μm)	2.8	3.1	4.0	3.0
Number of bilayer	6	6	8	7
Bilayer thickness, (nm)	450	500	500	400
Thickness ratio of the layers in bilayer	1:1	1:1	1:1	1:1

2.2. Characterization

The microstructure and morphology of the coating surfaces were determined using a scanning electron microscope (JEOL JSM-5500LV). An analysis of the composition was completed using energy dispersive X-ray spectroscopy, EDS (Oxford Link ISIS 300).

The surface roughness of the coatings was evaluated with a Hommel Werke T8000 profilometer. The test was performed five times for each sample.

To define the size of the macroparticles and number of macroparticles on the unit of area on the surface of the coatings, the metallographic microscope Nikon Eclipse MA200 equipped with NIS-Elements software was used. The measurement area was about 133×171 μm^2. The surfaces of the coatings were recorded at the same magnification ($400\times$), contrast, and sharpness. For each sample, five measurements in one line at equal distances from each other were carried out.

The coating thickness was determined using the ball-cratering method (Calotest), according to the EN 1071-2: 2002 European Standard.

The adhesion of the coatings was evaluated using a scratch test (Revetest Scratch Tester, CSEM) with a diamond indenter Rockwell C type with a 0.2 mm tip radius. The pre-defined parameters were: a scratch length of 10 mm, scratch speed of 10 mm/min, and normal load (100 N) progressively increasing at the speed of 100 N/min. The Lc$_1$ critical load was determined as the load at which the first cracks in the coating appeared, and the Lc$_2$ critical load as the load at which the total detachment of the coating from the substrate was observed. These loads were evaluated by observation using an optical microscope as the average of at least three measurements. These results were verified using the Daimler-Benz test [33], which qualitatively evaluates the adhesion of coatings as a size and types of failure modes of the coating formed as a result of a Rockwell indentation force of about 1500 N.

For the hardness measurements, the nanohardness tester FISCHERSCOPE HM2000 equipped with a Berkovivh indenter was used. The hardness measurement of the coatings deposited using cathodic arc evaporation was difficult because of the large number of macroparticles on the surface, resulting in a high roughness. Thus, meeting the requirements for a correct measurement of hardness, so that the depth of the indentation did not exceed 1/10 of the total coating thickness (in order to eliminate the effect of the substrate) and the depth of the indentation was not less than 20 × Ra (arithmetic mean roughness), was difficult. Therefore, the method described by Romero et al. [34] was used. For this reason, the coatings were polished with fine diamond powder (1 μm) in order to remove these surface defects. After this operation, significantly reducing the surface roughness of the coatings, the statistic of the measurements was highly improved. In total, 20 measurements were made for each sample. At a fixed indenter penetration depth of 0.2 μm, using the device software (WIN-HCU software), the average value of the coating hardness and its Young's modulus, as well as measurement uncertainties, were obtained.

The stress of the coating was determined using the Stoney formula [35]:

$$\sigma = \frac{E}{1 - v} \frac{t_s^2}{6t_f} \frac{1}{R} \tag{1}$$

Silicon wafers with a thickness of 0.3 mm, a length of 30 mm, and a width of 4 mm were used.

Studies carried out in the pin-on-disc system enable an estimation of the tribological properties of coatings, i.e., their coefficient of friction and wear rate. The parameters of the test were: a normal load of 20 N, sliding speed of about 0.2 m/s, distance of 1000 m, environment—dry friction in air at a temperature of 20 °C, and humidity of about 50%. Alumina balls with a diameter of 10 mm and roughness R_a of <0.03 μm were used as a counterpart. The wear rate was estimated as the wear volume (based on wear track profile—Hommel Werke T8000) divided by the normal load and the sliding distance [36].

Industrial tests were carried out using a Weinig H23 four-side planer machine equipped in 6 heads. The working parameters were: cutting speed—50 m/s, spindle speed—6000 min^{-1}, feed rate—73 m/min, and cutting depth—1 mm. The machined material was overdried pine wood. Under these conditions, industrial tests were carried out for 8 h. Their purpose was not to determine the actual increase in durability, but to evaluate the rake face of knives modified with the investigated coatings in the same time and under the same working conditions of the tools.

3. Results

3.1. Chemical Composition of the Coatings

The chemical composition of the investigated coatings is presented in Table 3. This composition applies to the entire coatings, not the individual layers.

3.2. Surface Morphology

The images of the PVD (Figure 1) coatings' surfaces revealed a large number of defects, macroparticles, or pinholes formed when removed, or as a shadow effect when it was applied. These defects resulted in an increase in the surface roughness of the coating and worsening of its quality. This is a negative phenomenon that is the subject of relatively few works [6,7,37,38].

Table 3. The chemical composition of the coatings (at.%).

	TiAlN/CrN	Mo$_2$N/CrN	CrCN/CrN
Ti	13.8	-	-
Al	25.7	-	-
Mo	-	29	-
Cr	6.8	23.5	50.2
C	-	-	4.9
N	53.7	47.5	44.9

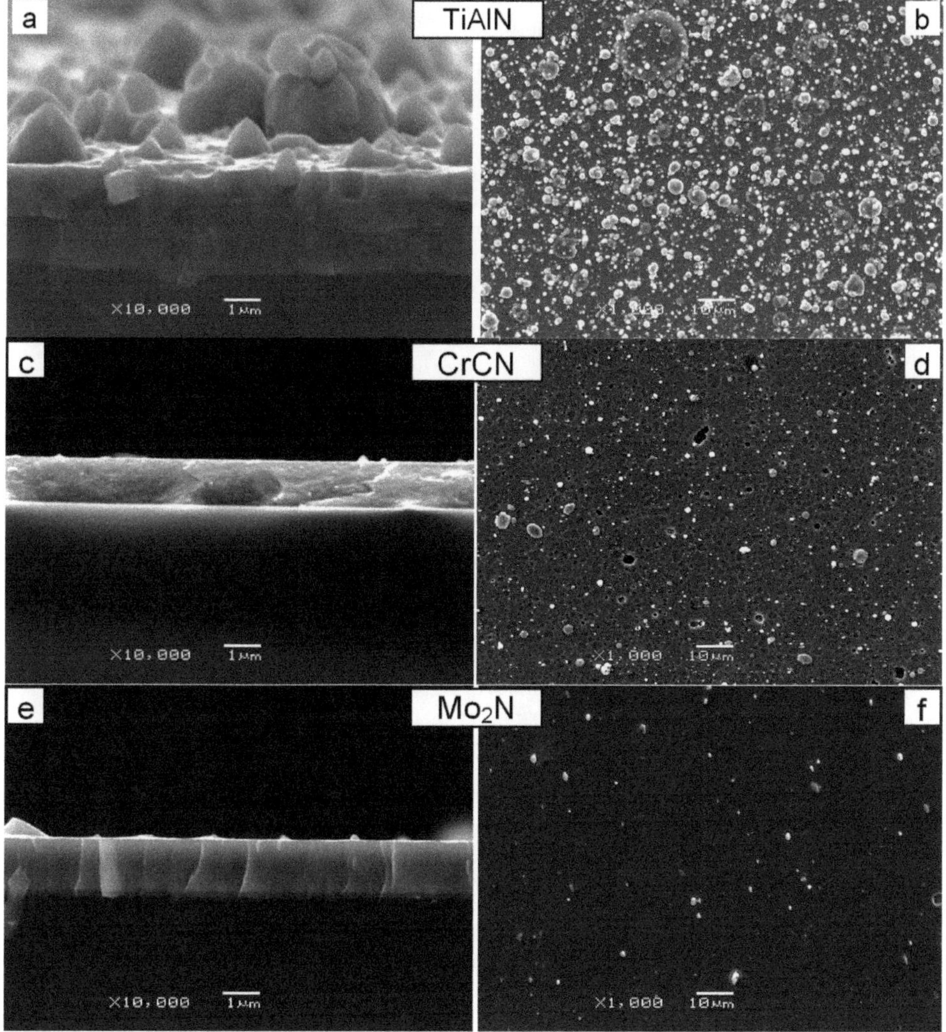

Figure 1. *Cont.*

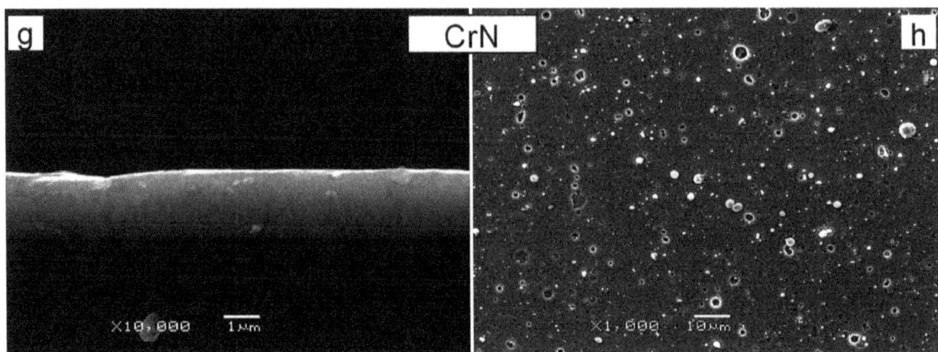

Figure 1. Images of cross-sections (**a,c,e,g**) and surface (**b,d,f,h**) of TiAlN (**a,b**), CrCN (**c,d**), Mo$_2$N (**e,f**), and CrN (**g,h**) monolayer coatings.

The top layer of the all the multilayer coatings investigated was a chromium nitride layer. Despite this, the amount of macroparticles on the coating surfaces varied. This was shown in two multilayer coatings: TiAlN/CrN and CrCN/CrN—Figure 2. The multilayer coating structure is clearly marked. The column growth was significantly limited, although possible.

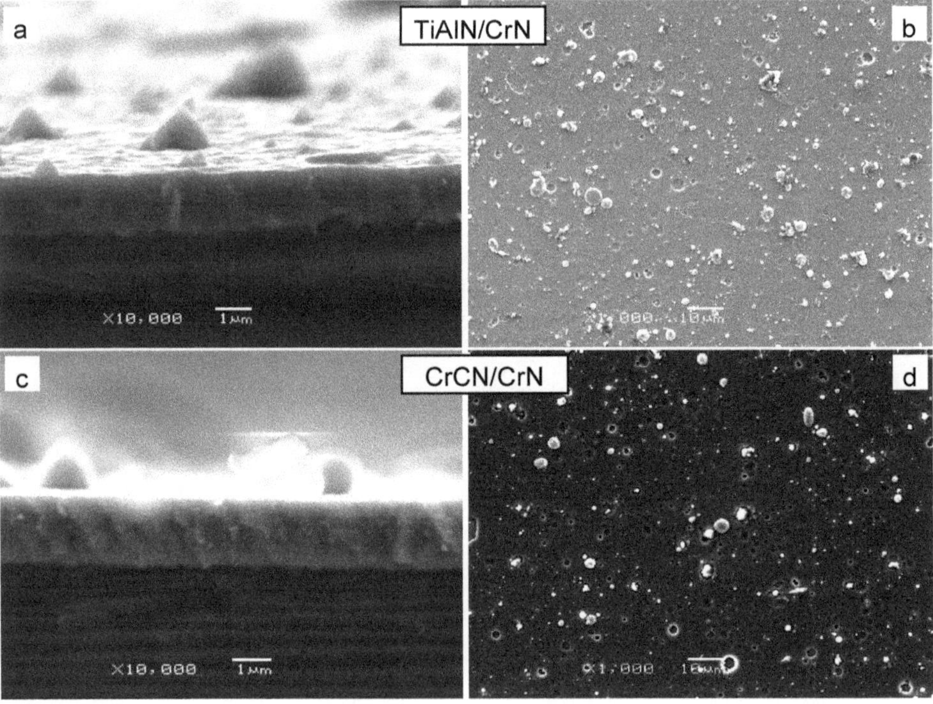

Figure 2. Images of cross-sections (**a,c**) and surface (**b,d**) of TiAlN/CrN (**a,b**) and CrCN/CrN (**c,d**) multilayer coatings.

In the cathodic arc evaporation method, the plasma jet contains not only electrons, ions, and atoms, but also droplets of the molten metal of the cathode. They are the biggest disadvantage of the arc method. Droplets of different sizes are deposited on the substrate. One can observe only a small part of them on the surface. Some of them extend from the substrate to the coating surface. Smaller droplets with spherical shapes and larger, more irregular droplets are often "anchored" in the coating.

3.3. Macroparticle Statistics

The measurement of the size and quantity of the macroparticles on the surface of the coating was performed at the same area of about 133×171 μm^2 at five randomly selected points. The number of registered particles and also the fraction of the sample area covered by the macroparticles, as well as the surface roughness Ra of the coating, are presented in Table 4.

Table 4. The number of macroparticles and fraction of sample area covered by macroparticles on investigated multilayer coatings formed using cathodic arc evaporation.

Type of the Coating	TiN/CrN	Mo$_2$N/CrN	TiAlN/CrN	CrCN/CrN
Area (μm^2)	2.27×10^4	2.27×10^4	2.27×10^4	2.27×10^4
Number of macroparticles	1281 ± 120	677 ± 40	3844 ± 720	612 ± 210
Fraction of sample area covered by macroparticles	15 ± 1	3 ± 1	41 ± 7	2.6 ± 0.9
Roughness Ra (μm)	0.20	0.09	0.32	0.08

Although the top layer of each multilayer coating was a CrN layer, one could observe very variable quantities of the macroparticles on their surfaces. The smallest amount (about 600–700 macroparticles) was observed on the CrCN/CrN and Mo$_2$N/CrN coatings, while much more was registered for the TiAlN/CrN coating. A large scatter of the fraction of the sample area covered by macroparticles was also found. For the Mo$_2$N/CrN and CrCN/CrN coatings, it was about 3%, while for the TiAlN/CrN coatings, it was more than 40%. These results correlated completely with the observations arising from Figures 1 and 2. Similar results were obtained by Wan et al. [37] for CrN coatings deposited using ion plating.

A dimension analysis of the macroparticles on the surfaces of the coatings tested indicated that about 80% of the objects were sized to 1.5 μm. The second-largest group of objects, about 10%, had dimensions in the range from 1.5 to 3.0 μm. It was interesting that part of each of the size fractions was similar for each of the coatings, regardless of their chemical composition—Figure 3.

A detailed analysis of the size distribution of the macroparticles was performed for two multilayer coatings: TiN/CrN and CrCN/CrN—Figure 4. The TiN/CrN multilayer coating had almost a two times greater macroparticles surface density compared to the more complex CrCN/CrN coating, while the dimensional distribution of the macroparticles in both coatings was similar. This may have been associated with the lower melting point of the titanium cathode (1660 °C) compared to the chromium cathode (1870 °C).

In multilayer coatings, their surface quality (e.g., expressed by the roughness parameter Ra) depends on the process conditions for the coating deposition of various cathodes and the kind of cathodes used. An analysis of the surfaces of the CrN and CrCN monolayer coatings [38] revealed that they had a surface roughness Ra, respectively, of (0.06 ± 0.01) μm and (0.10 ± 0.02) μm, and the CrCN/CrN multilayer coating had an intermediate value of Ra (0.08 ± 0.01) μm. Otherwise, it was with the Mo$_2$N/CrN multilayer coating, characterized by a roughness (Ra about 0,09 μm) higher than the coating component: Mo$_2$N, with a Ra about 0.04 μm [5]) and CrN, with a Ra about 0.06 μm [38].

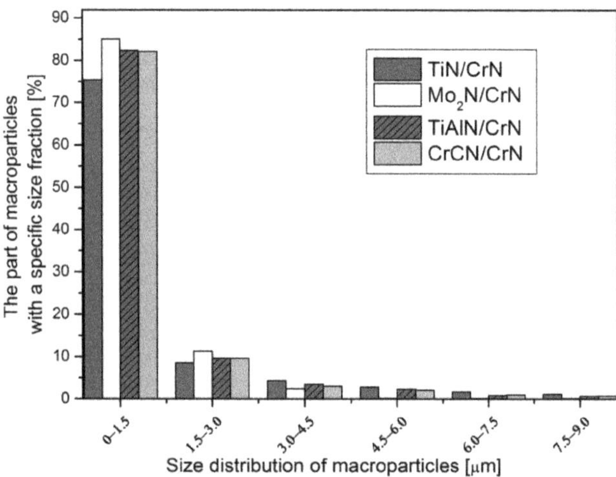

Figure 3. Part of macroparticles on the surface of multilayer coatings formed by cathodic arc evaporation, depending on the size distribution of macroparticles.

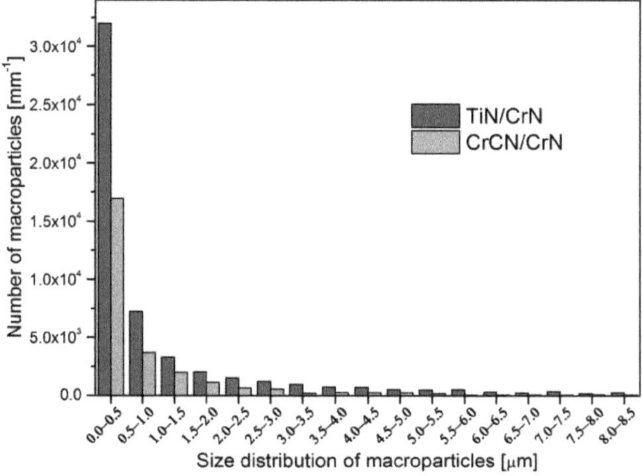

Figure 4. Size distribution of the macroparticles for TiN/CrN and CrCN/CrN multilayer coatings.

Surface roughness is defined as its irregularity associated with the occurrence of height and valleys in relation to a specific mean line. In many tribological applications, the maximum irregularity heights above the mean line are an important parameter, since they may damage the contact surfaces of the friction bodies. However, in the case of lubrication in the cavities, a lubricant reducing the friction between rubbing bodies may accumulate. Rough surfaces usually wear more quickly and have higher friction coefficients than smooth surfaces.

The surface roughness of coatings can be regulated by an arc current—Figure 5, but it should be remembered that a reduction in the arc current decreases the deposition rate of the coating, which increases the time of the deposition process.

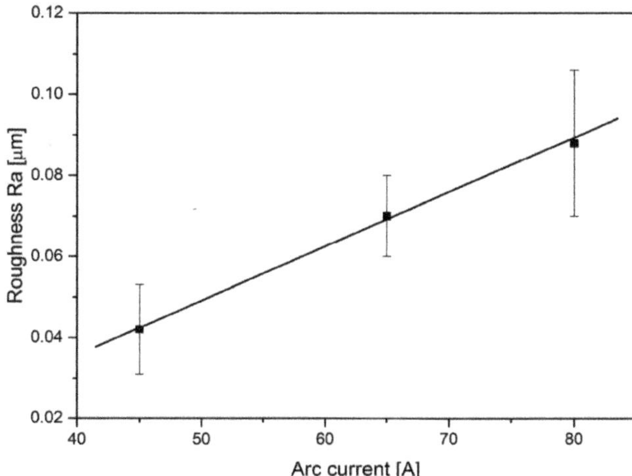

Figure 5. The relationship between surface roughness Ra of CrN monolayer coating obtained by cathodic arc evaporated depending to the arc current.

3.4. Hardness

The hardness of the coatings and other properties are summarized in Table 5. It should be noted that the hardness of TiAlN/CrN (21 ± 1 GPa) was lower than described in the literature [27,39–41], however, the thickness of the (TiAlN + CrN) bilayer was much smaller, at less than 50 nm. The hardness of the CrCN/CrN coating (24 ± 2 GPa) was higher than the hardness of both the chromium nitride (18 GPa) and chromium carbonitride (22 GPa). The Mo_2N/CrN coatings' hardness was comparable with Ref. [42].

Table 5. The mechanical properties of TiAlN/CrN, CrCN/CrN, and Mo_2N/CrN multilayer coatings.

Type of the Coating	TiAlN/CrN	CrCN/CrN	Mo_2N/CrN
Hardness H (GPa)	21 ± 1	24 ± 2	26 ± 2
Stress (GPa)	−1.0 ± 0.1	−1.5 ± 0.1	−1.1 ± 0.1
Young modulus E (GPa)	291 ± 15	258 ± 9	310 ± 12
H/E	0.072 ± 0.007	0.093 ± 0.011	0.085 ± 0.010
H^3/E^2 (GPa)	0.11 ± 0.03	0.21 ± 0.07	0.18 ± 0.06
Critical load Lc_2 (N)	82 ± 3	105 ± 2	59 ± 4
Coefficient of friction	0.72 ± 0.04	0.58 ± 0.04	0.40 ± 0.02
Wear rate of the coating ($mm^3N^{-1}m^{-1}$)	$(2.5 ± 0.2) \times 10^{-7}$	$(1.3 ± 0.6) \times 10^{-7}$	$(9.0 ± 0.6) \times 10^{-8}$

3.5. Adhesion

The multilayer coatings discussed were characterized by a good adhesion to the substrate, Figure 6. This amounted to 59 N for the Mo_2N/CrN coating, 82 N for the TiAlN/CrN coating, and about 105 N for the CrCN/CrN coating. These values are close to those presented in the literature, although some single-layer coatings have a lower critical load [18,28,43,44].

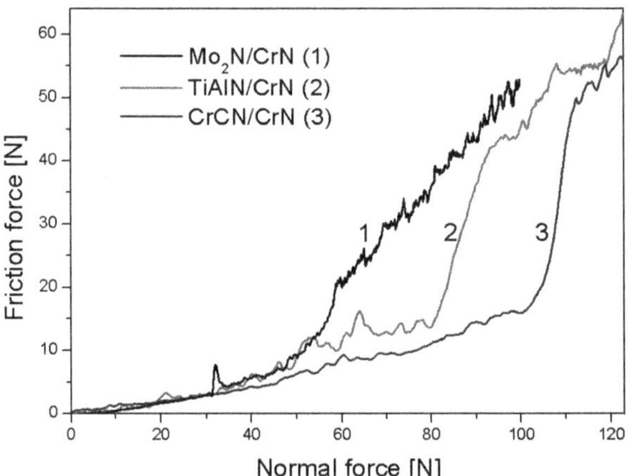

Figure 6. The friction force versus normal force in scratch test for coatings investigated.

Optical micrographs of the scratch section for the coatings tested at three loads are shown in Figure 7. The first is where the first coating failure was observed (Lc_1), the second shows the detachment of the coating from the substrate occurring (Lc_2), and the third is for an intermediate load. A relatively large critical force Lc_1 of the investigated coatings and a large difference between the loads Lc_1 and Lc_2 should be noted. This indicates a considerable resistance to the destruction of the coating. In the image of the crack formed after applying a load greater than Lc_1, there are numerous coating chips at the scratch border. This cracking of the coating and its partial delamination were related to the plastic deformation of the substrate and resulted from the different mechanical properties (hardness and Young's modulus) of the coating and the substrate.

In the scratch track, fine conformal hemispherical cracks perpendicular to the scratch direction are visible, indicating the high hardness and high adhesion of the coating. Minor chipping and cracking of the coating did not adversely affect its good properties. In the pictures of the scratches loaded by Lc_2, there are significant areas of coating losses.

The Daimler-Benz test is easy to apply, competent, and reliable, due to the large force applied indentation in the coating evaluation method for its adhesion and mechanisms for its destruction. The test allows for a visual assessment of the adhesion and durability of hard coatings. As the adhesion depends on the film thickness, hardness, and roughness of the substrate, a coating thickness of about 3–4 μm was used.

The Mo_2N/CrN coating adhesion test results are shown in Figure 8. Both distinct and circumferential and radial cracks are visible. Due to the pill-up of the substrate material around the indentation, there existed a high level of tensile stress. For brittle and hard materials, these stresses cause the formation of large circular cracks on the circumference of indentation [32].

In the TiAlN/CrN and CrCN/CrN coatings, probably due to their slightly lower hardness, only radial cracks occurred, which are frequent in the case of brittle coatings showing a good adhesion. In the TiAlN/CrN coatings, angular intersecting cracks also appeared, which may have contributed to an easier detachment of the coating from the substrate, as shown in Figure 8. The coatings tested were characterized by a good adhesion and they belonged to the group of HF1.

The results of the Daimler-Benz test and scratch test were comparable and showed that, among the tested coatings, Mo_2N/CrN showed a poorer adhesion.

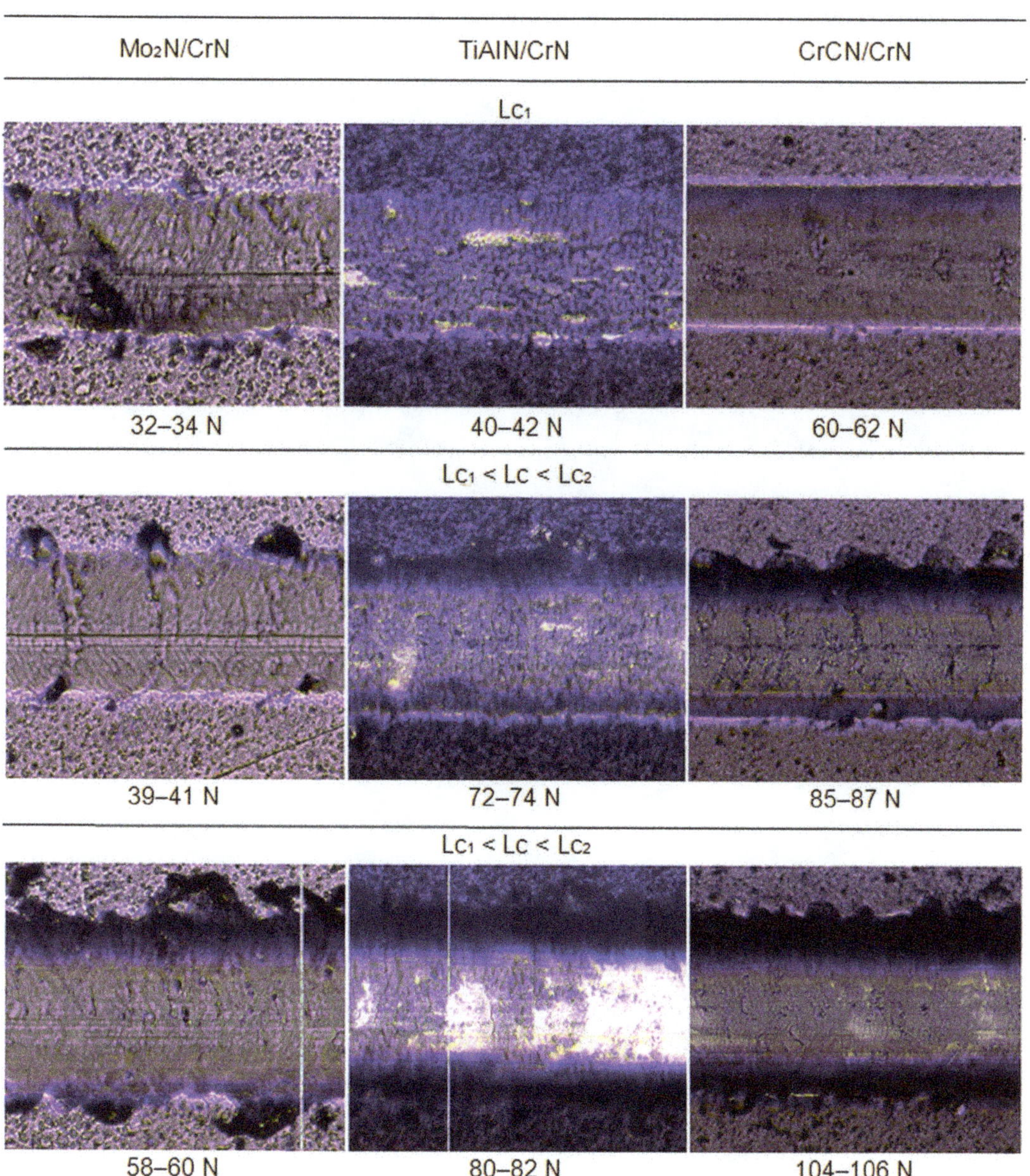

Figure 7. The scratch test and optical micrographs of failures at the critical load Lc_1, $Lc_1 < Lc < Lc_2$, and Lc_2 for coatings investigated.

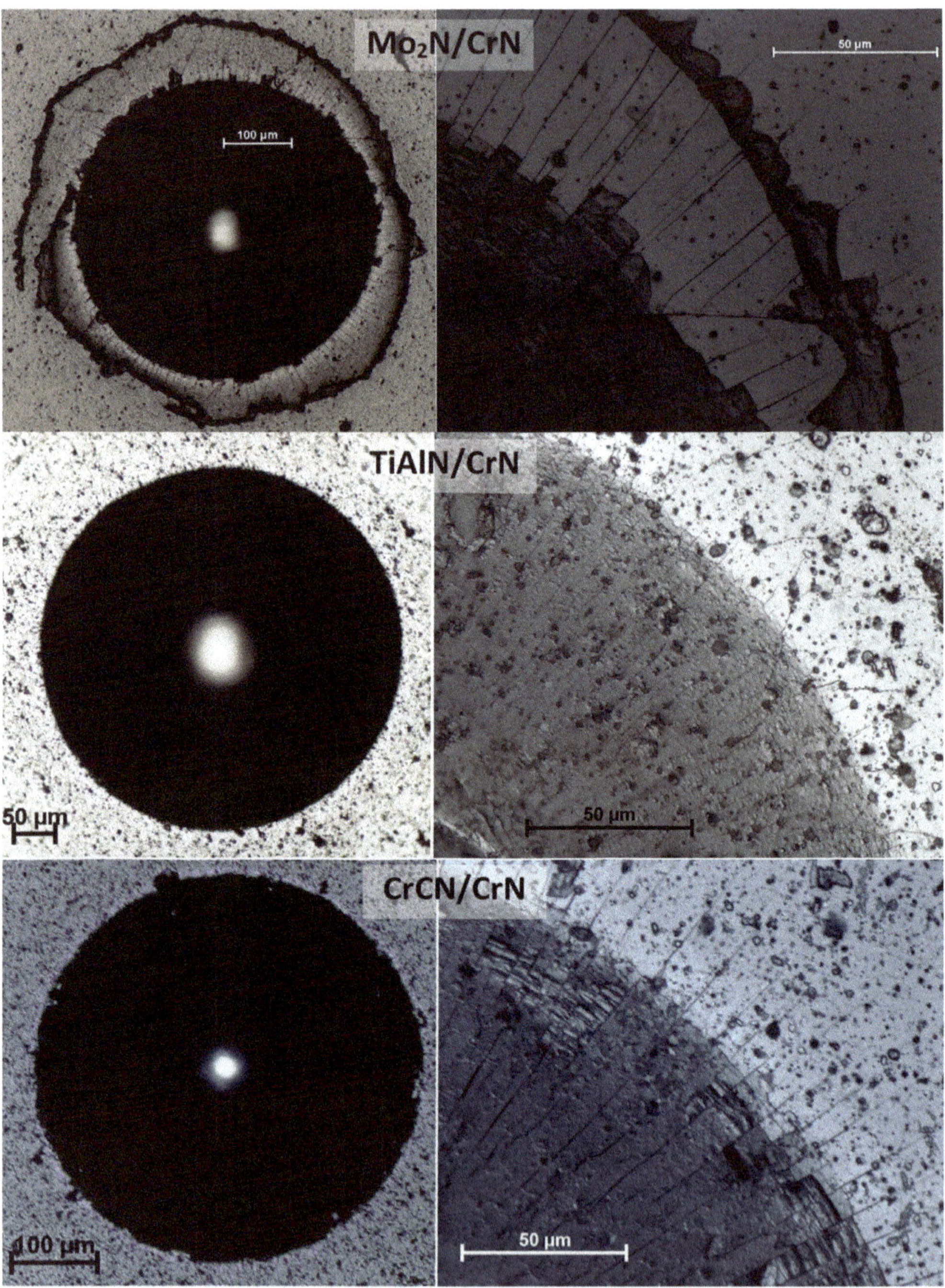

Figure 8. The picture of a Rockwell indentation in Mo_2N/CrN, TiAlN/CrN, and CrCN/CrN multi-layer coatings.

3.6. Wear

The chromium addition to the TiAlN coating significantly reduced the wear rate, and for the TiAlN/CrN multilayer coating, it was about 30 times smaller (2.5×10^{-7} mm^3/Nm) compared to that of the TiAlN coating (7.8×10^{-6} mm^3/Nm). It is interesting that CrCN/CrN coating was characterized by an almost two times lower wear rate (1.3×10^{-7} mm^3/Nm), and that of the Mo$_2$N/CrN coating was even lower—9×10^{-8} mm^3/Nm. It should be noted that the test of friction for all the coatings was carried out under identical conditions. The results of the coefficient of friction are presented in Table 4. They corresponded to the coating wear rates. The highest was for the TiAlN/CrN coating (0.72) and the lowest for Mo$_2$N/CrN (0.4), which is similar to that presented in Refs. [39,41,45].

The real test of the wear of coatings dedicated to woodworking tools is an industrial test. Due to the anisotropy of the physical properties and anomalies in wood structures, as well as the use of high cutting speeds (to 100 m/s) and feed rate ($20 \div 40$ m/min), this test is more reliable than the laboratory test with a low sliding speed and normal load. The above values are much higher than those for metal machining, respectively—to 2 m/s and to 1 m/min. The cooling of the tool and workpiece used in metal processing is not possible with woodworking, therefore, the temperature of the tool is higher, which contributes to more severe test conditions and increased tool wear. The photographs of the cutting edge and rake face of the steel (HS6-5-2) knives with the coatings tested are shown in Figure 9.

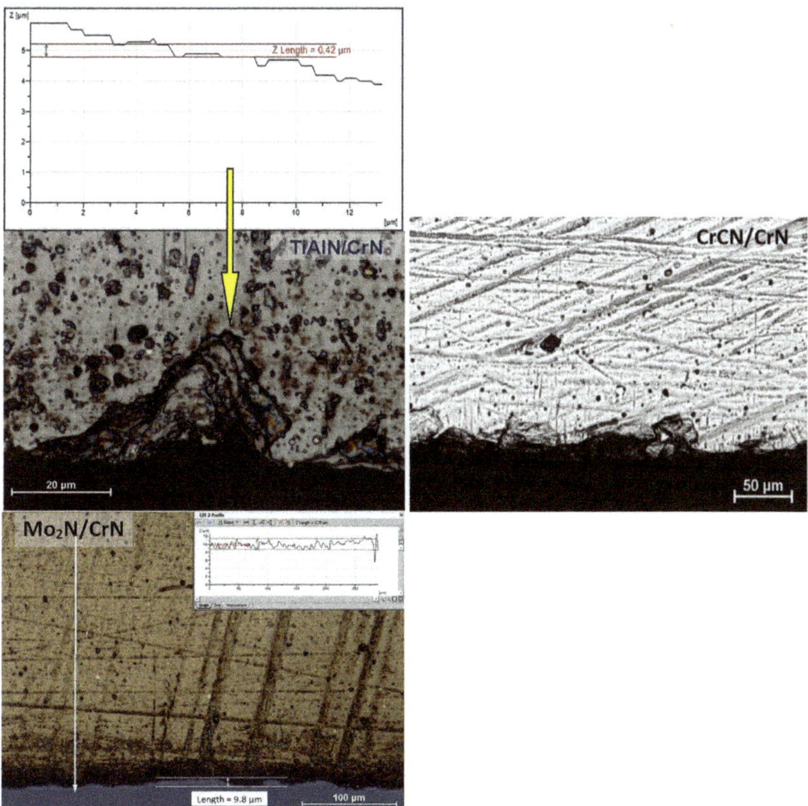

Figure 9. The images of the cutting edge and rake face of the planer knife made from HS6-5-2 steel with TiAlN/CrN, CrCN/CrN, and Mo$_2$N/CrN multilayer coatings.

On the edge of a planer knife coated with TiAlN/CrN, the chipping of the coating with a thickness of about 420 nm was observed. It corresponded to the thickness of the coating bilayer. The yellow arrow in Figure 9 indicates the direction of the profiling of the tool surface, perpendicular to the edge of the blade. For other coatings, there were only a few small chippings. The surface profile of the Mo_2N/CrN coating in the direction perpendicular to the surface of the blade edge, marked with a white arrow, is shown in the upper right part of the figure. It shows no significant abrasive wear.

The presented coatings were characterized by a good adhesion and they significantly improved the tool life. They exhibited an abrasive nature of wear, which meant that the temperature during machining did not exceed the tempering temperature of the blade material.

The wear of coatings is a complex process, in which one should take into account a number of possible changes in the chemical, physical, and mechanical properties both in the coating and interface of the coating–substrate. For this reason, the wear should be considered as a random process.

4. Discussion

The arc process of coating deposition is characterized by a high average kinetic energy of particles and a high degree of ionization of plasma. Almost all the particles (ions and atoms) and drops leaving around the cathode space reach the substrate. Hence, it follows a high deposition rate. The substrate bias voltage increases the energy of the particles and the substrate bombardment of high-energy particles results in the removal of macroparticles weakly bound to the substrate. A similar dimensional distribution for CrN monolayer coatings formed via cathodic arc evaporation was previously presented [38].

The disadvantage of TiAlN (Figure 1) coatings is their relatively high surface roughness, resulting from the low melting point of the TiAl cathode. The studies conducted by Munz et al. [6,7] suggested that, as the melting point of the cathode material increases, the amount of macroparticles of the cathode material deposited on a steel substrate decreases. Another negative feature is the decrease in the deposition rate as a result of the "target poisoning". In the case of TiAl cathodes with a relatively low melting point (about 1400–1500 °C, depending on the concentration of aluminum), it is particularly important because of the high melting points of TiN and TiAlN (about 3000 °C). The melting points of chromium and molybdenum amount to about 1900 °C and 2620 °C, respectively. The images of the surfaces of the TiAlN, CrCN, Mo_2N, and CrN coatings, deposited using cathodic arc evaporation, indicated a large amount of macroparticles on the TiAlN coating surface, less on the surfaces of the CrCN and CrN coatings, and an especially small amount on the surface of the Mo_2N coating, Figure 1.

The test results confirmed that the melting point of the cathode material had a large impact on the number of defects on the surface of the formed coating. The top layer of each tested multilayer coating was CrN. In spite of this fact, a significant difference was observed in the number of surface defects (Figure 2) or the roughness of the tested coatings, Table 4. The type of the second layer of the multilayer coating played an important role here. This was particularly evident in the case of the TiN/CrN and TiAlN/CrN coatings. In the former, the melting point of the cathode (1660 °C) was higher than that for the TiAl cathode, which directly translated into their roughness parameters Ra of 0.20 μm and 0.32 μm, respectively.

The chemical composition of the TiAlN/CrN coating was somewhat surprising, in fact, the Ti/(Ti + Al) ratio was about 0.35, which is more than that of the titanium concentration in the $Ti_{30}Al_{70}$ cathode. This was probably related to the resputtering effect of the deposited coatings and lower atomic weight of Al. This lower mass caused a greater scattering of the Al ions in collisions with nitrogen, which led to a lower vapor density and fewer of them reaching the substrate.

The hardness of the CrCN/CrN and Mo_2N/CrN coatings was close to the literature data, while the hardness of the TiAlN/CrN coating was slightly lower. This may have been due to the high concentration of aluminum in the coating, which reduced its hardness.

The wear of coatings can be considered in many areas. Often, the hardness property is considered to be decisive for their resistance to wear [46]. It is preferable, from the viewpoint of stability, to use the H/E ratio as an indicator of the durability of the coating, because this parameter generally is related to the ability of the elastic deformation of the coating material. The better adaptation of the elastic properties of the coating and substrate can improve its ability to adapt to the deformation of the substrate. The H/E ratio is a more important factor for wear resistance than high hardness. When considering the wear of the coating, the ability to deform both the substrate and the coated substrate and the ability to absorb deformation energy should be taken into account. A higher value for the H/E rate is a strong indication of the resistance to plastic deformation of the coating, but it does not always improve the ability of elastic deformation and fracture toughness. Both a high hardness of the coating and/or a low Young's modulus should cause a decrease in the effect of the deformation of the coating (and substrate) during scratch tests, reducing the coefficient of friction. The initiation of cracks in brittle material occurs with a lower energy than plastic material. When designing coatings, one should pay attention to the H/E ratio, because above a certain characteristic material value, its wear rate increases. For a lower wear, the ratio of the coating destruction mechanism changes from a brittle-to-ductile. A decrease in the H/E ratio due to a hardness decrease or Young's modulus increase results in the lowering of the coating's resistance to dynamic loads, and the coating becomes more plastic. For coatings with a low H/E (plastic materials), even low loads cause their strain and a rapid increase in stress, often exceeding the local strength. This results in the creation and propagation of microcracks that connect and lead to local delamination.

The H/E ratio was the lowest for the TiAlN/CrN coating and the highest for CrCN/CrN coating. The Mo_2N/CrN coating was characterized by a slightly lower value for the H/E ratio. It should be noted, however, that taking into account both the H/E value and measurement uncertainty, the CrCN/CrN and Mo_2N/CrN coatings had comparable values of this parameter. The same applied to their resistance to plastic deformation, i.e., the H^3/E^2 ratio. It was the lowest for the TiAlN/CrN coating, and comparable (within the limits of measurement uncertainty) for the other tested coatings. It should be noted that the roughness of the TiAlN/CrN coating was significantly higher than that of the other coatings, Table 4. The H/E and H^3/E^2 rates indicated the TiAlN/CrN coating as a coating with a lower wear resistance, and these were also the results of the wear measurement under laboratory tests, Table 5.

Significant effects on the wear are played by the surface conditions, its roughness, and the number of macroparticles on the surface of the coating. Macroparticles, in most cases, are the droplets of cathode material. They are soft in relation to the hardness of the coating and weakly bonded to it, and they can create a lubricating effect by grinding on the surface of the machined metal. However, even the small efficiency of a chemical reaction (e.g., nitridation) in the macroparticle sphere can cause their considerable hardening. Such detached particles can serve as an effective abrasive material, resulting in the faster processing of the material, but also accelerate the wear of the coating itself. Simultaneously, probably due to its hardness, a significant change in the dimension or degradation of the macroparticles does not occur. The fragmentation of the particles depends on their resistance to fracture toughness, and the higher the hardness, the higher the fracture toughness. Thus, the amount of macroparticles on the surface coating can be related to the wear of the coating during a friction test. In the case of the TiAlN/CrN coating, the number of macroparticles was the largest (also roughness parameter Ra); hence, the greater the wear rate of this coating. In the case of the CrCN/CrN and Mo_2N/CrN coatings, their quantities of macroparticles were similar (the roughness parameter Ra also) and wear rates were similar, although for the Mo_2N/CrN coating, these were slightly smaller.

The relatively low coefficient of friction of the CrN and Mo_2N coatings, Table 5, compared to TiN-based coatings, has also been the subject of many studies [47,48]. The physical and chemical processes occurring on the rubbing material surface favor a change in their chemical composition and phase. Metal oxides formed on the surface of the coating usually reduce the coefficient of friction and temperature in the frictional contact, and, consequently, reduce the wear of the coating [49]. Based on our previous research, we can conclude that, in the friction track of the CrCN coatings tested in laboratory conditions (normal force 20 N, sliding speed 0.2 m/s, and dry friction), an increased oxygen concentration is present [44]. The same effect was found in AlCrSiN coatings tested under the same conditions [50] and others.

It is known that the kind of wear products forming during friction determines the tribological behavior [51]. In particular, the chemical nature of the products formed during friction tests, i.e., the type of compound: simple, complex, and stoichiometry, has a significant effect on the phenomena taking place in the frictional contact. One of the theoretical approximations explaining the relationship between the chemistry of the friction products and lubrication is the approach proposed by Erdemir [49,52]. In this approach, the ionic potential plays an important role. Oxides or mixtures of metal oxides with a higher ionic potential φ, defined as the ratio of cation charge to the cation ionic radius, can decrease the coefficient of friction of the rubbing surfaces. A significant decrease in the coefficient of friction is observed for mixtures of oxides, in which the ionic potential difference is large.

The presence in friction pairs of oxides with a low ionic potential results in higher friction forces, which usually results in a higher wear. Metal oxides with a high ionic potential easily deform or are subjected to shear during friction. They are often called lubricious oxides. The test results of the products after the friction test carried out at room temperature revealed the existence of oxides produced by tribochemical reactions [48]. The ionic potentials of Cr_2O_3, CrO_2, TiO_2, Al_2O_3, MoO_3 were 4.0, 7.3, 5.8, 6.0, and 8.2, respectively [49]. The potential difference in ionic TiO_2 and Al_2O_3 is close to zero, which can result in a high coefficient of friction and wear, which is consistent with the approach proposed by Erdemir.

The Mo_2N/CrN coating case is a good example for considerations of wear. It was characterized by the smallest adhesion to the substrate, but also the highest wear resistance of the coatings tested. Because of the possibility of the formation of an oxide lubricious coating, for treatment without the use of coolants (elevated temperature), this coating can constitute an alternative to coatings based on TiAlN. In the Mo_2N/CrN coating at an elevated temperature (about 400 °C [53] or 680 °C [54]), Mo_2N transformed into MoO_3 molybdenum oxide via the reaction of [54]:

$$Mo_2N + 2O_2 = 2MoO_2 + \frac{1}{2}N_2 \tag{2}$$

$$MoO_2 + \frac{1}{2}O_2 = MoO_3 \tag{3}$$

Molybdenum oxide is characterized by a relatively low coefficient of friction [54]. The formation of oxides on the surface of the Mo_2N coating during friction is useful and can lead to a reduction in the coating wear. An abraded oxide surface layer is reproduced in the presence of oxygen in air. Even at room temperature, in wear products of Mo_2N coatings, the Raman spectroscopy reveals the presence of molybdenum oxide [48]. Such a mechanism of surface oxidation and the abrasion of the surface oxide layer have been described by Zhu [55].

CrN coatings are resistant to oxidation to about 800 °C, and MoN is likely to oxidize at about 500 °C [45]. The nano-layered structure of Mo_2N/CrN should be hard and lubricious at an elevated temperature. Chromium nitride CrN is a barrier against oxygen diffusion into the inner part of the coating and protects against the oxidation of the all the coating. Therefore, the inner Mo_2N layers of the multilayer coating do not oxidize, which, in view of a layered structure of molybdenum oxide MoO_3 with weak van der Waals bonds, could lead to a rapid degradation of the coating under load.

Industrial tests showed that the greatest abrasive wear, as well as the local chipping of the coating, occurred in the TiAlN/CrN coating. Despite the registered high adhesion, the cohesion of the coatings was weaker, which was observed by the local layered chipping of the coating, as shown in Figure 9. The other coatings showed only slight abrasive wear with a slight chipping of the coating on the edge of the blade. A correlation of the wear results in the laboratory and industrial tests was observed. These results confirmed the applicability of the wear prediction H/E and H^3/E^2 ratios.

5. Conclusions

TiAlN/CrN, CrCN/CrN and Mo_2N/CrN multilayer coatings were synthesized using cathodic arc evaporation. The research results can be gathered into three areas:

(1) An analysis of the surface defects. A surface analysis of the coatings formed was carried out, including the determination of the number and size of the coating surface defects and their roughness. It was found that the TiAlN/CrN coating was characterized by the highest roughness Ra, about 0.32 μm, and the Mo_2N/CrN and CrCN/CrN coatings had a similar roughness Ra, slightly below 0.1 μm. The dimensions of the macroparticles were different; most objects were small with dimensions of up to 0.5 μm—about 60%—and up to 1 μm total of about 73%. A few macroparticles which were larger than the thickness of the coating were also observed. The fraction of the sample area covered by the macroparticles increased monotonically with the amount of macroparticles. The melting point of the cathode material directly affected the number and size of the macroparticles and growth-related defects in the coatings. The amount of macroparticles and surface roughness of the coatings increased with the lowering of the melting point of the cathode material. The TiAlN/CrN coating was characterized by a high surface roughness, which was caused by the relatively low melting point intermetallic phase of TiAl.

(2) The mechanical properties of the coatings. The coatings were characterized by a high hardness, with the TiAlN/CrN coating having the lowest, at 21 GPa. The lower hardness of the Ti-based coating compared to the other tested coatings translated into the lowest H/E and H^3/E^2 ratios, suggesting a higher wear among other coatings. These ratios for the Mo-based coating (H = 26 GPa) were similar to those for CrCN/CrN coatings, but despite having the lowest adhesion (Lc = 56 N), they were characterized by the lowest wear rate, 9.0×10^{-8} mm^3/Nm. Coating roughness played an important role here. The titanium coating had more than three times the roughness of the others and this may have been the main reason for the lower wear resistance of these coatings. A phase transformation of $Mo_2N \rightarrow MoO_3$ was also possible. Molybdenum oxide is known as a solid lubricant and can reduce the coefficient of friction and wear rate.

(3) The results of industrial tests. Industrial comparative tests of the coatings showed greater damage to the TiAlN/CrN coating compared to other coatings. Here, as in the case of the other coatings, there was abrasive wear, but also the chipping of the coating layers on the edge of the blade. This chipping, with a thickness of about 200 nm, corresponded to the thickness of a single layer in the coating, rather indicating their weaker cohesion. The rake surface of the tools covered with other coatings showed abrasive wear and there were also only small chippings of the coating on the edge of the blade. This effect could have been associated with the high H/E and H^3/E^2 ratios and/or low surface roughness of the coatings.

Author Contributions: Conceptualization, B.W. and A.G.; methodology, B.W.; software, A.G.; validation, B.W. and A.G.; formal analysis, B.W.; investigation, A.G.; resources, A.G. and M.T.; data curation, A.G. and M.T.; writing—original draft preparation, B.W.; writing—review and editing, B.W. and A.G.; visualization, A.G. and M.T.; supervision, A.G.; project administration, A.G.; funding acquisition, A.G. All authors have read and agreed to the published version of the manuscript.

Funding: This research was funded by the National Center for Research and Development, Poland, grant number BIOSTRATEG3/344303/14/NCBR/2018.

Data Availability Statement: The data presented in this study are available on request from the corresponding author.

Conflicts of Interest: The authors declare no conflict of interest.

References

1. Muhammed, M.; Javidani, M.; Heidari, M.; Jahazi, M. Enhancing the Tribological Performance of Tool Steels for Wood-Processing Applications: A Comprehensive Review. *Metals* **2023**, *13*, 1460. [CrossRef]
2. Warcholinski, B.; Gilewicz, A. Surface Engineering of Woodworking Tools, a Review. *Appl. Sci.* **2022**, *12*, 10389. [CrossRef]
3. Munz, W.D. Titanium aluminum nitride films: A new alternative to TiN coatings. *J. Vac. Sci. Technol. A* **1986**, *4*, 2717–2725. [CrossRef]
4. Kong, Y.; Tian, X.; Gong, C. Enhancement of toughness and wear resistance by CrN/CrCN multilayered coatings for wood processing. *Surf. Coat. Technol.* **2018**, *344*, 204–213. [CrossRef]
5. Adamiak, S.; Bochnowski, W.; Dziedzic, A.; Szyller, Ł.; Adamiak, D. Characteristics of the Structure, Mechanical, and Tribological Properties of a Mo-Mo_2N Nanocomposite Coating Deposited on the Ti6Al4V Alloy by Magnetron Sputtering. *Materials* **2021**, *14*, 6819. [CrossRef]
6. Creasey, S.; Lewis, D.B.; Smith, L.J.; Munz, W.D. SEM image analysis of droplet formation during metal ion etching by a steered arc discharge. *Surf. Coat. Technol.* **1997**, *97*, 163–175. [CrossRef]
7. Munz, W.D.; Smith, L.J.; Lewis, D.B.; Creasey, S. Droplet formation on steel substrates during cathodic steered arc metal ion etching. *Vacuum* **1997**, *48*, 473–481. [CrossRef]
8. Harris, S.G.; Doyle, E.D.; Wong, Y.C.; Munroe, P.R.; Cairney, J.M.; Long, J.M. Reducing the macroparticle content of cathodic arc evaporated TiN coatings. *Surf. Coat. Technol.* **2004**, *183*, 283–294. [CrossRef]
9. Hogmark, S.; Jacobson, S.; Larsson, M. Design and evaluation of tribological coatings. *Wear* **2000**, *246*, 20–33. [CrossRef]
10. Stueber, M.; Holleck, H.; Leiste, H.; Seemann, K.; Ulrich, S.; Ziebert, C. Concepts for the design of advanced nanoscale PVD multilayer protective thin films. *J. Alloys Compd.* **2009**, *483*, 321–333. [CrossRef]
11. Holleck, H.; Schier, V. Multilayer PVD coatings for wear protection. *Surf. Coat. Technol.* **1995**, *76–77*, 328–336. [CrossRef]
12. Santana, A.E.; Karimi, A.; Derflinger, V.H.; Schütze, A. Microstructure and mechanical behavior of TiAlCrN multilayer thin films. *Surf. Coat. Technol.* **2004**, *177–178*, 334–340. [CrossRef]
13. Kot, M.; Rakowski, W.A.; Major, Ł.; Major, R.; Morgiel, J. Effect of bilayer period on properties of Cr/CrN multilayer coatings produced by laser ablation. *Surf. Coat. Technol.* **2008**, *202*, 3501–3506. [CrossRef]
14. Tranca, D.E.; Sobetkii, A.; Hristu, R.; Anton, S.R.; Vasile, E.; Stanciu, S.G.; Banica, C.K.; Fiorentis, E.; Constantinescu, D.; Stanciu, G.A. Structural and Mechanical Properties of CrN Thin Films Deposited on Si Substrate by Using Magnetron Techniques. *Coatings* **2023**, *13*, 219. [CrossRef]
15. Dinu, M.; Mouele, E.S.M.; Parau, A.C.; Vladescu, A.; Petrik, L.F.; Braic, M. Enhancement of the Corrosion Resistance of 304 Stainless Steel by Cr–N and Cr(N,O) Coatings. *Coatings* **2018**, *8*, 132. [CrossRef]
16. Abdallah, B.; Kakhia, M.; Alssadat, W.; Zetoun, W. Study of Power Effect on Structural, Mechanical Properties and Corrosion Behavior of CrN Thin Films Deposited by Magnetron Sputtering. *Prot. Met. Phys. Chem. Surf.* **2021**, *57*, 80–87. [CrossRef]
17. Faga, M.G.; Settineri, L. Innovative anti-wear coatings on cutting tools for wood machining. *Surf. Coat. Technol.* **2006**, *201*, 3002–3007. [CrossRef]
18. Djouadi, M.A.; Nouveau, C.; Beer, P.; Lambertin, M. Cr_xN_y hard coatings deposited with PVD method on tools for wood machining. *Surf. Coat. Technol.* **2000**, *133–134*, 478–483. [CrossRef]
19. Nouveau, C.; Djouadi, M.A.; Decès-Petit, C. The influence of deposition parameters on the wear resistance of Cr_xN_y magnetron sputtering coatings in routing of oriented strand board. *Surf. Coat. Technol.* **2003**, *174–175*, 455–460. [CrossRef]
20. Tritremmel, C.; Daniel, R.; Lechthaler, M.; Polcik, P.; Mitterer, C. Influence of Al and Si content on structure and mechanical properties of arc evaporated Al–Cr–Si–N thin films. *Thin Solid Films* **2013**, *534*, 403–409. [CrossRef]
21. Nouveau, C.; Labidi, C.; Ferreira Martin, J.P.; Collet, R.; Djouadi, M.A. Application of CrAlN coatings on carbide substrates in routing of MDF. *Wear* **2007**, *263*, 1291–1299. [CrossRef]
22. Nadolny, K.; Kapłonek, W.; Sutowska, M.; Sutowski, P.; Myśliński, P.; Gilewicz, A.; Warcholiński, B. Experimental tests of PVD AlCrN-coated planer knives on planing Scots pine (*Pinus sylvestris* L.) under industrial conditions. *Eur. J. Wood Prod.* **2021**, *79*, 645–665. [CrossRef]

23. Benlatreche, Y.; Nouveau, C.; Marchal, R.; Ferreira Martin, J.P.; Aknouche, H. Applications of CrAlN ternary system in wood machining of medium density fibreboard (MDF). *Wear* **2009**, *267*, 1056–1061. [CrossRef]
24. Benlatreche, Y.; Nouveau, C.; Aknouche, H.; Imhoff, L.; Martin, N.; Gavoille, J.; Rousselot, C.; Rauch, Y.J.; Pilloud, D. Physical and Mechanical Properties of CrAlN and CrSiN Ternary Systems for Wood Machining Applications. *Plasma Process. Polym.* **2009**, *6*, S113–S119. [CrossRef]
25. Batista, C.A.; Godoy, C.; Buono, V.T.L.; Matthews, A. Characterisation of duplex (Ti,Al)N and Cr-N PVD coatings. *Mater. Science Eng.* **2002**, *A336*, 39–51. [CrossRef]
26. PalDey, S.; Deevi, S.C. Single layer and multilayer wear resistant coatings of (Ti,Al)N: A review. *Mater. Science Eng.* **2003**, *A342*, 58–79. [CrossRef]
27. Barshilia, H.C.; Prakash, M.S.; Jain, A.; Rajam, K.S. Structure, hardness and thermal stability of TiAlN and nanolayered TiAlN/CrN multilayer films. *Vacuum* **2005**, *77*, 169–179. [CrossRef]
28. Kazmanli, M.K.; Ürgen, M.; Çakir, A.F. Effect of nitrogen pressure, bias voltage and substrate temperature on the phase structure of Mo-N coatings produced by cathodic arc PVD. *Surf. Coat. Technol.* **2003**, *167*, 77–82. [CrossRef]
29. Ürgen, M.; Eryilmaz, O.L.; Çakir, A.F.; Kayali, E.S.; Nilüfer, B.; Işik, Y. Characterization of molybdenum nitride coatings produced by arc-PVD technique. *Surf. Coat. Technol.* **1997**, *94–95*, 501–506. [CrossRef]
30. Sarioglu, S.; Demirler, U.; Kazmanli, M.; Urgen, M. Measurement of residual stresses by X-ray diffraction techniques in MoN and Mo$_2$N coatings deposited by arc PVD on high-speed steel substrate. *Surf. Coat. Technol.* **2005**, *190*, 238–243. [CrossRef]
31. Gulbiński, W.; Suszko, T. Thin films of MoO$_3$-Ag$_2$O binary oxides—The high temperature lubricants. *Wear* **2006**, *261*, 867–873. [CrossRef]
32. Rebholtz, C.; Ziegele, H.; Leyland, A.; Matthews, A. Structure, mechanical and tribological properties of nitrogen-containing chromium coatings prepared by reactive magnetron sputtering. *Surf. Coat. Technol.* **1999**, *115*, 222–229. [CrossRef]
33. Vidakis, S.; Antoniadis, A.; Bilalis, N. The VDI 3198 indentation test evaluation of a reliable qualitative control for layered compounds. *J. Mater. Process. Technol.* **2003**, *143–144*, 481–485. [CrossRef]
34. Romero, J.; Gómez, M.A.; Esteve, J.; Montalà, F.; Carreras, L.; Grifol, M.; Lousa, A. CrAlN coatings deposited by cathodic arc evaporation at different substrate bias. *Thin Solid Films* **2006**, *515*, 113–117. [CrossRef]
35. Stoney, G.G. Tension of metallic films deposited by electrolysis. *Proc. R. Soc. Lond.* **1909**, *A82*, 172–175.
36. Archard, J.F. Contact and rubbing of flat surfaces. *J. Appl. Phys.* **1953**, *24*, 981–988. [CrossRef]
37. Wan, X.S.; Zhao, S.S.; Yang, Y.; Gong, J.; Sun, C. Effects of nitrogen pressure and pulse bias voltage on the properties of Cr-N coatings deposited by arc ion plating. *Surf. Coat. Technol.* **2010**, *204*, 1800–1810. [CrossRef]
38. Warcholinski, B.; Gilewicz, A.; Ratajski, J.; Kuklinski, Z.; Rochowicz, J. An analysis of macroparticle-related defects in the CrCN and CrN coatings in dependence on the substrate bias voltage. *Vacuum* **2012**, *86*, 1235–1239. [CrossRef]
39. Chang, Y.Y.; Yang, S.J.; Wang, D.Y. Structural and mechanical properties of AlTiN/CrN coatings synthesized by a cathodic-arc deposition process. *Surf. Coat. Technol.* **2006**, *201*, 4209–4214. [CrossRef]
40. Chang, Y.Y.; Wang, D.Y.; Hung, C.Y. Structural and mechanical properties of nanolayered TiAlN/CrN coatings synthesized by a cathodic arc deposition process. *Surf. Coat. Technol.* **2005**, *200*, 1702–1708. [CrossRef]
41. Chang, C.L.; Jao, J.Y.; Ho, W.Y.; Wang, D.Y. Influence of bi-layer period thickness on the residual stress, mechanical and tribological properties of nanolayered TiAlN/CrN multi-layer coatings. *Vacuum* **2007**, *81*, 604–609. [CrossRef]
42. Koshy, R.A. Thermally Activated Self-lubricating Nanostructured Coating for Cutting Tool Applications. Ph.D. Thesis, Northwestern University, Evanston, IL, USA, December 2008.
43. Warcholinski, B.; Gilewicz, A. The properties of multilayer CrCN/CrN coatings dependent on their architecture. *Plasma Process. Polym.* **2011**, *8*, 333–339. [CrossRef]
44. Gilewicz, A.; Warcholinski, B. Tribological properties of CrCN/CrN multilayer coatings. *Tribol. Int.* **2014**, *80*, 34–40. [CrossRef]
45. Koshy, R.A.; Graham, M.E.; Marks, L.D. Synthesis and characterization of CrN/Mo$_2$N multilayers and phases of Molybdenum nitride. *Surf. Coat. Technol.* **2007**, *202*, 1123–1128. [CrossRef]
46. Leyland, A.; Matthews, A. On the significance of the H/E ratio in wear control: A nanocomposite coating approach to optimised tribological behaviour. *Wear* **2000**, *246*, 1–11. [CrossRef]
47. Luo, Q.; Rainforth, W.M.; Münz, W.D. TEM observations of wear mechanisms of TiAlCrN and TiAlN/CrN coatings grown by combined steered-arc/unbalanced magnetron deposition. *Wear* **1999**, *225–229*, 74–82. [CrossRef]
48. Öztürk, A.; Ezirmik, K.V.; Kazmanli, K.; Ürgen, M.; Eryilmaz, O.L.; Erdemir, A. Comparative tribological behaviors of TiN-, CrN- and MoN-Cu nanocomposite coatings. *Tribol. Int.* **2008**, *41*, 49–59. [CrossRef]
49. Erdemir, A. A crystal chemical approach to the lubrication by solids oxides. *Tribol. Lett.* **2000**, *8*, 97–102. [CrossRef]
50. Warcholinski, B.; Gilewicz, A.; Myslinski, P.; Dobruchowska, E.; Murzynski, D.; Kuznetsova, T.A. Effect of Silicon Concentration on the Properties of Al-Cr-Si-N Coatings Deposited Using Cathodic Arc Evaporation. *Materials* **2020**, *13*, 4717. [CrossRef]
51. Gulbiński, W.; Suszko, T.; Sienicki, W.; Warcholiński, B. Tribological properties of silver and copper-doped transition metal oxide coatings. *Wear* **2003**, *254*, 129–135. [CrossRef]
52. Erdemir, A. A crystal chemical approach to the formulation of self-lubricating nanocomposite coatings. *Surf. Coat. Technol.* **2005**, *200*, 1792–1796. [CrossRef]
53. Suszko, T.; Gulbiński, W.; Jagielski, J. The role of surface oxidation in friction processes on molybdenum nitride thin films. *Surf. Coat. Technol.* **2005**, *194*, 319–324. [CrossRef]

54. Solak, N.; Ustel, F.; Urgen, M.; Aydin, S.; Cakir, A.F. Oxidation behavior of molybdenum nitride coatings. *Surf. Coat. Technol.* **2003**, *174–175*, 713–719. [CrossRef]
55. Zhu, X.; Yue, D.; Shang, C.; Fan, M.; Hou, B. Phase composition and tribological performance of molybdenum nitride coatings synthesized by IBAD. *Surf. Coat. Technol.* **2013**, *228*, S184–S189. [CrossRef]

 *lubricants*

Article

Improving the Surface Integrity and Tribological Behavior of a High-Temperature Friction Surface via the Synergy of Laser Cladding and Ultrasonic Burnishing

Nan Xu [1], Xiaochen Jiang [1,*], Xuehui Shen [1,*] and Hao Peng [2]

[1] School of Mechanical and Electrical Engineering, Shandong Jianzhu University, Jinan 250101, China; chenqu_sd@163.com
[2] School of Mechanical Engineering, Qilu University of Technology (Shandong Academy of Sciences), Jinan 250353, China; qilu_hao@163.com
*   Correspondence: jiangxiaochen20@sdjzu.edu.cn (X.J.); xuehuishen@163.com (X.S.)

Abstract: Quite a lot of engineering friction components serve at high temperatures, and are thus required to have excellent friction and wear resistance. The said study aims to fabricate high-wear-resistance coating on ordinary low-cost materials, achieving the low-cost manufacturing of some high-end friction components that are usually made with expensive solid alloys. The coating was prepared via laser cladding with a sort of widely used Fe-based self-fluxing alloy powder. The chosen substrate material was forged 42CrMo, which is popular in high-temperature friction engineering applications. In order to achieve the best possible high-temperature friction and wear properties, the prepared coating was turned and then ultrasonic burnished. Three samples, i.e., the substrate sample, the cladded sample without burnishing, and the cladded sample with burnishing, were prepared. For the three samples, the surface characteristics and friction properties at a 200 °C temperature were compared and investigated. According to the results, the cladded sample with burnishing exhibited the best surface finishing and friction behavior. Ultrasonic burnishing after cladding led to a further hardness improvement of 15.24% when compared with the cladded sample without burnishing. Therefore, ultrasonic burnishing is an effective low-cost post-treatment method for a wearable coating serving at a high temperature.

Keywords: hybrid surfacing; wear resistance; additive manufacturing; surface modification

Citation: Xu, N.; Jiang, X.; Shen, X.; Peng, H. Improving the Surface Integrity and Tribological Behavior of a High-Temperature Friction Surface via the Synergy of Laser Cladding and Ultrasonic Burnishing. *Lubricants* **2023**, *11*, 379. https://doi.org/10.3390/lubricants11090379

Received: 8 August 2023
Revised: 3 September 2023
Accepted: 5 September 2023
Published: 7 September 2023

1. Introduction

42CrMo is widely used to manufacture engineering components owing to its excellent strength, toughness, and hardenability [1]. However, 42CrMo has low hardness and insufficient wear resistance [2,3]. Therefore, under high-temperature and pressure friction conditions, surface strengthening is necessary for this sort of material to meet some specific service requirements, for instance, a high wear resistance. Engine pistons, which serve at a max instantaneous temperature of about 2500 K, is a very typical case. In this case, the adhesion between friction pairs easily occurs.

Surface coating is a commonly applied surface modification method. There exists several coating preparation techniques for varying application cases, such as thermal spraying [4], electroplating [5], physical vapor deposition [6], etc. However, these techniques have varying defects including low coating and substrate bonding strength, a thin thickness, and environmental pollution [7]. Thus, these coating preparation methods are not very suitable in some cases similar to engine pistons.

Metal arc welding [8] and laser cladding [9] are two popular coating additive manufacturing techniques. Meanwhile, in recent years, laser cladding has been earning more engineering applications. Laser cladding utilizes an extremely high-energy-density laser beam to melt a layer of powder material on a substrate, and therefore, the cladding coating is metallurgically bonded with the substrate. This technique can repair and remanufacture many high-end engineering parts with a much lower cost and less resource consumption [10]. Therefore, it has been earning more and more attention. Feng [11] prepared an Fe-based coating on a 42CrMo substrate via laser cladding and achieved apparent wear resistance improvement. Liu and Shi [12] used laser cladding to prepare a Ni-based self-lubricating coating on a 42CrMo substrate. They found that at a laser scanning speed of 300 mm/s, the sample obtained the least wear rate of 6.824×10^{-6} mm^3/N·m within the experimental scope. Kumar et al. fabricated a self-lubricating coating on a stainless steel substrate and achieved a 60% friction reduction at room temperature and a 40% friction reduction at 400 °C in comparison to the unmodified alloy [13].

However, in spite of their high bonding strength, cladding coatings still suffer from issues such as a rough surface finishing, tensile residual stress, and pores. These issues are mainly attributed to the difference of the thermal expansion coefficients between the cladding material and substrate as well as the rapid heating and cooling of the coating material. Therefore, in most cases, coatings need to be post-treated after cladding.

Ultrasonic burnishing (UB) is developed from traditional burnishing without ultrasonic vibration. This technique not only reduces the surface roughness, but also introduces compressive residual stress and refines the grain size, thus improving the mechanical properties of the treated materials [14]. Currently, ultrasonic burnishing technology is applied to various materials to improve different service performances [15]. Furthermore, the excellent process effect of UB has been established [16–18]. In terms of the post-treatment of cladding coatings, a few studies on UB could be found. For example, Hao [19] prepared an Inconel 625 multi-layer cladding coating on an H13 steel substrate. The coating was milled and then subjected to ultrasonic burnishing (UB). As a result, it was found that the treated coating had a surface roughness of less than 0.12 μm, and had better wear performance with less wear scars than the sample without UB. Ye et al. [20] applied ultrasonic burnishing on a laser-cladded Cr-Ni coating, and reported obvious fracture toughness improvement. Zhang et al. prepared Fe-based laser cladding coatings on carbon steel, and then treated the coatings via UB at room temperature and at a medium temperature of 200°, respectively. They reported a roughness decrease of 82.72% as well as a wear rate drop of 49.9% in comparison to the turned sample [21].

A few previous works have explored UB as a post-treatment of cladding coating. However, the effect of this technique crucially depends on the material [22]. Moreover, most available reports on the friction behaviors of cladding coatings were performed at room temperature, and investigations were rarely conducted at high temperatures. In this study, laser cladding and ultrasonic burnishing treatment was employed to address the shortcomings of 42CrMo friction surface serving at a high temperature, laying the foundation for meeting higher engineering requirements.

2. Experimental Details

2.1. Material and Sample Preparation

The substrate was a Ø50 × 30 mm forged 42CrMo bar. One end of the bar was turned to a roughness of Ra1.1μm and set as the cladding target surface. The composition of the substrate material was as follows: 0.43 wt % C, 0.31 wt % Si, 0.67 wt % Mn, 0.014 wt % P, 0.007 wt % S, 1.09 wt % Cr, and 0.20 wt % Mo and Fe margin. Before cladding, the substrate was cleaned in an ultrasonic acetone bath for 15 min.

The cladding material was a commercial Spherical Fe-based alloy powder with a grain diameter range of 10–30μm. The chemical composition was as follows: 0.43 wt % C, 0.31 wt % Si, 0.67 wt % Mn, 0.014 wt % P, 0.007 wt % S, 1.09 wt % Cr, and 0.20 wt % Mo and Fe margin. The left image in Figure 1 is the SEM photo of the chosen powder.

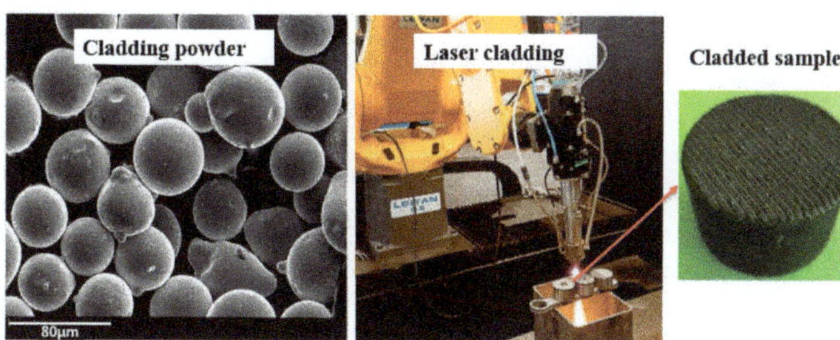

Figure 1. Cladding material and process.

Figure 1 shows photos of the laser cladding process and a cladded sample. The operation was performed on a robot laser cladding workstation (LYRF-4000) with an integrated control system (LYRF1500). The parameters of laser cladding were as follows: the laser power was 2000 W, spot diameter was 3.2 mm, scanning speed was 1500 mm/min, overlap rate was 30%, and powder feeding rate was 28.8 g/min. The distance of the spot from the surface being treated was about 15 mm. Argon served as protective gas with a flow rate of 5 L/min.

Self-developed ultrasonic burnishing equipment was used to post-treat the prepared cladding coatings. Tables 1 and 2 list the chosen parameters. Figure 2 shows the UB setup and illustrates the process principle. The detailed working principle of the designed UB device can be found in a previous work [23] and will not be further described here. After cladding, one sample was only hard turned (referred to as HT-treated sample), and the other was hard turned and then ultrasonic burnished (referred to as UB-treated sample). In addition, a forged 42CrMo sample was hard turned and set as the control.

Table 1. Turning parameters.

Cutting Depth (mm)	Spindle Speed (r/min)	Feed (r/mm)
0.1	300	0.06

Table 2. Ultrasonic burnishing parameters.

Frequency (kHz)	Amplitude (μm)	Feed (mm/min)	Static Load (N)	Spindle Speed (rpm)
28	7	10	320	160

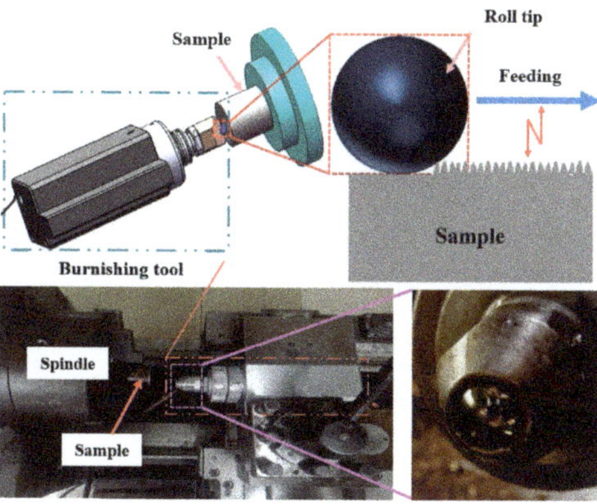

Figure 2. Schematic diagram and experimental setup of ultrasonic burnishing.

2.2. Microstructure Characterization

A scanning electron microscope (SEM, Phenom ProX, Eindhoven, the Netherlands) and its attached energy dispersion analyzer (EDS) were used to observe and measure the surface morphology and chemical composition of the samples. The sample surface morphology and roughness were observed and measured using a white light interferometer (Contour Elite K, Bruker Technology, Billerica, MA, USA). The samples were cut via WEDM and etched with aqua regia (a solution of HCl: HNO_3 = 3:1) to observe the microstructure. The sample microstructure was observed under a three-dimensional digital microscope (VHX-5000). XRD (D8-ADVANCE, Bruker, Germany) was used to determine the phase pattern. The chosen diffraction angle of the XRD was 20~85°, and the scanning speed was 5°/min. The samples were irradiated via Cu-kα ray (λ = 0.1542 nm) at 45 kV and 40 mA. Prior to XRD analysis, the samples were cleaned in an ultrasonic anhydrous ethanol bath for 10 min to remove surface impurities. The blind hole method was used to measure the surface residual stress. The porosity was calculated using ImageJ software from the SEM image. The microhardness was measured using a Vickers hardness tester (HXD-1000TMC). The section hardness was tested under a load of 500 GF and a holding time of 10 s. For each surface character parameter, five measurements were averaged to avoid random errors.

2.3. Wear

A ball-on-plate sliding wear experiment was conducted on a tribometer (HT-1000). The test temperature was set at 200 °C according to the usual working condition of piston pin-hole friction pair. The samples tested were shaped as blocks with dimensions of 20 mm × 10 mm × 5 mm as determined via WEDM. When being tested, the samples were slightly polished to remove oxides and then cleaned in an acetone bath for 10 min.

The counterbody was a Si_3N_4 ball with a 6 mm diameter and HV2200 hardness. The applied load was 150 N, and the mean and maximum contact pressures were 1.46 GPa and 2.19 GPa, respectively. The one-way sliding distance was 5 mm, the reciprocating frequency was 5 Hz, and the test holding time was 30 min. For each sample, three tests were performed to avoid gross error. The wear morphology was observed via SEM. Meanwhile, a white light interferometer and a 3D Super Depth Digital Microscope (VHX-5000) were used to measure the wear loss.

3. Results and Discussion

3.1. Surface Morphology and Porosity

Figure 3 shows the surface SEM images and the three-dimensional morphologies of the three prepared samples. In Figure 3a,b, for the control and HT-treated samples, there exists clear cutting marks with obvious peaks and valleys. In contrast, as shown in Figure 3c, the UB sample has a flat surface finishing without a visible cutting trace. According to the measurements, the roughness values of the three samples are Ra1.19, Ra1.12, and Ra0.17. Compared with the HT-treated sample, UB led to a 79.2% roughness reduction. The smoothing surface is a usual case for UB. During UB, the burnishing tool tip extruded the near-surface materials of the sample being treated to flow from peaks to valleys, and thus flattened the surface [24].

Another observation from Figure 3 is that, in all cases, there exists distributed black dots on the samples' surfaces. According to the EDS analysis, the content of the carbon element was relatively high at the black dots. It was analyzed that these black dots were pores and carbides generated in laser cladding. From Table 3, for the HT-treated and UB-treated samples, the contents of the major elements were not shown to be distinct except for carbon. UB smooths and strengthens the surface via material plastic deformation, and therefore, it could not cause a composition change in a usual case [23]. Here, for the UB-treated sample, the reduction in the C element was mainly attributed to the detachment of carbides [25]. Meanwhile, owing to the material flow from UB, some pores in near surfaces produced via laser cladding were filled [20], and thus, much fewer black dots were found in the UB-treated sample than in the HT-treated one.

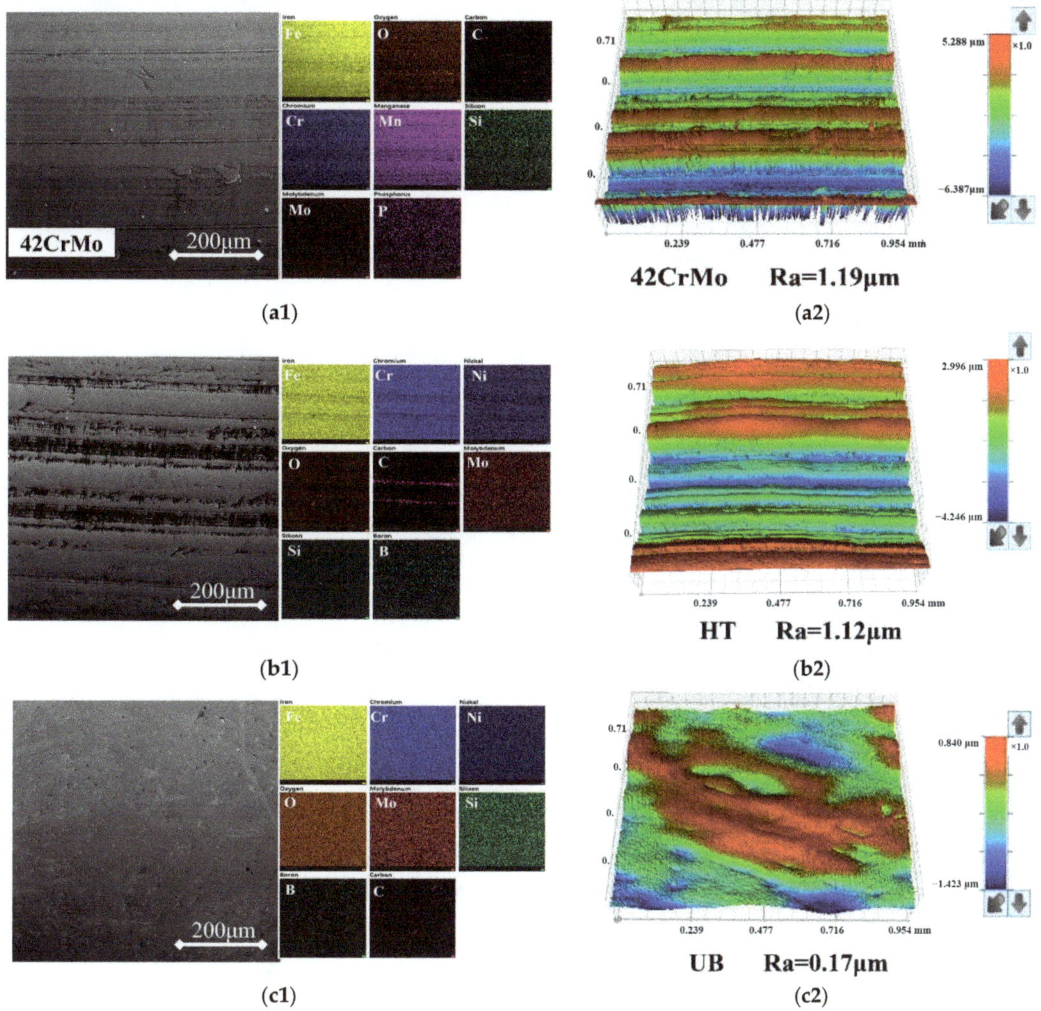

Figure 3. Surface SEM morphology. (**a1**) 42CrMo, (**b1**) HT samples, (**c1**) UB samples, and three-dimensional morphology: (**a2**) 42CrMo, (**b2**) HT samples, and (**c2**) UB samples. The EDS spectrum of UB samples is on the right side of (**a1,b1,c1**), respectively.

Table 3. Main element content of samples (wt%).

Samples	C	Si	Cr	Mo	Fe	Ni	O
Powder	0.43	0.31	1.09	0.20	97.28	-	-
Control	5.09	0.82	1.09	0.15	87.47	-	4.58
HT treated	11.89	1.59	14.41	1.18	49.88	5.44	12.23
UB treated	0.68	1.71	16.29	1.15	57.18	6.33	12.45

As shown in Figure 4, the porosity values of the control and HT-treated samples were 4.09% and 3.97%, respectively. The pores were generated from impurities and from gas generation during the melting and solidification of the powder. From Figure 4, turning alone could not improve the porosity inside the coating. In contrast, the UB treatment led to an 85.6% porosity drop in comparison to the control. The significant pore elimination effect was attributed to the dynamic load from UB. In the experiment, the prepared cladding coat had a one-layer structure with about a 1 mm thickness. During UB, the applied high-frequency vibration could cause a dynamic stress wave propagating along an in-depth direction inside the coating [23]. Therefore, UB had a more strengthening effect than that of conventional burnishing without vibration.

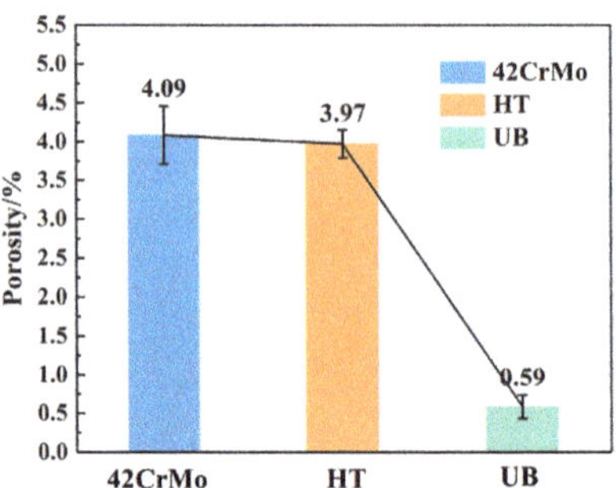

Figure 4. Porosity measurements.

3.2. Surface Microstructure

Figure 5 shows the surface microstructure of the samples with two magnifications, which were observed under a digital microscope after etching. From Figure 5(a1,a2), for the control sample, it can be seen that the 42CrMo material was mainly composed of tempered sorbite and ferrite. In Figure 5(b1,b2), there exists coarse strip-shaped and cellular dendrites. The shape of the dendrite structure is irregular and uneven. It is known that this sort of coarse microstructure is not conducive to mechanical performance. In contrast, from Figure 5(c1,c2), for the UB-treated sample, it can be seen that the dendrites were refined into equiaxed grains. Moreover, the sizes and shapes of the equiaxed crystals were uniform, making it easier to maintain a stable structure and thus have better mechanical properties. Grain refinement is a usual case in UB treatment, which is commonly explained as an outcome of dislocation slip, accumulation, interaction, annihilation, and rearrangement from the co-action of ultrasonic vibration and static load [26].

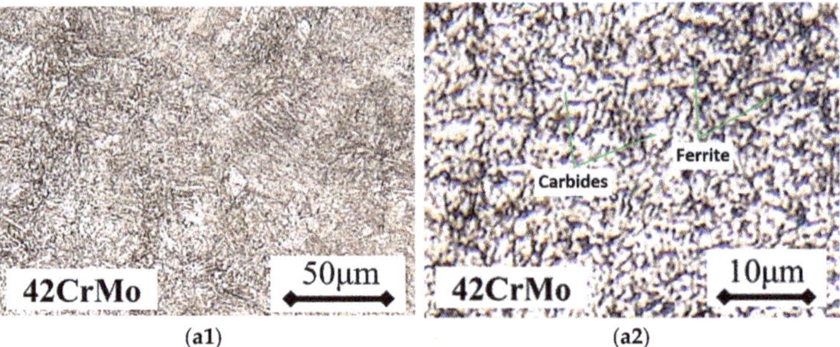

Figure 5. *Cont.*

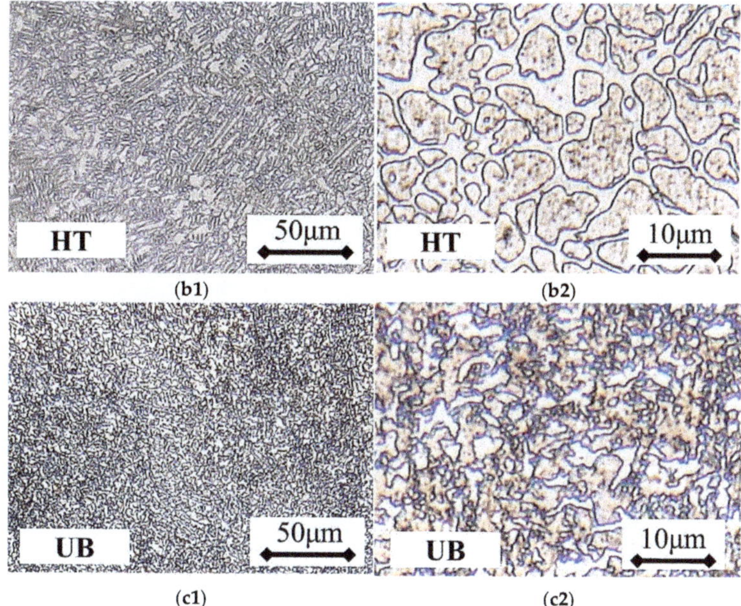

Figure 5. Surface microstructure: (**a1,a2**) control, (**b1,b2**) HT-treated, and (**c1,c2**) UB-treated samples.

3.3. XRD Analysis

The XRD pattern of the tree-tested samples are shown in Figure 6a. It can be observed that there was no clear element distribution variation for the HT-treated and UB-treated samples [25]. Figure 6a shows an enlarged view of the XRD's main peak. It can be seen that the peak intensity of the UB-treated sample was greatly weakened compared with the HT-treated one. This phenomenon was attributed to the grain reorientation, lattice distortion, and micro-strain resulting from UB [26]. Moreover, it is noted that the main peak of the XRD shifted to the right in the UB case, indicating an increase in the diffraction angle, a decrease in the crystal plane spacing, and an intensification of the micro-strain [27]. Figure 6b shows the FWHM measurements. Such values are 0.43 deg, 0.55 deg, and 0.74 deg, for the control, the HT-treated sample, and the UB-treated sample, respectively. According to the Scherrer equation, the broadening of the diffraction peak and the increase in the FWHM meant there was a decrease in the grain size [28,29], which agrees well with the microstructure above.

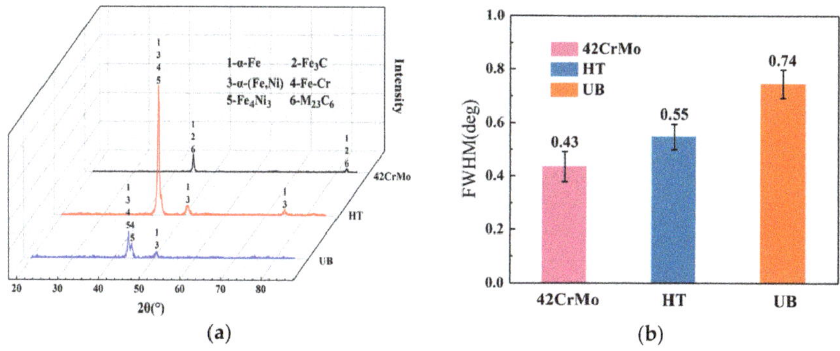

Figure 6. XRD measurements ((**a**) XRD spectrum; (**b**) FWHM measurements).

3.4. Residual Stress

Figure 7 shows the residual stress measurements. The residual stress values of the control and HT-treated samples are 404.48 $\pm$ 27.45 MPa and 571.48 $\pm$ 36.25 MPa, respectively, indicating tensile stress. Tensile residual stress introduction is the usual case for laser cladding. Two aspects are responsible for that. One is the great thermal expansion coefficient difference between the substrate and cladding material. The other is the rapid heating and cooling characteristics of the cladding process. It is known that tensile stress could facilitate crack initiation and propagation [30,31], and thus harm the mechanical property of the material.

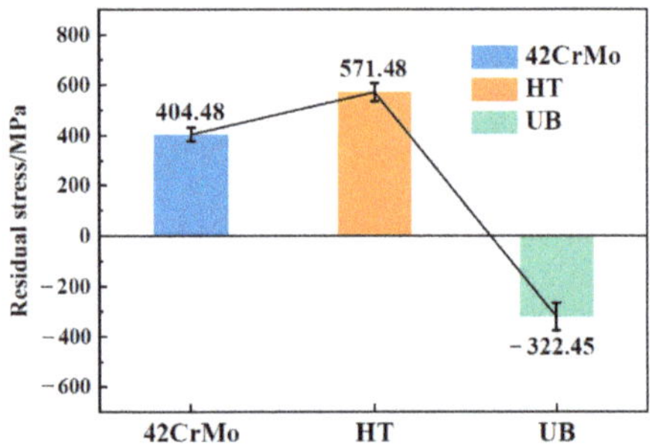

Figure 7. Residual stress measurements.

Compressive residual stress introduction is a key process effect of UB [32]. According to the results, the UB-treated coating had a compressive stress of about 322.45 MPa. Similarly, Korzynski et al. [33] also reported that ultrasonic burnishing could greatly improve the residual stress of the coating. The compressive stress introduction effect of UB could be explained from two aspects. On one hand, under the hammering of the burnishing tool tip, the near-surface materials flowed and deformed. Therefore, a layer of the near-surface material was compressed. On the other hand, UB introduced a large strain into the treated surface, generating high-density dislocations. The dislocation movement formed dislocation walls and entanglements, resulting in grain refinement [20,34]. Microstress was stored inside the refined grains, generating compressive stress. By combining Figures 6 and 7, it can be found that the residual stress measurements agree well with the diffraction peak shift.

3.5. Microhardness

The microhardness is the key to assess the wear behaviors of materials. As shown in Figure 8a, the surface microhardness values of the control and HT-treated coatings are 328.3 $HV_{0.5}$ and 416.6 $HV_{0.5}$, respectively. Compared with the control, the microhardness of HT-treated sample increased by 26.9%. The surface microhardness of the UB-treated sample was 480.1 $HV_{0.5}$, showing an increase of 15.24% compared to the HT-treated sample.

Figure 8b shows the varying in-depth hardness. The microhardness of the HT-treated sample was relatively stable in the coating region, while it sharply decreased at the coating–substrate interface. In contrast, the UB-treated sample exhibited a gradient hardness variation along the in-depth direction. During UB, the top surface material deformed the most, and the farther away it was from the top surface, the less deformation occurred [35,36]. Thus, UB could produce a gradient structure inside the coating, which was beneficial to the coating's wear resistance. Grain refinement was one main reason for the coating hardening. In the UB treatment, due to the material deformation, the crystal plane space and the distance between the atoms both decreased, and thus, the hardness was increased. In addition, the compressive residual stress introduction, the dislocation strengthening, and the work hardening were also responsible for the hardness improvement.

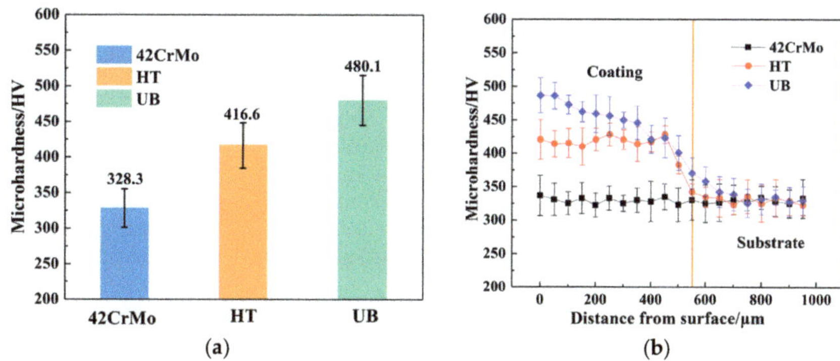

Figure 8. Microhardness measurements: (**a**) surface microhardness and (**b**) cross-sectional microhardness variation.

3.6. High-Temperature Friction Behavior

Figure 9a shows the friction coefficient varying for the three tested samples. The friction coefficient of all the samples increased rapidly at the beginning, and then decreased sharply, and finally became stable. Generally, the wear process includes three periods: running in, stable wear, and severe wear. Here, the running-in period was about 400 s. At this stage, the initial friction coefficients of the control and HT-treated samples were between 0.25 and 0.3, while such value for the UB-treated one was more than 0.4. Above all, according to the classic friction theory, it can be determined that all samples were in a boundary lubrication state. The larger friction coefficient of the UB-treated sample was mainly caused by lubricant loss from the smooth surface. In the control and HT-treated cases, there existed cutting marks, which could be served as lots of small lubricant reservoirs. In contrast, for the UB-treated sample, the smooth surface finishing facilitated lubricant loss.

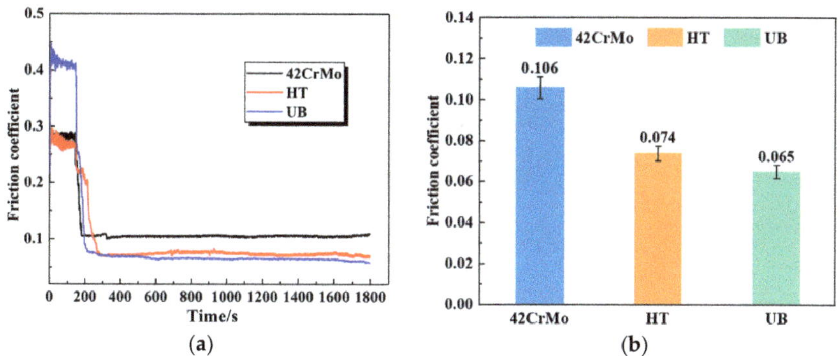

Figure 9. Friction coefficient measurements ((**a**) varying frictional coefficient; (**b**) average value).

At the stable wear period (400–1800 s), the contact area between the counterbody and the sample surfaces increased, and thus, the influence of the surface roughness greatly decreased. Meanwhile, a layer of oil film formed on the friction surface, which served to decrease the friction coefficient fluctuation. At this stage, the friction coefficient of the UB-treated sample was slightly lower than that of the HT-treated sample. The friction coefficient data in the stable friction period were averaged and are shown in Figure 9b. Compared to the value of the control sample, which was 0.106, the friction coefficient of the HT-treated sample decreased by 31.13%. Such value for the UB-treated sample was 0.065, which was further reduced by 10.96% compared to that of the HT-treated one. At this stage, for all the three samples, the friction coefficients were in a range of 0.01–0.11, which indicated a hybrid of boundary lubrication and fluid lubrication. For the UB-treated sample, the decrease in the friction coefficient mainly resulted from surface improvement, including small roughness, high hardness, and the induced compressive residual stress [37,38].

Figure 10 shows the worn zone morphologies of the three samples. As shown in Figure 10a, in the control case, the sample surface suffered from severe wear along the sliding direction. There existed many debris, furrows, and pits, indicating abrasive and adhesion wear mechanism. At a high temperature, under the press load, the surface asperities slightly melted and welded with the counterbody instantaneously, and then with the sliding going on, peeling and cracks occurred.

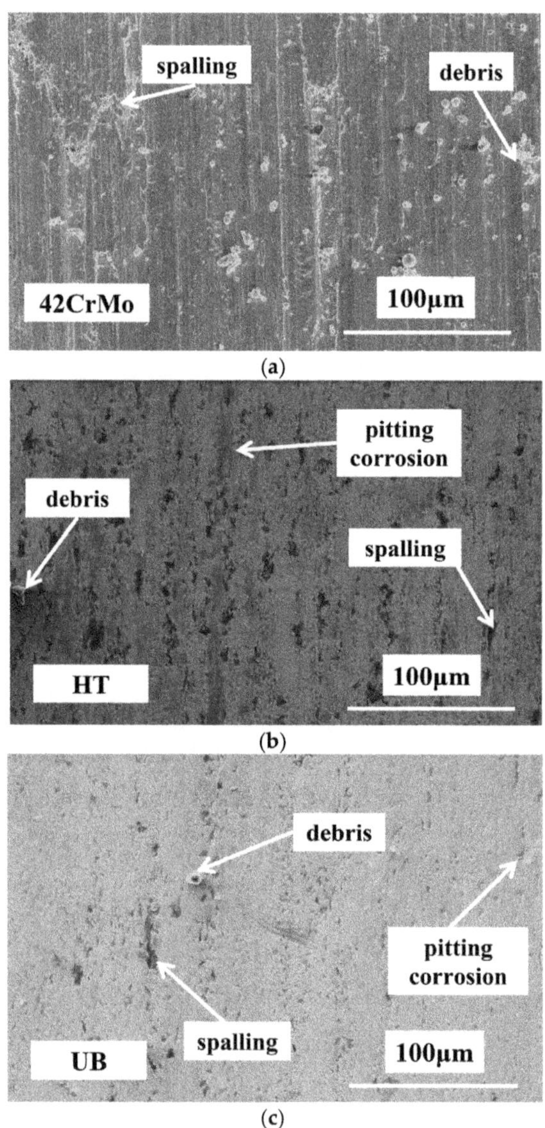

Figure 10. Worn zone images ((**a**) control; (**b**) HT-treated sample; (**c**) UB-treated sample).

In contrast, as seen in Figure 10b, the HT-treated surface had almost no plowing marks, and only a few wear debris were observed. However, there existed some pitting. As seen in Figure 10c, the worn morphologies of the HT-treated and UB-treated surfaces were similar. However, the latter had much less worn marks, exhibiting better wear resistance. The high hardness was mainly responsible for the wear resistance improvement.

The cross-sectional wear scar profiles of the three tested samples are shown in Figure 11a. Clearly, the control sample had the maximum worn depth and width, while the UB-treated one had the least worn dimension. The wear loss together with the wear rate were calculated and are compared in Figure 11b. The wear loss values of the control, HT-treated sample, and UB-treated sample were 0.071 mm³, 0.024 mm³, and 0.017 mm³, respectively. In comparison to the control case, the wear rates of the two treated samples had a decrease of 66.62% and 76.06%, respectively. According to Archard's law, the wear loss is inversely proportional to the hardness. Here, the hardness and wear loss measurements conformed to the law.

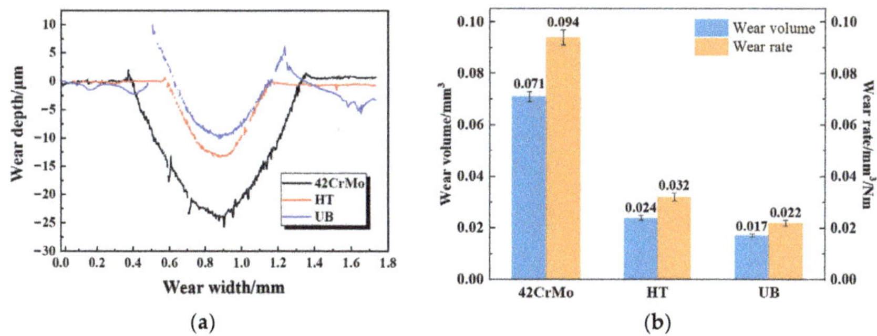

Figure 11. Worn zone measurements ((**a**) cross-sectional profile; (**b**) wear loss and wear rate).

The mechanism for the wear resistance improvement of the UB-treated surface could be explained from micro and macro aspects. At the micro level, as stated above, by UB, the grain was refined, and the amount and strength of the grain boundary both increased, making boundary slip difficult during sliding friction. At the macro level, UB caused work hardening and thus made the material difficult to deform. In addition, the compressive residual stress introduced from UB could serve to reduce the maximum shear stress and equivalent stress, and thus prevent surface crack initiation and propagation. Furthermore, the severe plastic deformation of the material from UB introduced high-density grain boundaries and dislocations, which could boost oxygen atom adsorption and thus form a dense oxide layer [39].

4. Conclusions

To improve the high-temperature wear resistance of some engineering parts, Fe-based alloy cladding coatings were prepared on a 42CrMo substrate via a laser and then post-treated via turning alone (HT-treated) or turning with an ultrasonic burnishing chain (UB-treated). The forged 42CrMo was set as the control for the comparative analysis.

The main conclusions are as follows:

(1) In comparison to the control, the hardness of the HT-treated sample was raised by 26.90%. However, turning alone could not improve the roughness or change the tensile residual stress.

(2) In comparison to the HT-treated sample, the UB-treated sample had a 79.2% roughness reduction, a further 15.2% increase in the hardness, and an 85.1% decrease in the porosity. Moreover, UB introduced compressive residual stress in the near-surface material of the coating.

(3) The two coated samples had better high-temperature wear resistance than the control. When compared with the HT-treated sample, although the UB-treated sample showed a larger friction coefficient in the running-in friction period, it had a lower friction coefficient and much less wear loss.

(4) A wearable coating serving at a high temperature could be manufactured and remanufactured via a laser and then post-treated via an HT-UB chain in a low-cost manner.

Author Contributions: Conceptualization, N.X. and X.S.; methodology, N.X. and X.S.; validation, N.X., H.P. and X.J.; investigation, X.J. and H.P.; writing—original draft preparation, N.X. and X.J.; writing—review and editing, N.X. and X.S. All authors have read and agreed to the published version of the manuscript.

Funding: This study was supported by the Shandong Provincial Natural Science Foundation (grant number ZR2023ME104).

Data Availability Statement: The data are contained within the article.

Conflicts of Interest: The authors declare no conflict of interest.

References

1. Qi, K.; Yang, Y.; Hu, G.; Lu, X.; Li, J. Thermal expansion control of composite coatings on 42CrMo by laser cladding. *Surf. Coat. Technol.* **2020**, *397*, 125983. [CrossRef]
2. Yang, Z.; Li, S.; Zhang, J.; Zhang, J.; Li, G.; Li, Z.; Hui, W.; Weng, Y. The fatigue behaviors of zero-inclusion and commercial 42CrMo steels in the super-long fatigue life regime. *Acta Mater.* **2004**, *52*, 5235–5241. [CrossRef]
3. Lin, Y.; Chen, M.; Zhong, J. Microstructural evolution in 42CrMo steel during compression at elevated temperatures. *Mater. Lett.* **2008**, *62*, 2132–2135. [CrossRef]
4. Tejero-Martin, D.; Rad, M.; McDonald, A.; Hussain, T. Beyond traditional coatings, a review on thermal sprayed functional and smart coatings. *J. Therm. Spray Technol.* **2018**, *28*, 598–644. [CrossRef]
5. Wu, H.; Wu, Y.; Yan, M.; Tu, B.; Li, Y. Microstructure and mechanical properties of surface coating prepared on grade 2 titanium by Ni-B composite electroplating and laser cladding. *Opt. Laser Technol.* **2023**, *164*, 109498. [CrossRef]
6. Jansons, E.; Lungevics, J.; Kanders, U.; Leitans, A.; Civcisa, G.; Linins, O.; Kundzins, K.; Boiko, I. Tribological and Mechanical Properties of the Nanostructured Superlattice Coatings with Respect to Surface Texture. *Lubricants* **2022**, *10*, 285. [CrossRef]
7. Xu, J.; Zhang, X.; Xuan, F.; Wang, Z.; Tu, S. Rolling contact fatigue behavior of laser cladded WC/Ni composite coating. *Surf. Coat. Technol.* **2014**, *239*, 7–15. [CrossRef]
8. Kołodziejczak, P.; Bober, M.; Chmielewski, T. Wear Resistance Comparison Research of High-Alloy Protective Coatings for Power Industry Prepared by Means of CMT Cladding. *Appl. Sci.* **2022**, *12*, 4568. [CrossRef]
9. Barr, C.; Rashid, R.; Palanisamy, S.; Watts, J.; Brandt, M. Examination of steel compatibility with additive manufacturing and repair via laser directed energy deposition. *J. Laser Appl.* **2023**, *35*, 022015. [CrossRef]
10. Rashid, R.; Nazari, K.; Barr, C.; Palanisamy, S.; Orchowski, N.; Matthews, N.; Dargusch, M. Effect of laser reheat post-treatment on the microstructural characteristics of laser-cladded ultra-high strength steel. *Surf. Coat. Technol.* **2019**, *372*, 93–102. [CrossRef]
11. Feng, Y.; Pang, X.; Feng, K.; Feng, Y.; Li, Z. Residual stress distribution and wear behavior in multi-pass laser cladded Fe-based coating reinforced by M 3 (C, B). *J. Mater. Res. Technol.* **2021**, *15*, 5597–5607. [CrossRef]
12. Liu, J.; Shi, Y. Microstructure and wear behavior of laser-cladded Ni-based coatings decorated by graphite particles. *Surf. Coat. Technol.* **2021**, *412*, 127044. [CrossRef]
13. Kumar, R.; Torres, H.; Aydinyan, S.; Antonov, M.; Varga, M.; Hussainova, I.; Ripoll, M. Tribological behavior of Ni-based self-lubricating claddings containing sulfide of nickel, copper, or bismuth at temperatures up to 600 °C. *Surf. Coat. Technol.* **2023**, *456*, 129270. [CrossRef]
14. Yin, M.; Cai, Z.; Zhang, Z.; Yue, W. Effect of ultrasonic surface rolling process on impact-sliding wear behavior of the 690 alloy. *Tribol. Int.* **2020**, *147*, 105600. [CrossRef]
15. Huang, H.; Wang, Z.; Lu, J.; Lu, K. Fatigue behaviors of AISI 316L stainless steel with a gradient nanostructured surface layer. *Acta Mater.* **2015**, *87*, 150–160. [CrossRef]
16. Zhao, J.; Liu, Z.; Wang, B.; Cai, Y.; Song, Q. Analytical Prediction and Experimental Investigation of Burnishing Force in Rotary Ultrasonic Roller Burnishing Titanium Alloy Ti-6Al-4V. *J. Manuf. Sci. Eng.* **2020**, *142*, 031004. [CrossRef]
17. Shi, Y.; Shen, X.; Xu, G.; Xu, C.; Wang, B.; Su, G. Surface integrity enhancement of austenitic stainless steel treated by ultrasonic burnishing with two burnishing tips. *Arch. Civ. Mech. Eng.* **2020**, *20*, 79. [CrossRef]
18. Teimouri, R.; Amini, S. Analytical modeling of ultrasonic burnishing process: Evaluation of active forces. *Measurement* **2019**, *131*, 654–663. [CrossRef]
19. Hao, J.; Hu, F.; Le, X.; Liu, H.; Han, J. Microstructure and high-temperature wear behaviour of Inconel 625 multi-layer cladding prepared on H13 mould steel by a hybrid additive manufacturing method. *J. Mater. Process. Technol.* **2020**, *291*, 117036. [CrossRef]

20. Ye, H.; Zhu, J.; Liu, Y.; Liu, W.; Wang, D. Microstructure and mechanical properties of laser cladded Cr-Ni alloy by hard turning (HT) and ultrasonic surface rolling (USR). *Surf. Coat. Technol.* **2020**, *393*, 125806. [CrossRef]
21. Zhang, C.; Shen, X.; Wang, J.; Wang, J.; Xu, C.; He, J.; Bai, X. Improving surface properties of Fe-based laser cladding coating deposited on a carbon steel by heat assisted ultrasonic burnishing. *J. Mater. Res. Technol.* **2021**, *12*, 100–116. [CrossRef]
22. Amanov, A. Effect of local treatment temperature of ultrasonic nanocrystalline surface modification on tribological behavior and corrosion resistance of stainless steel 316L produced by selective laser melting. *Surf. Coat. Technol.* **2020**, *398*, 126080. [CrossRef]
23. Shen, X.; Gong, X.; Zhang, J.; Su, G. An investigation of stress condition in vibration-assisted burnishing. *Int. J. Adv. Manuf. Technol.* **2019**, *105*, 1189–1207. [CrossRef]
24. Su, H.; Shen, X.; Xu, C.; He, J.; Su, G. Surface characteristics and corrosion behavior of TC11 titanium alloy strengthened by ultrasonic roller burnishing at room and medium temperature. *J. Mater. Res. Technol.* **2020**, *9*, 8172–8185. [CrossRef]
25. Gong, L.; Pan, Y.; Peng, C.; Fu, X.; Jiang, Z.; Jiang, S. Effect of ultrasonic surface rolling processing on wear properties of Cr12MoV steel. *Mater. Today Commun.* **2022**, *33*, 104762. [CrossRef]
26. Liu, D.; Liu, D.; Zhang, X.; Liu, C.; Ao, N. Surface nanocrystallization of 17-4 precipitation-hardening stainless steel subjected to ultrasonic surface rolling process. *Mater. Sci. Eng.-A Struct.* **2018**, *726*, 69–81. [CrossRef]
27. Su, Y.; Zhu, Y.; Zhang, B.; Zhou, H.; Li, J.; Wang, F. Spectral response of polarization properties of fiber Bragg grating under local pressure. *Opt. Fiber Technol.* **2015**, *25*, 15–19. [CrossRef]
28. Krzysztof, T.; Halina, G. Manufacturing of nanostructured titanium Grade2 using caliber rolling. *Mater. Sci. Eng. A Struct.* **2019**, *739*, 277–288.
29. Marcello, C. Minimum necessary strain to induce tangled dislocation to form cell and grain boundaries in a 6N-A1. *Mater. Sci. Eng. A Struct.* **2020**, *770*, 138420.
30. Suárez, A.; Amado, J.; Tobar, M.; Yanez, A.; Frag, E.; Peel, M. Study of residual stresses generated inside laser cladded plates using FEM and diffraction of synchrotron radiation. *Surf. Coat. Technol.* **2010**, *204*, 1983–1988. [CrossRef]
31. Shen, X.; He, X.; Gao, L.; Su, G.; Xu, H.; Xu, N. Study on crack behavior of laser cladding ceramic-metal composite coating with high content of WC. *Ceram. Int.* **2022**, *48*, 17460–17470. [CrossRef]
32. Ravil, K.; Young, S.; Pyun, C.; Min, S.; Riichi, M. Mechanical and fatigue characteristics of Ti-6Al-4V extra low interstitial and solution-treated and annealed alloys after ultrasonic nanocrystal surface modification treatment. *J. Nanosci. Nanotechnol.* **2014**, *14*, 9430–9435.
33. Korzynski, M.; Pacana, A.; Cwanek, J. Fatigue strength of chromium coated elements and possibility of its improvement with slide diamond burnishing. *Surf. Coat. Technol.* **2009**, *203*, 1670–1676. [CrossRef]
34. Ye, C.; Telang, A.; Gill, A.; Suslov, S.; Idell, Y.; Zweiacker, K.; Wiezorek, J.; Zhou, Z.; Qian, D.; Mannava, S.; et al. Gradient nanostructure and residual stresses induced by Ultrasonic Nano-crystal Surface Modification in 304 austenitic stainless steel for high strength and high ductility. *Mater. Sci. Eng. A Struct.* **2014**, *613*, 274–288. [CrossRef]
35. Fernandes, F.; Mahesh, K.; Silva, R.; Gurau, C.; Gurau, G. XRD study of the transformation characteristics of severely plastic deformed Ni-Ti SMAs. *Phys. Status Solidi C* **2010**, *7*, 1348–1350. [CrossRef]
36. Qin, W.; Li, J.; Liu, Y.; Yue, W.; Wang, C.; Mao, Q.; Li, Y. Effect of Rolling Strain on the Mechanical and Tribological Properties of 316 L Stainless Steel. *J. Tribol.* **2019**, *141*, 021606. [CrossRef]
37. Amanov, A.; Urmanov, B.; Amanov, T.; Pyun, Y. Strengthening of Ti-6Al-4V alloy by high temperature ultrasonic nanocrystal surface modification technique. *Mater. Lett.* **2017**, *196*, 198–201. [CrossRef]
38. Konyashin, I.; Ries, B.; Hlawatschek, D.; Zhuk, Y.; Park, D. Wear-resistance and hardness, Are they Directly related for nanostructured hard materials? *Int. J. Refract. Met. Hard* **2015**, *49*, 203–211. [CrossRef]
39. Amanov, A. Improvement in mechanical properties and fretting wear of Inconel 718 superalloy by ultrasonic nanocrystal surface modification. *Wear* **2020**, *446*, 203208. [CrossRef]

Article

Thermodynamic Analysis Based on the ZL205A Alloy Milling Force Model Study

Jing Cui, Xingquan Shen *, Zhijie Xin, Huihu Lu, Yanhao Shi, Xiaobin Huang and Baoyu Sun

School of Mechanical Engineering, North University of China, Taiyuan 030051, China; s202202018@st.nuc.edu.cn (J.C.); 13935173390@163.com (Z.X.); luhuihu@nuc.edu.cn (H.L.); shiyanhao1209@163.com (Y.S.); hxbgys@163.com (X.H.); 15247673446@163.com (B.S.)
* Correspondence: m15535395030@163.com

Abstract: The ZL205A aluminum alloy is mostly used in automobiles, aircraft, aerospace, and other mechanical components, but now, it focuses on the study of its casting performance, and there is still a lack of research on its cutting performance. In this paper, the milling ZL205A aluminum alloy was milled for testing and simulation analysis. The milling test showed that the impact of the axial cutting depth, radial cutting depth, feed, and cutting speed on the milling force was successively reduced. A thermodynamic analysis model is proposed to evaluate the cutting force and tool design in milling. The model considers the front angle and friction angle of the tool, in which the friction angle is adjusted by the friction coefficient, the variable is the cutting parameter, the constant is fitted through the milling experiment, and the effectiveness of the model is verified to predict the milling force. The pre-grinding test was carried out before the experiment, and the stability of the test was proved by observing the macroscopic shape of the chip and the wear of the cutting edge. The model comprehensively considers the tool angle and quickly calculates the minimum load on the milling cutter based on the optimal geometric parameters, which can be used to optimize the milling cutter structure and provide a theoretical basis for the preparation of ZL205A aluminum alloy mechanical components.

Keywords: ZL205A aluminum alloy; milling forces; thermomechanical behavior; analytical modeling; parametric study

Citation: Cui, J.; Shen, X.; Xin, Z.; Lu, H.; Shi, Y.; Huang, X.; Sun, B. Thermodynamic Analysis Based on the ZL205A Alloy Milling Force Model Study. *Lubricants* **2023**, *11*, 390. https://doi.org/10.3390/lubricants11090390

Received: 17 August 2023
Revised: 5 September 2023
Accepted: 6 September 2023
Published: 11 September 2023

1. Introduction

Since the 20th century, aluminum–copper cast aluminum alloys have been widely used in automotive, aircraft, and aerospace mechanical components [1,2]. Because of its good room temperature performance and high temperature resistance, ZL205A is an aluminum alloy with a tensile strength of up to 500 MPa and an elongation of up to 10% [3,4]. It has excellent comprehensive properties, such as mechanical processing, electroplating, stress corrosion resistance, etc. It is suitable to produce large-forced structural parts [5]. Due to the complex content of ZL205A material alloy elements, and with a wide crystallization range and various uncertainties in the casting process, various defects often occur, resulting in organizational performance that does not reach the service requirements [6,7]. Most researchers have carried out research on how to improve the strength of aluminum alloys [8,9]. However, the component has plane, hole, and other cutting structures, resulting in a direct impact on the service performance of the component in the application process [10,11]. Therefore, it is of great practical significance and engineering value to carry out research on the machinability of aluminum alloys.

There are numerous problems in milling, such as the breakage of the milling cutter, the vibration of milling, etc., which are largely caused by the action of milling force [12]. In the milling process, the milling force will directly affect the generation of cutting heat and have an impact on tool wear, machining accuracy, and the workpiece-processed surface quality [13]. Therefore, it is of great significance to study and establish a mathematical model of the cutting force of the Z205A aluminum alloy. The research on milling force is mainly divided into empirical models based on experimental verification [14], finite element models [15], and analytical models [16].

The American scholar Merchant [17] established a cutting theory with shear slip as the main feature of right-angle free cutting as the basic model. E.H. Lee and B.W. Shaffer [18] deduced the calculation formula of the shear angle by constructing a sliding line field model of right-angle free cutting. P.L.B. Oxley [19] studied the strain rate and strain hardening of the material in the cutting process and proposed a cutting model with a parallel quadrilateral cutting surface. E.D. Doyle [20] established a cutting model considering knife-chip friction. Zhang, X. [21] discussed the influence of tool geometry parameters on milling force by establishing an empirical model considering quadratic polynomial factors. Zhu, S. [22] proposed a milling force prediction model based on the Taylor factor to study the influence of different milling parameters on the evolution of surface organization. Chen, Y.H. [23] proposed a tool radial jump milling model considering the static jump of the tool for batch precision machining. Su, X. [24] proposed a milling force model applied to complex contour milling cutters. Over the years, most scholars have focused on analyzing milling models from multiple cutting angles, but this inevitably increases the complexity of prediction, so this study is committed to proposing a fast and effective milling force model. In addition to the influence of friction on material properties [25–27], friction in milling cannot be ignored [28], so this study adjusted the friction angle by adjusting the friction coefficient.

This study proposes a milling force model in the ZL205A aluminum alloy based on thermodynamic analysis and verifies the effectiveness of the model through milling tests. The model considers material properties and tool structure (tool front angle, friction angle), where the friction angle is calculated by the friction coefficient. This research provides a theoretical basis for optimizing the tool structure of the milling cutter and provides a theoretical basis for the preparation of mechanical components for automobiles, aircraft, and aerospace with the ZL205A aluminum alloy, making the use of the material more convenient.

2. Experimental Woke

The milling test was carried out to analyze the cutting process of the ZL205A aluminum alloy. The following will introduce the workpiece materials, milling cutters, NC machine tools, dynamometers, and the tests that were carried out.

2.1. Workpiece Materials

The mechanical characteristics of the ZL205A aluminum alloy are shown in Table 1. The Johnson–Cook constitutive model and Johnson–Cook damage model were established for quasi-static tensile, Hopkinson rod, and high-temperature tensile tests. Considering the impact of adiabatic temperature rise, parameter C was corrected. The model was verified by quasi-static smooth stretching, Hopkinson rod, and high-temperature tensile tests, and the maximum error of the model did not exceed 4%. The chemical composition of the ZL205A aluminum alloy is shown in Table 2.

Table 1. Mechanical characteristics of ZL205A aluminum alloy.

ρ (g/cm^3)	H (HRC)	$R_{p0.2}$ (T°: Ambient) (MPa)	$R_{p0.2}$ (T°: 623 K) (MPa)	R_{pm} (T°: Ambient) (MPa)
2.8	55	297.94	111.38	417.76

2.2. Milling Test

The CNC machine tool adopts HCZK1340 machine tool designed by Zhengdang Precision Machinery Co., LTD., Shenzhen City, Guangdong Province, China; as shown in

Figure 1a, the tool handle adopts WINIO Winon BAP300R 050T4-22 CNC cutter head from Taiwan Province, China; as shown in Figure 1c, the tool adopts APKT11T308-LH milling cutter of Diamond Factory, Zhuzhou City, Hunan Province, China; as shown in Figure 1d, the dynamometer adopts the Swiss KISTLER four-part measurement system by installing the KISTLER four-part dynamometer on the CNC machine tool, using a fixture to connect the dynamometer to the workpiece, as shown in Figure 1b. Water soluble cutting fluid from Dr. You of Shenzhen City, Guangdong Province, China was used in the test.

Table 2. Chemical composition of ZL205A aluminum alloy (wt.%).

Element	Cu	Mn	Ti	Zr	Cd	B	V	Al
Cont.	4.95	0.4	0.25	0.125	0.2	0.035	0.175	Bal.

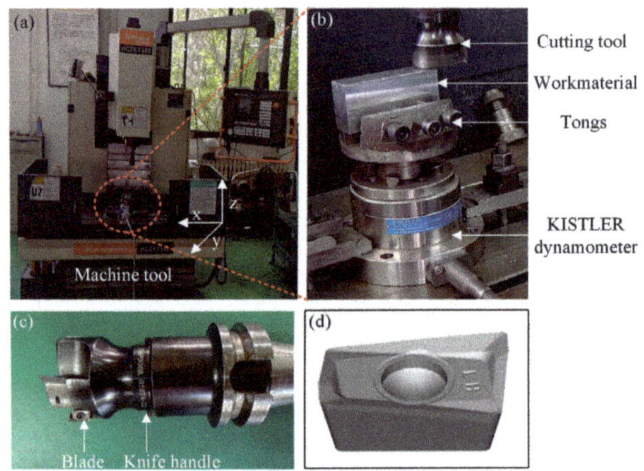

Figure 1. Test equipment diagram. (**a**) Machining center; (**b**) KISTLER dynamometer; (**c**) tool handle; (**d**) blade.

2.3. Measure the Axial Force

The purpose of this experiment is to identify the relatively stable cutting range in the milling process of a given tool/workpiece material. The milling test is carried out by changing the cutting parameters of the cutting speed, feed speed, axial cutting depth, and radial cutting depth. The cutting parameters are designed according to the cutting conditions recommended by the tool manufacturer. For the ZL205A aluminum alloy, the cutting speed, feed per tooth, axial cutting depth, and radial cutting depth are 200 m/min, 0.3 mm/r, 2 mm, and 15 mm, respectively. The milling tests carried out are shown in Table 3. Each set of tests is repeated 4 times, and new blades are used each time. No tool acceleration or early grinding was detected during all testing processes, and conventional chips were formed. Therefore, the milling test can be regarded as stable cutting.

The axial cutting force was recorded by the KISTLER four-component dynamometer. Figure 2 shows the variation trend of the axial cutting force with cutting speed, feed rate per tooth, axial cutting depth, and radial cutting depth. The axial milling force increases with the increase in cutting speed, feed rate, axial cutting depth, and radial cutting depth.

2.4. Cutting Performance

By observing the chip shape and tool wear, the stability of the milling process in each milling experiment is verified using Japan's KEYENCE ultra-depth-field microscope to observe chips. Figure 3 shows the change in chip macroscopic morphology with the change in parameters, and the number in the figure corresponds to the experimental number in Table 3.

Table 3. Milling test design of ZL205A aluminum alloy.

	Test No.	Cutting Speed (m/min)	Feed Rate (mm/r)	Axial Depth of Cut (mm)	Radial Depth of Cut (mm)
1st Phase	1	100	0.3	2	15
	2	200	0.3	2	15
	3	300	0.3	2	15
	4	400	0.3	2	15
2nd Phase	5	200	0.1	2	15
	6	200	0.2	2	15
	7	200	0.3	2	15
	8	200	0.4	2	15
3rd Phase	9	200	0.2	1	15
	10	200	0.2	2	15
	11	200	0.2	3	15
	12	200	0.2	4	15
4th Phase	13	200	0.2	2	3
	14	200	0.2	2	9
	15	200	0.2	2	15
	16	200	0.2	2	21

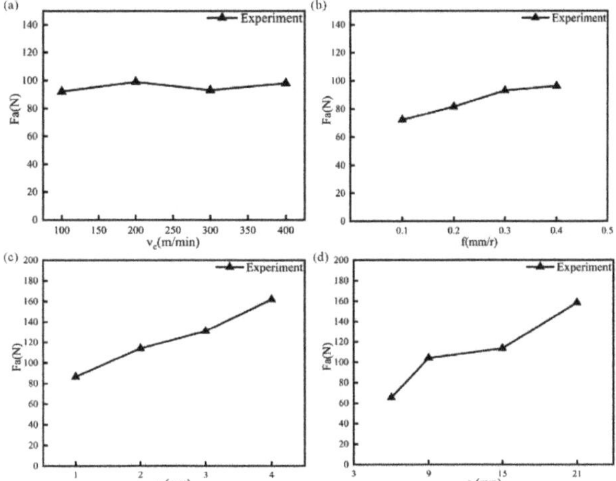

Figure 2. Measuring the axial force of milling. (**a**) The change in the axial force of milling with the cutting speed. (**b**) The change in the axial force of milling with the feed of each tooth. (**c**) The change in milling axial force with axial cutting depth. (**d**) The change in milling axial force with radial cutting depth.

The influence of the radial cutting depth, axial cutting depth, feed volume, and cutting speed on the chip is successively reduced. The ZL205A aluminum alloy will generate continuous chips under the cutting parameters of the test design. Even if the cutting speed increases, the macroscopic change in the chips is not obvious (Figure 3a). Increasing the feed (Figure 3b) and cutting depth (Figure 3c,d) will increase the cutting speed in the unit and the removal amount. As a result, the extension length of the chip in the cutting area increases so the chip length becomes longer, and the length deformation coefficient lessens (Figure 3b). When the axial cutting depth increases, it causes greater cutting force and cutting heat, which will affect the cutting force distribution and chip formation mode in the cutting area.

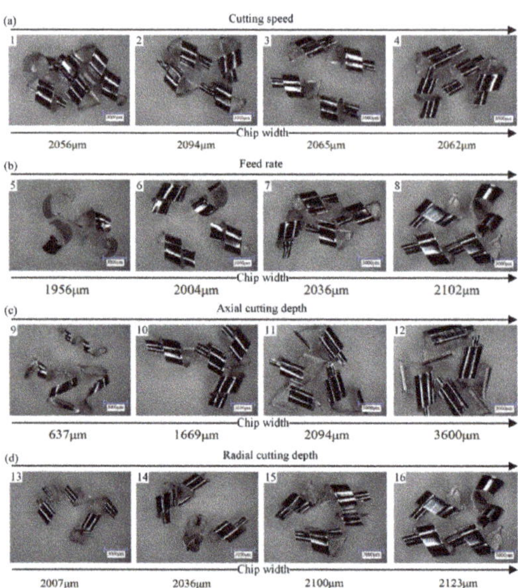

Figure 3. Macroform of chip. (**a**) Change in chips with cutting speed. (**b**) Change in chips with feed. (**c**) The change in chip with axial cutting depth. (**d**) The change in chip with the radial cutting depth.

Figure 4 shows tool wear. The new blade (Figure 4a) is compared with the blade used in the milling test (Figure 4b). The four blades are all stable.

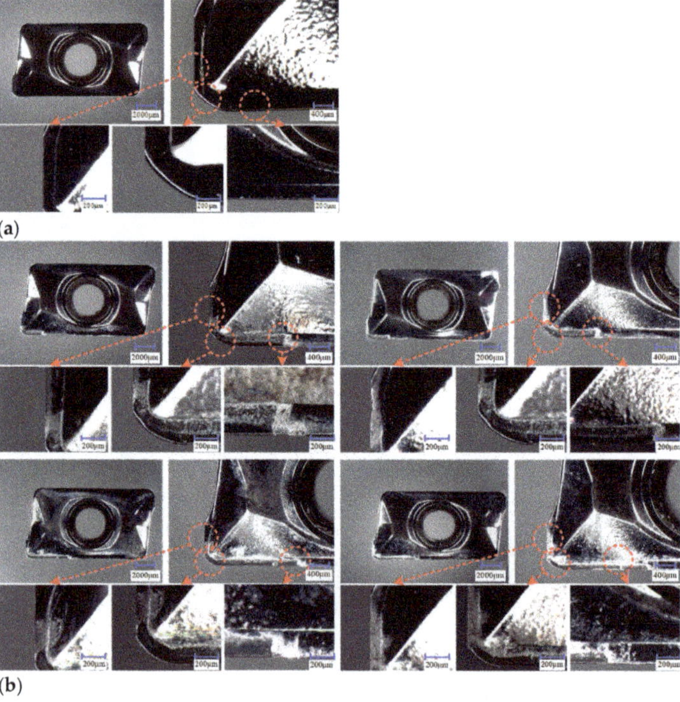

Figure 4. Tool wear. (**a**) New blade. (**b**) Milling cutter blade four blades.

3. Thermoviscoelastic Model for Predicting Milling Forces

The mathematical model of cutting force is often limited by the workpiece material, tool, etc., and is effective within the range of the available data. Therefore, the verification of the mathematical model of cutting force should be completely based on the test data. The thermodynamic analysis model established in this paper, combined with the material performance parameters, tool structure, and cutting parameters, can quickly evaluate the cutting force of the milling ZL205A aluminum alloy. The analysis method involves the material performance and the law of contact/friction, and the equation is established based on certain assumptions.

3.1. Modeling of Cutting Forces

During the milling process, the cutting layer and internal workpiece undergo elastic and plastic deformation resistance, chip outflow, and friction between the workpiece movement and the tool. These combined forces acting on the tool can be decomposed into three parts. It is assumed that lubrication during milling is indirectly considered by adjusting the friction coefficient in the mathematical model of milling forces.

By discrediting the milling cutter, the local coordinate system Oxyz of the tool is established, with the center of the circle on the base of the milling cutter as the coordinate origin, the direction of the milling cutter axis as the z-axis and upward as the positive direction, and the direction of the intersection of the cutting edge on the base circle of the milling cutter from the origin pointing to a certain cutting edge as the y-axis. Based on the idea of discrete milling edges, the spiral milling edge cutting into the workpiece part is divided axially into a certain number of micro-element cutting edges, i.e., the dz section in the diagram. The cutter teeth are numbered j. The cutting edge with the starting point on the y-axis is recorded as blade number 1 and is numbered sequentially in the direction of the helix, and the cutter teeth micro-elements are numbered i and are numbered sequentially in the direction of the helix, with a maximum number of N.

The force applied to the workpiece is derived from the force applied to each micro-element of the workpiece. The component of the combined force applied to the workpiece is calculated by calculating the component of the combined force for each micro-element. Figure 5 illustrates the micro-element of the milling cutter edge and the applied force.

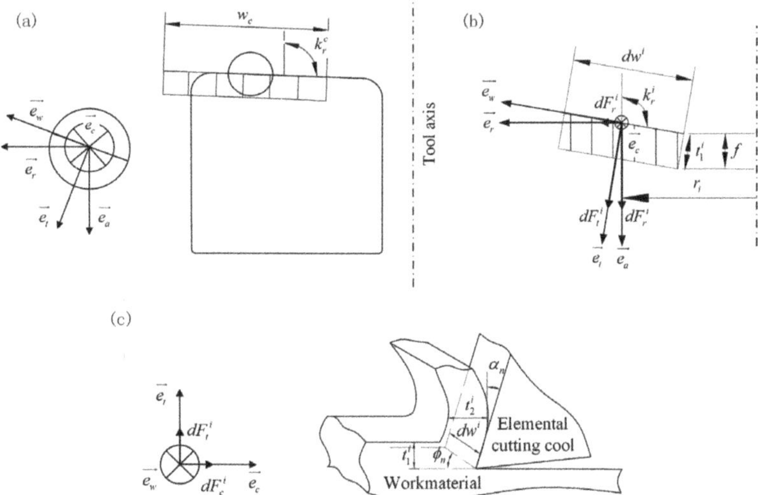

Figure 5. Illustration of (**a**) the resultant component of the milling force acting on the milling cutter; (**b**) the elemental component of the milling force acting on the cutting edge; and (**c**) shows the right-angle cutting corresponding to each micro-element.

The results applied to the workpiece can be expressed in terms of tangential milling forces, radial milling forces, and axial milling forces as follows:

$$F_c = \sum_{j=1}^{4} F_c^j, F_r = \sum_{j=1}^{4} F_r^j, F_a = \sum_{j=1}^{4} F_a^j \tag{1}$$

where F_c, F_r, F_a are tangential, radial, and axial forces, respectively.

Where the components of the milling synthesis force (F_c^j, F_r^j, F_a^j) can be calculated by integrating the components of the milling force. This is shown below:

$$F_c^j = \int dF_c, F_a^j = \int dF_a, F_r^j = \int dF_r \tag{2}$$

By discrediting the edges of the milling cutter, the synthetic components of each cutting edge are summed as follows:

$$F_c^j = \sum_{i=1}^{N} \Delta F_c^j, F_r^j = \sum_{i=1}^{N} \Delta F_r^j = \sum_{i=1}^{N} \Delta F_r^j\left(\Delta F_f^i, \Delta F_c^i\right), F_a^j = \sum_{i=1}^{N} \Delta F_a^j = \sum_{i=1}^{N} \Delta F_a^j\left(\Delta F_f^i, \Delta F_c^i\right) \tag{3}$$

Equation (3) is used to predict the combined force component for each cutting edge, which is then substituted into Equation (2). However, here, it is necessary to calculate ΔF_f^i and ΔF_c^i. This is performed in the analytical thermodynamic model for right-angle cutting, which is described in the next section.

3.2. Elemental Cutting Force Modeling

During milling, for radial (Figure 5), an infinitesimal milling micro-element, given by the milling length, produces a chip from a right-angle and is modeled using thermal methods. The cutting conditions corresponding to the milling micro-element are therefore given by the following:

$$t_1 = f \sin k_r, \ v_c = wr, \ \alpha_n \text{ and } dw \tag{4}$$

where t_1 (mm), f (mm/r), w (rad/s), v_c (mm/min), $\alpha_n(^\circ)$ and dw (mm) are the cutting thickness, feed, angular speed, cutting speed, tool front angle, and cutting width.

For chip formation under right-angle cutting, the shear zone is defined as a shear zone of constant thickness within which the chip deforms, characterized by a planar shear angle perpendicular to the cutting edge. The secondary shear zone at the tool–chip and the flow of the complex material at the edge of the tool material are not considered. The analysis is restricted to stationary flow and has no time dependence, and the material flow within the primary shear zone is modeled using a one-dimensional approach. Therefore, all variables are dependent only on the normal coordinates along this range. These variables are determined by the thermodynamic behavior of the material, the cutting conditions, the thickness shear angle, and the average friction coefficient of the front tool face [11]. Furthermore, the shear in the band is adiabatic, and the normal shear angle is determined by minimizing the minimum of the basic cutting energy.

Metal cutting deformation is a large strain non-linear elastic–plastic deformation; although the cutting process is accompanied by elastic deformation, compared with the amount of plastic deformation, it can be neglected. High-speed cutting materials have a high temperature, a large strain, and a large strain rate for elastic–plastic deformation. Fast cutting materials into chips in a very short time, reflecting the cutting area at each point of the strain, strain rate, and temperature on the material dynamic stress of the intentional equation in the numerical simulation calculation analysis, which is extremely critical. On balance, the Johnson–Cook model was chosen as the intrinsic model to describe the deformation behavior of the material [29], where the workpiece material should be isotropic, rigid, and viscoelastic, and it is described by the Johnson–Cook law as follows:

$$\tau = \frac{1}{\sqrt{3}}\left[A + B\left(\frac{\gamma}{\sqrt{3}}\right)^n\right]\left[1 + mIn\left(\frac{\dot{\gamma}}{\dot{\gamma_0}}\right)\right]\left[1 - \left(\frac{T - T_r}{T_m - T_r}\right)^v\right] \tag{5}$$

where A is the yield strength of the material (MPa), B is the strain reinforcement factor (MPa), n is the strain reinforcement index, m is the strain rate reinforcement factor, γ is the current shear strain of the mass, $\dot{\gamma}$ is the current shear strain rate of the mass (1/s), $\dot{\gamma_0}$ is the reference strain rate (1/s) and takes the value of 0.001/s, T is the current temperature of the mass (K), T_r is the room temperature (K), T_m is the melting temperature of the material (K) and the melting point, v is the thermal softening index, and τ is the shear stress of the mass (MPa). The three square brackets in Equation (5) successively represent the strain strengthening term, the strain rate term, and the temperature term. When a set of strain, strain rate, and temperature data is given, the corresponding mass flow stress can be obtained from Equation (5). The Johnson–Cook flow stress parameters for the ZL205A aluminum alloy are shown in Table 4.

Table 4. JC flow stress parameters for ZL205A aluminum alloy.

A (MPa)	B (MPa)	n	C	m	T_r (K)	T_m (K)	$\dot{\varepsilon_0}$
297.94	735.56	0.66	0.00672	1.30282	293	862	10^{-3}

In the main shear zone, the distributions of shear stress (γ) and temperature (T) are obtained from the solutions of the equations of motion and heat (via a one-dimensional formulation assuming smooth and adiabatic conditions), respectively. Note that the details of the equations are shown in [30,31], and only the solutions are as follows:

$$\tau = \rho(v_c \sin \phi_n)^2 \gamma + \tau_0 \tag{6}$$

$$T = T_w + \frac{\beta}{\rho c}\left(\rho(v_c \sin \phi_n)^2 \frac{\gamma^2}{2} + \tau_0 \gamma\right) \tag{7}$$

where β, c, ρ, ϕ_n, and τ_0 are the Taylor–Quincy coefficient, heat capacity, material density, shear angle, and flow stress at the start of the cut.

From the Johnson–Cook principal structure, Equation (5), it follows that

$$\begin{cases} \dot{\gamma} = \dot{\gamma_0} \exp(g(\gamma)) \\ g(\gamma) = \dfrac{\tau\sqrt{3}}{m\left(A + B(\gamma/\sqrt{3})^n\right)\left(1 - ((T - T_r)/(T_m - T_r))^v\right)} - \dfrac{1}{m} \end{cases} \tag{8}$$

Combining Equations (5)–(7), we can note that the shear strain rate is a function of shear strain and shear stress. Therefore, Equation (8) can be written as follows:

$$\dot{\gamma} = \dot{\gamma}(\gamma, \tau_0) \tag{9}$$

The distribution of this strain is controlled by the following first-order differential equation:

$$\dot{\gamma} = \frac{D\gamma}{Dt} = v_c \sin \phi_n \frac{d\gamma}{dy} \tag{10}$$

where $D\gamma/Dt$ is the derivative of the strain rate in the shear zone, and y is the coordinate along the axial direction.

The shear band bandwidth is h, and shear strain occurs within this range resulting in the following [32]:

$$\int_0^{\gamma_h} \frac{v_c \sin \phi_n}{\dot{\gamma}(\gamma, \tau_0)} d\gamma - h = 0 \tag{11}$$

The shear stress can be calculated using Equation (11). It is important to focus on the fact that the shear strain only occurs within the shear zone, and therefore, the boundary conditions for this shear strain are as follows:

$$\gamma_h = \frac{\cos \alpha_n}{\sin \phi_n \cos(\phi_n - \alpha_n)} \tag{12}$$

The micro-element cutting forces dF_c^i and feed forces dF_f^i are determined by the following equations:

$$\begin{aligned}
dF_c &= \frac{(\cos \alpha_n + \tan \lambda \sin \alpha_n)}{\sin \phi_n (\cos(\phi_n - \alpha_n) - \tan \lambda \sin(\phi_n - \alpha_n))} dw t_1 \tau_h \\
dF_f &= \frac{(- \sin \alpha_n + \tan \lambda \sin \alpha_n)}{\sin \phi_n (\cos(\phi_n - \alpha_n) - \tan \lambda \sin(\phi_n - \alpha_n))} dw t_1 \tau_h
\end{aligned} \tag{13}$$

Substituting Equation (12) into Equation (6) gives the following:

$$\tau_h = \rho(v_c \sin \phi_n)^2 \frac{\cos \alpha_n}{\sin \phi_n \cos(\phi_n - \alpha_n)} + \tau_0 \tag{14}$$

where α_n is a normal rake angle.

In Equation (4), the milling length is defined by the microtitration (by the specified geometrical parameters), where the feed to the cutting layer and the approach angle are related as follows:

$$t_1 = f \sin k_r^j \tag{15}$$

The shear and front angles and friction angles are related as follows:

$$\phi_n = 35° + \frac{1}{2}(\alpha_n - \lambda) \tag{16}$$

Elemental cutting forces are as follows:

$$\begin{aligned}
\Delta F_c^i &= \frac{(\cos \alpha_n + \tan \lambda \sin \alpha_n)}{\sin\left(35° + \frac{1}{2}(\alpha_n - \lambda)\right)\left(\cos\left(35° - \frac{1}{2}(\alpha_n + \lambda)\right) - \tan \lambda \sin\left(35° - \frac{1}{2}(\alpha_n + \lambda)\right)\right)} \Delta w f \sin k_r^j \tau_h \\
\Delta F_f^i &= \frac{(- \sin \alpha_n + \tan \lambda \sin \alpha_n)}{\sin\left(35° + \frac{1}{2}(\alpha_n - \lambda)\right)\left(\cos\left(35° - \frac{1}{2}(\alpha_n + \lambda)\right) - \tan \lambda \sin\left(35° - \frac{1}{2}(\alpha_n + \lambda)\right)\right)} \Delta w f \sin k_r^j \tau_h
\end{aligned} \tag{17}$$

The calculation can be made by combining the tool front angle, the friction angle, and the material parameters. The combined component of the milling force can be obtained by integrating the components in the radial direction, given by Equation (3).

3.3. Thermodynamic Analysis Process

The analysis steps of calculating the milling force component of the milling ZL205A aluminum alloy are shown in Figure 6. First, design the cutting parameters and tool structure required for the test. Secondly, to micro-localize the cutting edge, it is necessary to understand the length of the cutting edge, the cutting position of each micro-element, the local cutting angle, and the local cutting conditions. Then, the axial force, tangent force, and radial force of each microfilm are calculated. Finally, the axial force, tangent force, and radial force, as well as the total axial force, tangent force, and radial force of each blade are predicted, respectively.

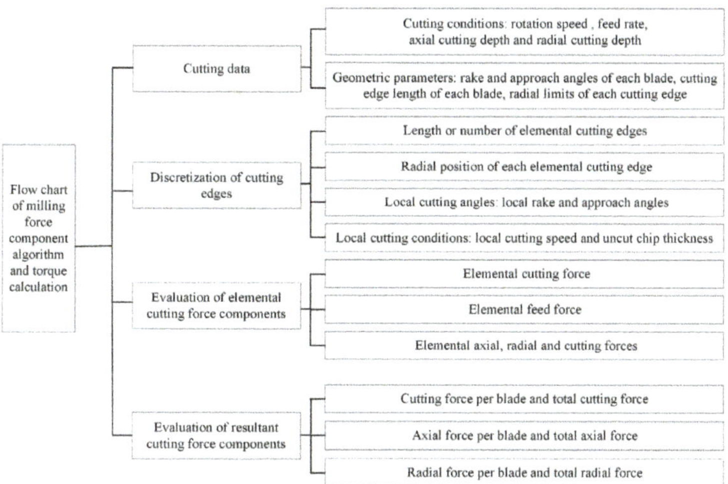

Figure 6. Thermodynamic analysis flow chart.

4. Results and Discussion

4.1. Milling Parameter Fitting

Orthogonal milling tests were carried out with the milling parameters, cutting speed (from 100 to 450 m/min), feed (from 0.1 to 0.45 mm/r), axial depth of cut (from 0.5 to 4.5 mm), and radial depth of cut (from 3 to 24 mm). This range includes the range of parameters designed for milling tests to verify machining stability (see Table 3). Figure 7 shows the average of the tangential and radial forces from the force gauge measurements.

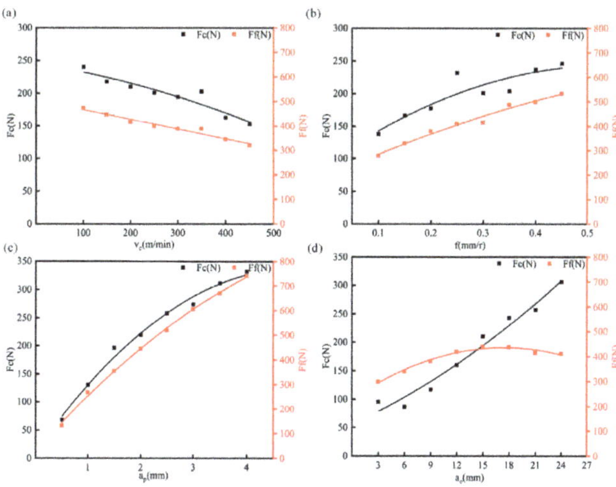

Figure 7. Tangential and radial forces for different cutting parameters. (**a**) The change in the tangential and radial forces of milling with the cutting speed. (**b**) The change in the tangential and radial forces of milling with the feed of each tooth. (**c**) The change in milling tangential and radial forces with axial cutting depth. (**d**) The change in milling tangential and radial forces with radial cutting depth.

Figure 7 shows the average value of the tangent force and radial force measured by the dynamometer. The horizontal coordinate of the figure is the cutting parameter; the left vertical coordinate is the tangent force; and the right coordinate is the feed force. With the increase in cutting speed, both the tangent force and the feed force are reduced (Figure 7a).

The increase in the cutting speed reduces the friction coefficient, increases the shear angle, reduces the deformation coefficient, and indirectly reduces the cutting force. On the other hand, the melting point of the aluminum alloy is low, and the cutting force is greatly affected by the temperature. With the increase in the cutting speed, the cutting temperature also increases. so that the strength and hardness of the aluminum alloy are reduced, thus reducing the cutting force, but the temperature has less impact on the cutting force. With the increase in the cutting speed, the contact length also decreases, and the aluminum alloy has good ductility. With the increase in the cutting temperature, the shear strength of the retention layer decreases. As a result, the cutting force decreases with the increase in the cutting speed.

Figure 7b shows the change in tangential force and radial force with the feed. Figure 7c shows the change in tangential force and radial force with axial cutting depth. Figure 7d shows the change in tangential force and radial force with the radial cutting depth. Both the tangential force and the radial force increase with the increase in the cutting parameters, but the degree of influence varies. For the tangential force, the impact of the axial cutting depth, radial cutting depth, and feed is reduced in turn. For the radial force, the impact of the axial cutting depth, feed, and radial cutting depth is reduced in turn. The increase in the cutting depth and feed directly increases the cutting area, thus increasing the deformation resistance, so the cutting force also increases. However, the increase in the feed volume also proportionally increases the cutting thickness. The increase in the cutting thickness reduces the deformation coefficient and the friction coefficient, thus reducing the cutting force. As a result of the positive and negative effects, the cutting force increases with the increase in the feed, but the impact of the feed on the cutting force is less than the impact of the cutting depth on the cutting force. The correctness of this conclusion is also verified by the test and prediction results.

For each set of tests, a coefficient of friction (COF) can be calculated, which can be adjusted by the friction angle to better predict the milling forces using the mathematical model of milling forces (see Figure 7). The corresponding coefficient is obtained by fitting the equation:

$$\mu = a_1 f + a_2 a_e + a_3 a_p + a_4 v_c$$
$$\lambda = \tan^{-1}(\mu) \tag{18}$$

The results of the fitting were substituted into the mathematical model of milling forces 17, where the cutting edge of each micro-element has a corresponding cutting speed, feed, axial depth of cut, and radial depth of cut. The coefficients of the friction law Equation (18) are shown in Table 5.

Table 5. Determining friction law parameters.

	a_1	a_2	a_3	a_4
Set 1	0.5407	2.8527×10^{-5}	5.3830×10^{-6}	1.1303×10^{-6}

Figures 8 and 9 show the radial and tangential force test values compared to the predicted values, and the friction coefficients were fitted by radial and tangential forces, so the average error between the two forces was small. The average error for the radial force is 4.5%, and the average error for the tangential force is 9.92%.

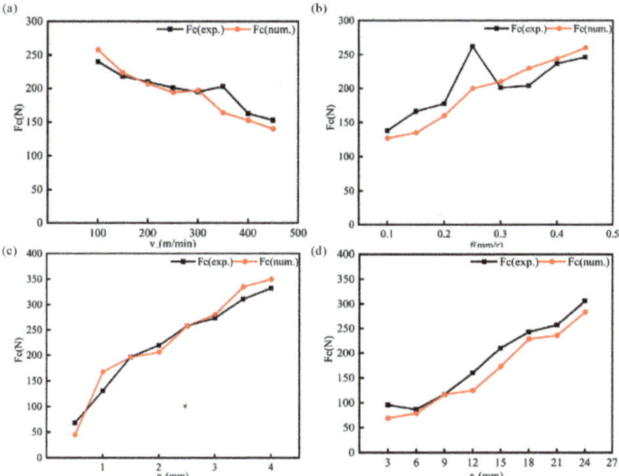

Figure 8. Orthogonal milling: experimental and predicted tangential. (**a**) The change in the tangential force of milling with the cutting speed. (**b**) The change in the tangential force of milling with the feed of each tooth. (**c**) The change in milling tangential force with axial cutting depth. (**d**) The change in milling tangential force with radial cutting depth.

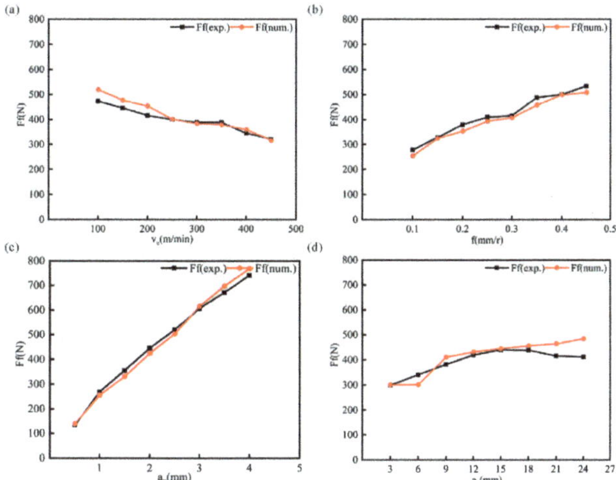

Figure 9. Orthogonal milling: experimental and predicted radial forces. (**a**) The change in the radial force of milling with the cutting speed. (**b**) The change in the radial force of milling with the feed of each tooth. (**c**) The change in milling radial force with axial cutting depth. (**d**) The change in milling radial force with radial cutting depth.

4.2. Milling Model Verification

Through milling force analysis, the calculation data of the thermodynamic model are compared with the measurement results of the milling data, and the axial force is used to verify the feasibility of the model. Within the milling range, the friction parameter Set 1 is also a good way to predict the milling cutting force data, with an average error of 17% (Figure 10). The influence of the radial cutting depth, axial cutting depth, feed volume, and cutting speed axial force is reduced in turn. With the increase in the cutting speed, the friction heating in the milling process increases, which softens the chips and increases the deformation rate of the material. The high deformation rate will lead to changes in the stress state and deformation characteristics of the material, so the axial force decreases with the increase in the cutting speed. However, increasing the radial cutting depth, axial cutting depth, and feed directly increases the cutting area, so that the deformation resistance increases and the internal friction force increases, and so the axial force also increases, but the impact of the feed on the axial force is far less than the impact of the cutting depth on the axial force. Therefore, when milling the ZL205A aluminum alloy, to improve the milling efficiency, improve the processing quality, and extend the tool life, the processing purpose of CNC milling can be achieved by adopting the cutting process parameters of large cutting width, small cutting depth, and appropriate feed volume.

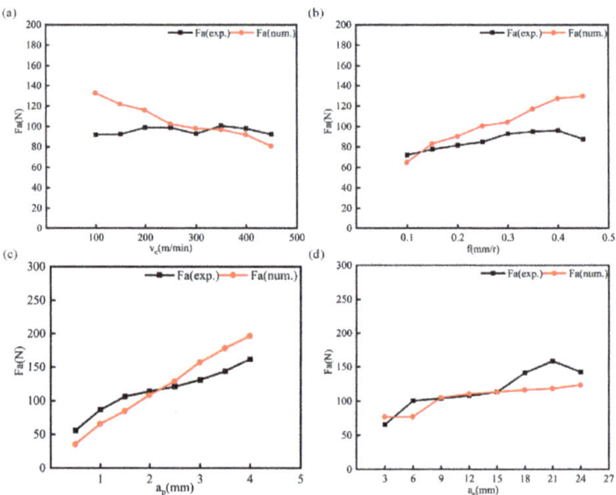

Figure 10. Axial force prediction and test. (**a**) The change in the axial force of milling with the cutting speed. (**b**) The change in the axial force of milling with the feed of each tooth. (**c**) The change in milling axial force with axial cutting depth. (**d**) The change in milling axial force with radial cutting depth.

5. Conclusions

Combined with the cutting parameters, material mechanical characteristics, and tool structure, using the material performance and contact/friction law analysis method, the friction coefficient in the milling model is fitted with the test results, and a thermodynamic analysis model of milling force is established.

(1) According to the test results, the influence of the axial cutting depth, radial cutting depth, feed, and cutting speed on the milling force is successively reduced. Based on the comprehensive analysis of roughness and milling force, the better milling parameters are 350 m/min cutting speed, 3.5 mm axial cutting depth, 15 mm radial cutting depth, and 0.15 mm/r feed.

(2) By improving the thermodynamic analysis model, the ZL205A aluminum alloy milling force prediction model based on thermodynamic analysis is established. The model predicts that the radial force and tangents are more sensitive, with an average error of 4.5% and a tangent force error of 9.92%. However, the sensitivity of axial force to cutting speed is relatively low, with an average error of 17%. The experimental results show that the milling force can be predicted by adjusting the friction coefficient.

(3) Based on the optimal cutting angle and cutting conditions, the milling model is used to predict the blade load, and the smaller load is used for cutting to improve the tool life, which can also be used to optimize the milling cutter structure. It provides a theoretical basis for the preparation of mechanical components with the ZL205A aluminum alloy.

Author Contributions: Methodology, Z.X. and X.H.; milling test, J.C. and Y.S.; data analysis, J.C. and B.S.; writing—original draft, formal analysis, data curation, J.C.; writing—review and editing, X.S. and H.L. All authors have read and agreed to the published version of the manuscript.

Funding: This work was financially supported by the National Natural Science Foundation of China (Grant Nos. 52105408 and 52075503).

Data Availability Statement: The original contributions presented in the study are included in the article, further inquiries can be directed to the corresponding author.

Conflicts of Interest: The authors declare no conflict of interest.

References

1. Wang, H.M.; Yi, Y.P.; Huang, S.Q. Investigation of quench sensitivity of high strength 2219 aluminum alloy by TTP and TTT diagrams. *J. Alloys Compd.* **2017**, *690*, 446–452. [CrossRef]
2. Wang, R.J.; Wu, S.P.; Chen, W. Mechanism of burst feeding in ZL205A casting under mechanical vibration and low pressure. *Trans. Nonferrous Met. Soc. China* **2018**, *28*, 1514–1520. [CrossRef]
3. Ren, L.; Wang, Z.; Wang, S.; Li, C.; Wang, W.; Ming, Z.; Zhai, Y. The Effect of Cu Content on the Microstructure and Properties of the Wire Arc Additive Manufacturing Al-Cu Alloy. *Materials* **2023**, *16*, 2694. [CrossRef] [PubMed]
4. Jiang, H.; Zhang, L.; Zhao, B.; Sun, M.; He, M. Microstructure and Mechanical Properties of ZL205A Aluminum Alloy Produced by Squeeze Casting after Heat Treatment. *Metals* **2022**, *12*, 2037. [CrossRef]
5. Shaga, A.; Shen, P.; Xiao, L.G.; Guo, R.F.; Liu, Y.B.; Jiang, Q.C. High dam-age-tolerance bio-inspired ZL205A/SiC composites with a lamellar-interpenetrated structure. *Mater. Sci. Eng. A* **2017**, *708*, 199–207. [CrossRef]
6. Guo, T.B.; Sun, Q.Z.; Li, K.Z.; Huang, D.W.; Wang, J.J.; Tai, X.Y. Effect of Temperature Gradient and Cooling Rate on Solidification Structure and Properties of ZL205A Alloy. *Rare Metal Mater. Eng.* **2022**, *51*, 2400–2408.
7. Luo, L.; Luo, L.S.; Su, Y.Q.; Su, L.; Wang, L.; Guo, J.J.; Fu, H.Z. Optimizing microstructure, shrinkage defects and mechanical performance of ZL205A alloys via coupling travel-ling magnetic fields with unidirectional solidification. *J. Mater. Sci. Technol.* **2021**, *74*, 246–258. [CrossRef]
8. Zhang, Z.H.; Liu, J.H.; Chen, J.M.; Wen, F.L.; Jia, R.; Ma, Q.X. Optimization of residual stress in low-pressure casting of ZL205A alloys. *J. Manuf. Process.* **2023**, *99*, 338–350. [CrossRef]
9. Luo, L.; Xia, H.Y.; Luo, L.S.; Su, Y.Q.; Cai, C.J.; Wang, L.; Guo, J.J.; Fu, H.Z. Eliminating shrinkage defects and improving mechanical performance of large thin-walled ZL205A alloy castings by coupling travelling magnetic fields with sequential solidification. *Trans. Nonferrous Met. Soc. China* **2021**, *31*, 865–877. [CrossRef]
10. Xu, H.Y.; Gong, Q.M.; Zhou, X.X.; Yang, F.W.; Han, B. Influence of the assisted kerf depth on cracks pattern and cutting performance of TBM cutter. *Int. J. Rock Mech. Min. Sci.* **2023**, *170*, 105516. [CrossRef]
11. Prajapati, P.K.; Biswas, P.; Singh, B.K.; Bapanapalle, C.O.; Ghosh, R.; Mandal, N. Rein-forcing potential of MWCNTs on mechanical and machining performance of hot-pressed ZTA-MgO ceramic cutting inserts. *Diam. Relat. Mater.* **2023**, *138*, 110202. [CrossRef]
12. Yuan, X.; Wang, S.T.; Mao, X.Y. Forced vibration mechanism and suppression method for thin-walled workpiece milling. *Int. J. Mech. Sci.* **2022**, *230*, 107553. [CrossRef]
13. Liu, T.; Wang, Q.; Wang, W. Micro-Milling Tool Wear Monitoring via Nonlinear Cutting Force Model. *Micromachines* **2022**, *13*, 943. [CrossRef]
14. Duan, Z.; Li, C.; Ding, W. Milling Force Model for Aviation Aluminum Alloy: Academic Insight and Perspective Analysis. *Chin. J. Mech. Eng.* **2021**, *34*, 18. [CrossRef]
15. Zang, P.; Yue, X.J.; Han, S.F. Experiment and simulation on the high-speed milling mechanism of aluminum alloy 7050-T7451. *Vacuum* **2020**, *182*, 109778.
16. Sun, Y.; Sun, J.; Wang, G. A modified analytical cutting force prediction model under the tool crater wear effect in end milling Ti6Al4V with solid carbide tool. *Int. J. Adv. Manuf. Technol.* **2020**, *108*, 3475–3490. [CrossRef]
17. Merchant, E. Basic mechanics of the metal cutting process. *Appl. Mech. Trans. ASME* **1944**, *66*, 160–168. [CrossRef]
18. Lee, E.H.; Shaffer, B.W. The theory of plasticity applied to a problem of machining. *Int. J. Appl. Mech.* **1951**, *18*, 405–413. [CrossRef]
19. Oxley, P.L.B.; Humphreys, A.G.; Larizadeh, A. The influence of rate of strain-hardening in machining. *Proc. Inst. Mech. Eng.* **1961**, *175*, 881–891.
20. Doyle, E.D.; Horne, J.G.; Tabor, D. Frictional interactions between chip and rake face in continuous chip formation. Proceedings of the Royal Society of London. *Ser. A Math. Phys. Sci.* **1979**, *366*, 173–183.
21. Zhang, X.; Zhang, J.; Zhou, H.; Ren, Y.; Xu, M. A novel milling force model based on the influence of tool geometric parameters in end milling. *Adv. Mech. Eng.* **2018**, *10*, 9. [CrossRef]
22. Zhu, S.; Zhao, M.; Mao, J.; Liang, S.Y. A Ti-6Al-4V Milling Force Prediction Model Based on the Taylor Factor Model and Microstructure Evolution of the Milling Surface. *Micromachines* **2022**, *13*, 1618. [CrossRef]
23. Chen, Y.H.; Lu, J.; Deng, Q.L. Modeling study of milling force considering tool runout at different types of radial cutting depth. *J. Manuf. Process.* **2022**, *76*, 486–503. [CrossRef]

24. Su, X.; Wang, G.; Yu, J. Predictive model of milling force for complex profile milling. *Int. J. Adv. Manuf. Technol.* **2016**, *87*, 1653–1662. [CrossRef]
25. Zhou, Q.; Luo, D.W.; Ye, W.T.; Li, S.; Zou, Q.G.; Chen, Z.Q.; Wang, H.F. Design and characterization of metallic glass/graphene multilayer with excellent nanowear prop-erties. *Friction* **2022**, *10*, 1913–1921. [CrossRef]
26. Ren, Y.; Huang, Z.B.; Wang, Y.C.; Zhou, Q.; Yang, T.; Li, Q.K.; Jia, Q.; Wang, H.F. Fric-tion-induced rapid amorphization in a wear-resistant $(CoCrNi)_{88}Mo_{12}$ dual-phase medium-entropy alloy at cryogenic temperature. *Compos. Part B-Eng.* **2023**, *263*, 110833. [CrossRef]
27. Ye, W.T.; Xie, M.D.; Huang, Z.B.; Wang, H.M.; Zhou, Q.; Wang, L.; Chen, B.; Wang, H.F.; Liu, W.M. Microstructure and tribological properties of in-situ carbide/CoCrFeNiMn high entropy alloy composites synthesized by flake powder metallurgy. *Tribol. Int.* **2023**, *181*, 108295. [CrossRef]
28. Emine, S. Evaluation of tribological performance of MQL technique combined with LN_2, CO_2, N_2 ecological cooling/lubrication techniques when turning of Hastelloy C22 superalloy. *Tribol. Int.* **2023**, *188*, 108786.
29. Haddag, B.; Nouari, M.; Moufki, A. Experimental analysis of the BTA deep drilling and a new analytical thermomechanical model for assessment of cutting forces and BTA drill design. *Int. J. Adv. Manuf. Technol.* **2020**, *106*, 455–469. [CrossRef]
30. Moufki, A.; Dudzinski, D.; Molinari, A.; Rausch, M. Thermoviscoelastic modelling of oblique cutting: Forces and chip flow predictions. *Int. J. Mech. Sci.* **2000**, *42*, 1205–1232. [CrossRef]
31. Moufki, A.; Devillez, A.; Dudzinski, D.; Molinari, A. Thermomechanical modelling of oblique cutting and experimental validation. *Int. J. Mach. Tool. Manu.* **2004**, *44*, 971–989. [CrossRef]
32. Chen, Z.J.; Qian, L.Y.; Ji, B.P. Investigation into thermodynamic behavior of LA103Z MgLi alloy during turning based on modified Johnson—Cook model. *J. Manuf. Process.* **2023**, *87*, 260–272. [CrossRef]

Article

Research on Wear of Micro-Textured Tools in Turning GH4169 during Spray Cooling

Jingshu Hu, Jiaxuan Wei, Xinmin Feng * and Zhiwei Liu

Key Laboratory of Advanced Manufacturing Intelligent Technology, Ministry of Education,
Harbin University of Science and Technology, Harbin 150080, China; hujingshu@hrbust.edu.cn (J.H.);
2120110105@stu.hrbust.edu.cn (J.W.); lzw1361894040@163.com (Z.L.)
* Correspondence: fxmin7301@hrbust.edu.cn; Tel.: +86-13633640821

Abstract: In this study, the wear resistance of micro-textured tools was explored. Micro-textured tools with different morphologies were used in turning GH4169 during spray cooling. The tool wear on the rake face of the different micro-texture morphologies was investigated through simulation and experiments. Firstly, based on the existing research on bionics, micro-textures with five different morphologies were designed on the rake face of carbide tools. A simulation model of cutting GH4169 during spray cooling was established, and the tools with designed micro-textures were used in it. The influence of the different micro-texture morphologies on the tool wear was analyzed. Secondly, the designed micro-textured tools with five different morphologies were produced using a femtosecond laser. Cutting experiments were conducted using the micro-textured tools during spray cooling. The wear area of the rake face was measured based on the infinitesimal method, and the optimal morphology with the best anti-wear ability was obtained. This study provides technical support for the design and development of micro-textured tools with an improved cutting performance, and contribute to the promotion and application of micro-textured tools.

Keywords: micro-texture morphology; GH4169; tool wear; spray cooling

Citation: Hu, J.; Wei, J.; Feng, X.; Liu, Z. Research on Wear of Micro-Textured Tools in Turning GH4169 during Spray Cooling. *Lubricants* **2023**, *11*, 439. https://doi.org/10.3390/lubricants11100439

Received: 6 September 2023
Revised: 27 September 2023
Accepted: 28 September 2023
Published: 12 October 2023

1. Introduction

Nickel-based superalloy GH4169 is a kind of alloy that features high hardness, high strength, and excellent corrosion resistance and heat resistance, and can maintain good mechanical stability even in very harsh environments [1]. Because of its excellent performance, it has been widely used in the aerospace, automotive, and nuke industries, as well as in petrochemical and many other frontiers, in recent years. It is because of the above material properties that GH4169 is a difficult-to-process material. In the processing of GH4169, serious wear occurs in ordinary tools, which brings about a short lifespan and fast failure. How to improve the cutting performance of GH4169 and reduce tool wear is a key problem to be solved in the processing of GH4169 [2].

There are many reasons for tool wear in the cutting process, and some scholars have conducted relevant research on the influence of the cutting parameters and cooling and lubrication conditions on tool wear. Wu et al. carried out a superalloy turning test using a PCBN tool under high pressure cooling, and analyzed the tool wear under different cutting speeds and cooling pressures. The experimental results showed that with the increase in the cutting speed, the degree of tool wear first increases, then decreases, and subsequently increases, and the influence of the cooling pressure on tool wear is that the wear first decreases and then increases [3]. Liu et al. conducted cutting experiments on AISI 304 stainless steel using cemented carbide tools in minimum quantity cooling lubrication (MQCL) conditions, and studied the influence of the cutting parameters on tool wear under minimum quantity cooling lubrication. The experimental results show that the cutting speed has the greatest impact on tool wear, followed by the feed rate and cutting depth. At the same time, MQCL achieves a better surface quality and lower tool wear than MQL and dry cutting [4]. Sharma et al. studied the effect of turning a Ti6Al-4V alloy with soybean oil in minimum quantity lubrication (MQL). The performance of the turning operation in dry and MQL environments was compared. The research results show that, compared with dry cutting, the tool wear rate is lower in the MQL cutting environment [5]. M. Ukamanal et al. studied the machinability of AISI 316 stainless steel turning in a dry cutting and spring impingement cooling environment through turning experiments. The test results show that, compared with dry machining, the processing under spray cooling technology produces a better surface finish, lower tool wear, and lower cutting temperature, making SIC an effective cooling technology in turning operations [6]. Zhang et al. carried out tool wear tests for the high-speed milling of 300 M steel under dry and cooling minimum quantity lubrication (CMQL) conditions. The experimental results show that, compared to dry cutting, CMQL technology can effectively reduce abrasive wear of the blade, bond wear on the blade surface, oxidative wear at the flank boundary, and mild diffusion wear [7].

In recent years, micro-textured tools based on the bionics theory have provided a new direction for tool wear reduction design and for the efficient cutting of difficult-to-machine materials such as superalloys [8,9]. A micro-textured tool is a tool in which micro-structures—such as convex, bulge, pit, groove, scale, and so on—are machined on its rake/flank face [10,11]. These micro-textured tools are more conducive to the penetration, storage, and film formation of lubrication medium, achieve a better friction reduction effect, slow down tool wear, and extend the service life of tools [12]. The high-performance-surface texture enables good friction reduction, adhesion resistance, and improved wear resistance. Cao Tongkun et al. used micro-textured tools to cut AISI 1045 steel and studied the friction coefficient, tool–chip contact length, and tool wear. The test results indicate that, in comparison to traditional pan-lubricated cutting, micro-textured tools provide a better cutting performance than other tools [13]. Using finite element analysis, Liu Wei et al. designed three different micro-textures on the rake face of cemented carbide tools—namely, a semicircular concave micro-texture, semicircular convex micro-texture, and trapezoidal groove micro-textured—and studied the influence of the micro-textured tools on the friction by changing the diameter or width of the micro-textures, the spacing of the micro-textures, and the covering length of the micro-textures. By analyzing the simulation data, we can see that, for the three tested micro-textured tools, the friction in cutting can be reduced by 23.0%, 27.7%, and 21.9%, respectively, using the optimal micro-texture parameters. In the range of the tool–chip contact length, the longer the micro-textured covering length, the better the anti-wear ability [14]. Yang G. et al. carried out internal-cooling-texture treatment on TNMG160404-SG XL7020 cemented carbide inserts using electric spark and laser technology, so that the rake face and the main flank face were internally cooled. Compared with non-woven tools, woven tools can reduce the tool–chip contact area, improve the anti-friction performance, and reduce the average friction coefficient of the interface between the tool and chip by about 2–8.6% [15]. Wu Ze et al. used elliptical micro-textured self-lubricating carbide tools and traditional tools to carry out a comparative test of dry cutting 45 steel. Compared with traditional tools, micro-textured self-lubricating tools can effectively reduce the rake face wear [16]. Huang Ke et al. used a nanosecond laser to process grooves on the tool surface, and designed dry friction and wear tests for micro-textured tools with different area occupancy rates. The experimental results indicate that the micro-textured tools can store the debris generated by friction, reduce the roughness of the tool surface, and fully reflect the influence of the micro-texture on the anti-wear ability [17]. Long Yuanqiang et al. used a micro-textured self-lubricating tool, micro-textured tool, and traditional carbide tool to carry out dry cutting tests on 0Cr18Ni9 austenitic stainless steel. Through analyzing the experimental data, it was concluded that micro-textured self-lubricating tools can effectively improve the adhesive wear of tools [18]. Li Liang et al. studied the anti-wear ability of the surface micro-textured tool under the condition of minimal lubrication and no lubricant through an orthogonal cutting experiment of titanium alloy. Based on the experimental data, it can be concluded that surface micro-grooves can effectively improve the friction of the tool chips under the condition of lubricant in order to achieve an improved anti-wear ability [19]. In order to verify the anti-wear ability of micro-textured tools, Tu Chunjuan used micro-pit and micro-groove texture self-lubricating tools to turn hardened steel. The experimental results show that the micro-pit-textured tools show a better wear reduction performance than the micro-groove-textured tools [20].

Some research results have also been obtained on the influence of different micro-textured arrangements on the wear: Arulkirubakaran D et al. conducted numerical simulation and turning experiments on a Ti-6Al-4V alloy using surface-textured tungsten carbide tools. The tool has micro-grooves parallel, perpendicular, and intersecting with the direction of chip flow. The numerical simulation results were validated by the experimental results, which include the cutting temperature, machining force, and chip morphologies. Cutting tools with surface textures processed in a direction perpendicular to that of the chip flow exhibited a larger reduction in the cutting force, temperature generation, and reduced tool wear [21]. Tatsuya Sugihara developed a CBN micro-textured tool with a micro-groove on the flank face, and the groove micro-textured arrangement was perpendicular to the cutting edge; the cutting test was carried out on Inconel 718. The results indicate that the micro-textured tool significantly extends the service life of CBN tools for high-speed dry cutting Inconel 718 [22]. Fang Zhenglong et al. produced five different shapes of micro-textures on the flank face, and the arrangement of the micro-textures was mainly parallel or vertical to the cutting edges. The wear of a micro-textured tool and a non-textured tool in the longitudinal turning process of an Inconel 718 nickel-based superalloy was compared in the condition of high-pressure jet assisted cooling. It was seen that, in comparison to non-textured tools, tools with micro-textures have lower flank face wear and crescent wear, and the arrangement of the micro-textures will affect the tool wear rate. In particular, tools with micro-textures of plate-fins showed a better performance than those with pin-fins [23]. Based on the energy loss method, Zhang Yan et al. established the micro-textured tool continuous wear model in the process of micro-cutting Ti6Al4V titanium alloy material using a carbide tool in ABAQUS6.14-5, and designed single and multiple micro-texture models at different positions of the rake face to analyze the influence of the micro-textures on the rake face depression wear and the flank face wear. The analysis shows that the quantity and position of the micro-textures on the rake face have a marked impact on the wear of tools with no texture [24]. Duan et al. carried out cutting experiments on Inconel 718 using three kinds of micro-textured CBN tools; namely, pit, straight line, and sine. The cutting force, tool wear, and machined surface roughness of the different micro-textured tools were compared and analyzed. The experimental results show that the existence of surface micro-texture enhances the wettability of the tool surface. It effectively reduces the cutting force, the wear of the rake face, and the surface roughness of the machined parts. Compared with other morphologies, the sinusoidal micro-textured tool shows the best cutting performance [25].

Based on the above studies, it can be seen that micro-textures have a significant effect on the improvement of the cutting performance of tools, especially in the reduction in tool wear. The existing studies of materials for workpiece cutting using micro-textured tools mainly focus on titanium alloy, aluminum alloy, stainless steel, hardened steel, and other difficult-to-cut materials. However, few studies have been conducted on the tool wear of micro-textured tools used to cut GH4169, and there are even fewer studies on the properties of micro-textured tools during spray cooling. In this article, experiments of turning a nickel-based GH4169 superalloy using micro-textured tools during spray cooling were carried out to investigate the influence of different micro-texture morphologies on the wear of tools. This study aims to provide technical reference for the design and application of micro-textured tools.

2. Design and Processing of Micro-Textured Tools

2.1. Distribution of Micro-Textures

The distribution of micro-textures takes two factors into account: the tool–chip contact length and the wear area of the rake face. Based on a series of cutting experiments and simulation studies, it was found that the maximum tool–chip contact length of the cemented carbide tool in turning GH4169 was about 500–700 μm from the nose of the tool, and the wear area on the rake face was within 500 μm of the tool nose [26].

2.2. Morphology of Micro-Textures

Studies have shown that the micro-texture morphology has an important impact on its anti-wearability. A micro-texture with an unreasonable shape and size not only fails to achieve the purpose of wear reduction, but will aggravate the wear of the tool and reduce the strength of the tool [27].

Based on the research of bionics, the surface characteristics of natural creatures were observed under microscope and studied in terms of anti-friction performance. The body surface structure of the creatures is shown in Figure 1.

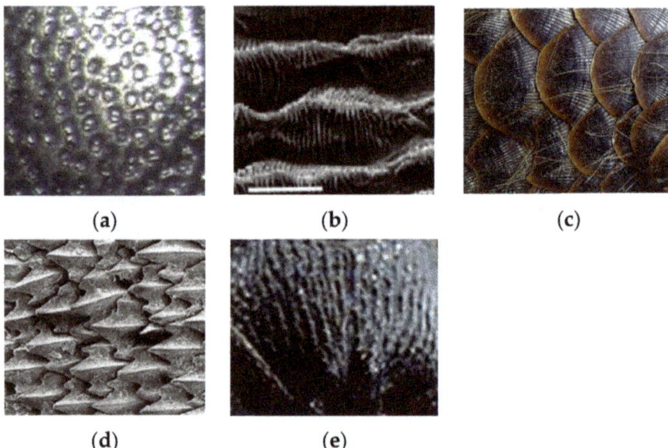

Figure 1. Body surface structure of creatures: (**a**) pit-structure on dung beetle shells; (**b**) waveform grooves on the shells of scapharca broughtonii; (**c**) straight and curved grooves on the shells of pangolin scales; (**d**) arrowtooth-shaped textures on shark skin; (**e**) straight grooves on the shells of ant.

In this work, five micro-texture morphologies were designed according to the study on the surface characteristics of natural creatures; namely, micro-pit, micro-wave-groove, micro-arc-groove, micro-vertical-groove, and micro-elliptic-groove. The schematic diagram of the micro-texture morphologies is shown in Figure 2.

The femtosecond laser method was used to process the micro-textures of different morphologies on the rake face of the cemented carbide tool, as shown in Figure 3. The distance of the micro-texture from the tool tip was 120 μm and the depths processed in this work were all 30 μm.

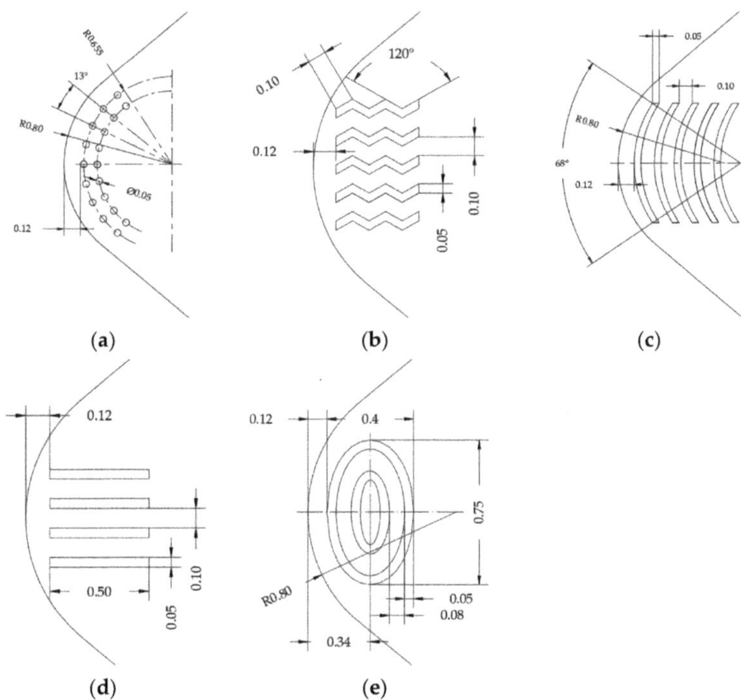

Figure 2. Schematic diagram of micro-texture morphologies: (**a**) micro-pit; (**b**) micro-wave-groove; (**c**) micro-arc-groove; (**d**) micro-vertical-groove; (**e**) micro-elliptic-groove.

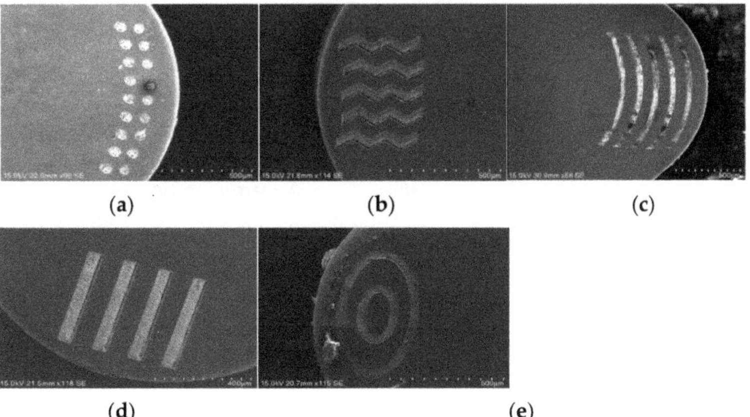

Figure 3. Partial larger views of the nose of the micro-textured tools: (**a**) micro-pit; (**b**) micro-wave-groove; (**c**) micro-arc-groove; (**d**) micro-vertical-groove; (**e**) micro-elliptic-groove.

3. Tool Wear Simulation Analysis

3.1. Establishment of Cutting Simulation Model

(1)　Construction of tool geometric model

The geometric models of the micro-textured tools with five different morphologies, established in the proportion of 1:1, are shown in Figure 4.

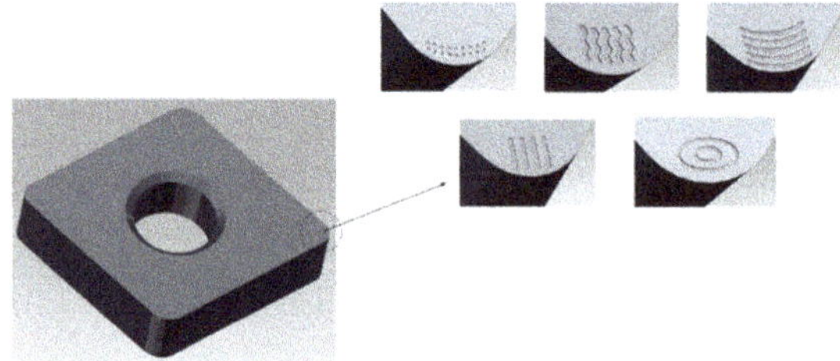

Figure 4. Geometric model and enlarged drawing of micro-textured tool.

The rake face used in this study was designed with micro-textures, which need to be refined in the mesh. In order to conduct simulation analysis efficiently, the cylindrical workpiece is selected as part of the thin-walled ring rotary workpiece, and the tool is simplified into the tool nose part. The simplified cutting model is shown in Figure 5.

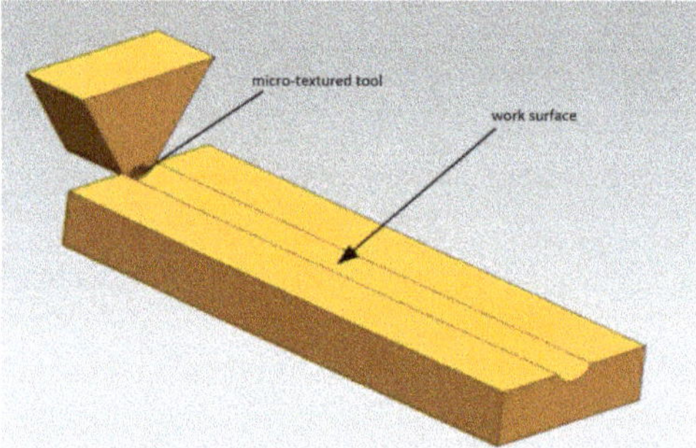

Figure 5. Simplified cutting model.

(2)   Workpiece material model

The Johnson–Cook constitutive model considers the strain hardening, strengthening, and thermalization of the material, which are closer to the actual phenomenon. The Deform software (Deform 11.0) comes with a rich material library. The simulation workpiece material is a nickel-based superalloy. In this simulation, the existing Inconel718 (performance similar to GH4169) in the Deform material library was selected, and the mathematical expression of the Johnson–Cook constitutive model was selected for the constitutive model. The material model expression is shown in Equations (1) and (2):

$$\overline{\sigma} = \overline{\sigma}(\overline{\varepsilon}, \dot{\overline{\varepsilon}}, T) = (A + B\overline{\varepsilon}^n)(1 + C \ln \dot{\overline{\varepsilon}})(1 - T^{*m}) \tag{1}$$

$$T^* = \frac{T - T_{room}}{T_{melt} - T_{room}} \tag{2}$$

where $\overline{\sigma}$ is the flow stress; $\overline{\sigma}$ is the equivalent plastic strain; $\dot{\overline{\varepsilon}}$ is the equivalent plastic strain rate; T is the deformation temperature; A is the yield stress; B is the pre-exponential coefficient; C is the strain rate sensitivity coefficient; m is the temperature sensitivity coefficient; n is the hardening coefficient; T^* is the relative temperature; T_{melt} and T_{room} are the melting temperature and room temperature of the workpiece material, respectively.

The J–C constitutive model parameters of the workpiece material are shown in Table 1.

Table 1. Johnson–Cook constitutive model parameters of Inconel 718.

$\rho/(kg/m^3)$	$C_P/J\cdot kg^{-1}\cdot K^{-1}$	E/GPa	A/MPa	B/MPa
8240	435	200	450	1700
C	**m**	**n**	$\dot{\varepsilon}/s^{-1}$	$T_{melt}/°C$
0.017	1.3	0.65	0.001	1300

(3)   Tool–chip friction model

Because the simulation is the process of cutting the nickel-based superalloy using a micro-textured tool, the tool nose is heated faster in the rake face during cutting, and the temperature difference between the tool nose and the rake face is generated, which reduces the normal stress ratio, and sliding friction occurs between the two cutting tools. In order to ensure the accuracy of the simulation and to accurately select the type of friction, the mixed-slip friction model was chosen in this simulation. The model expression is shown in Equation (3):

$$\begin{cases} T_f = \mu\sigma_n(\text{Sliding zone}, \mu\sigma_n \leq \tau_n) \\ T_f = \tau_n(\text{Bonding zone}, \mu\sigma_n \geq \tau_n) \end{cases} \tag{3}$$

where T_f is the friction stress; σ_n is the normal stress; μ is the friction coefficient, this simulation takes μ as 0.6; τ_n is the ultimate shear stress of the workpiece.

(4)   Mesh division and wear calculation model

When the tool model is divided into meshes, the micro-textured part needs to be locally refined. The mesh division method is absolute mesh division, the total number of meshes is 55,000, the minimum mesh size is 0.015 mm, and the mesh refinement ratio is 6. Taking the micro-arc-groove as an example, the divided mesh model is shown in Figure 6.

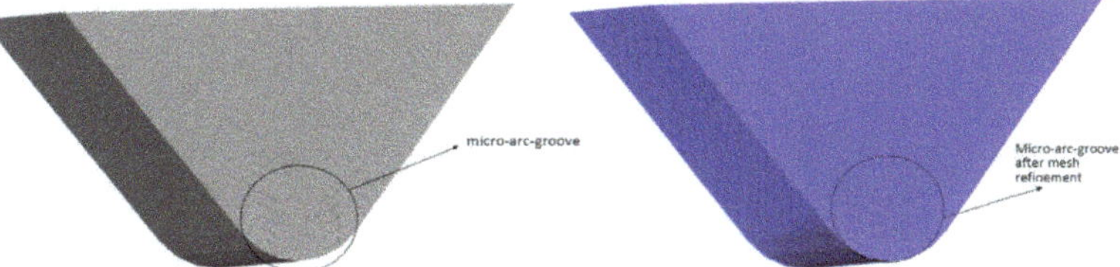

Figure 6. Grid division of micro-textured tool with arc groove.

As the simulation mainly observes the wear of the tool, the wear model needs to be selected during the setting. In Deform-3D preprocessing, there is a Usui wear model, and users can also set the wear model according to their own needs. In order to fit the actual cutting conditions and meet the existence of harsh conditions, such as a high strain rate, high temperature, and high pressure, the Usui calculation model is selected. Its mathematical expression is shown in Equation (4).

$$w = \int apVe^{-\frac{b}{T}}dt \tag{4}$$

where w is the wear depth; p is the chip interface stress; V is the slip velocity; T is the interface temperature; dt is the time increment; a and b are constants, according to the material properties of the tool, a is set to 1×10^{-6} and b is set to 855.

3.2. Simulation of Spray Cooling

As the simulation conducts a cutting test in the cutting environment of spray cooling, there is no direct setting of the spray cooling parameter conditions in the Deform software preprocessing; therefore, the research group uses the ANSYS simulation software (ANSYS

16.0) to simulate spray cooling, and analyzes the influence of different spray flow rates and pressures on the cutting temperature and convective heat transfer coefficient through the simulation and test. By means of the range analysis method, the optimal combination of spray parameters is obtained, with the lowest cutting temperature as the target. When the spray pressure is 0.3 Mpa and the flow rate 3.16 L/h, the corresponding convective heat transfer coefficient is 3.13737 w/m^2-k [28]. The convective heat transfer coefficient under the optimal parameters is imported into the Deform software, which can be equivalent to cutting under the condition of spray cooling.

In order to ensure the comparability of the simulation results and to reduce the influence of other condition settings on the simulation results, the analysis step size was set to 1500, step size increment to 0.01, ambient temperature to 20 °C, and friction factor to 0.6 for the five designed micro-textured tools.

3.3. Analysis of Simulation Results

A simulation of cutting GH4169 using the micro-textured tools was conducted with the same cutting parameters: the cutting speed V was set to 80 m/min; the feed rate was 0.2 mm/r; and the depth of cut a_p 0.2 mm. The finite element analysis of the tool wear obtained after cutting is shown in Figure 7.

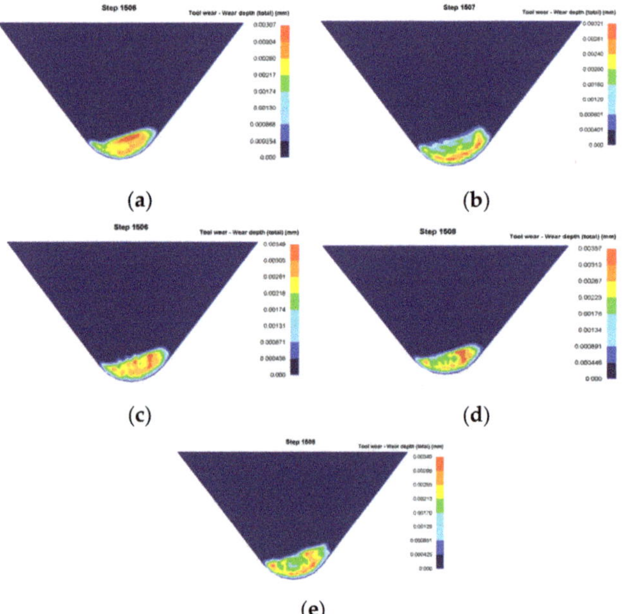

Figure 7. Finite element analysis of tool wear: (**a**) micro-pit; (**b**) micro-arc-groove; (**c**) micro-vertical-groove; (**d**) micro-wave-groove; (**e**) micro-elliptic-groove.

By observing Figure 7, it is found that deep wear bands appear on the rake face of the five kinds of micro-textured tools near the arc of the tool nose. In order to explore the changing trend of the wear depth of the five kinds of micro-textured tools with different morphologies, at the wear band near the radius of the arc of the tool nose, the same coordinate point is taken as the unit of 200 simulation step lengths and the simulation step length is changed. The tool wear depth of the five kinds of micro-textures is shown in Figure 8.

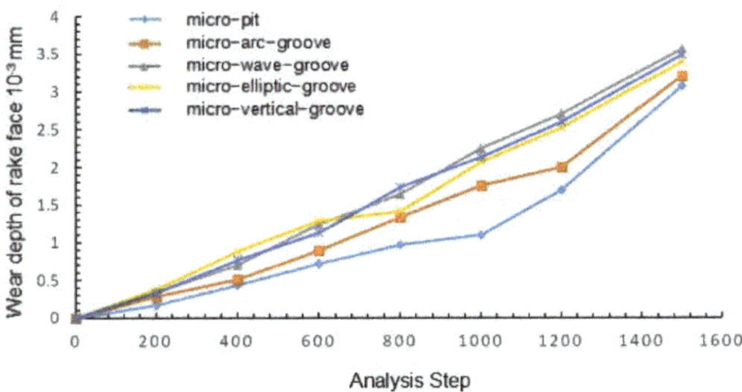

Figure 8. The wear depth of five kinds of micro-textured tools.

The wear depth of the rake face is shown in Table 2.

Table 2. The wear depth of rake face of micro-textured tools with different morphologies.

Micro-Textured Morphology	Minimum Wear Depth of the Rake Face (10^{-3} mm)	Maximum Wear Depth of the Rake Face (10^{-3} mm)
micro-pit	0.354	3.07
micro-arc-groove	0.401	3.21
micro-vertical-groove	0.436	3.49
micro-wave-groove	0.446	3.57
micro-elliptic-groove	0.425	3.4

It can be seen from Table 2 that under the same coordinates, the wear depth of the micro-pits' morphology changed the least with the simulation steps. By combining the cloud map of the wear changes of the five micro-textured tools, it can be found that the wear distribution of the micro-pit-textured tool on the rake face is more dispersed, and the wear depth value had the smallest variation range. Moreover, the maximum wear depth of the micro-pit-textured tools had a smaller distribution near the tool tip, and the maximum wear depth is more dispersed than the other tools. It can be seen that the micro-pit-textured tool had the smallest wear depth on the rake face, and its anti-wear capacity was better than the other micro-textured tools.

4. Cutting Experiment during Spray Cooling

4.1. Conditions of Experiment

Micro-textured tools with different morphologies were used in a turning experiment during spray cooling. The workpiece size was Φ 120 × 300 mm; the tool shank was Sandvik DCLNR 2525 M12; the insert was CNMA120408-KR 3225; and the micro-texture was processed on the rake face, as shown in Figure 5. The adopted cooling equipment was produced by Anmolin using the OoW129S model and the working pressure was in the range of 0.3 MPa to 0.8 MPa. The angle between the spray nozzle and rake face was 45 degrees, and the spray nozzle was 20 mm away from the tool nose. The nozzle outlet pressure was 0.3 MPa, and the flow rate was 3 L/h; the main components of the cutting fluid are mixed with Master and Castrol. The cutting parameters were as follows: v was 80 m/min; f was 0.2 mm/r; a_p was 0.2 mm; and cutting time was 30 s. The experiment setup is shown in Figure 9.

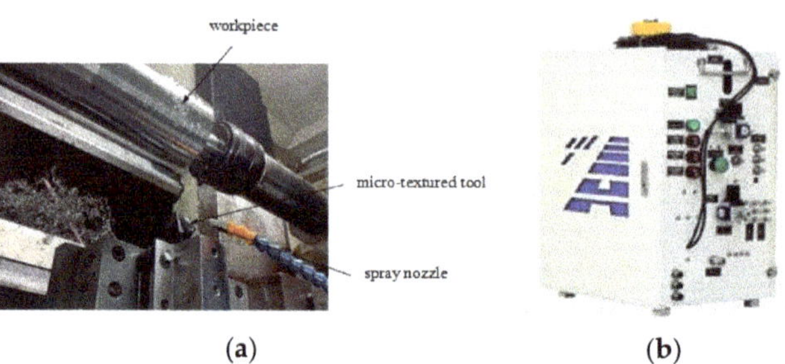

Figure 9. Experiment setup: (**a**) spray cooling site; (**b**) spray cooling equipment.

4.2. Analysis of Experiment Results

The rake faces of the micro-textured tools with different morphologies were observed through scanning electron microscopy, as shown in Figure 10.

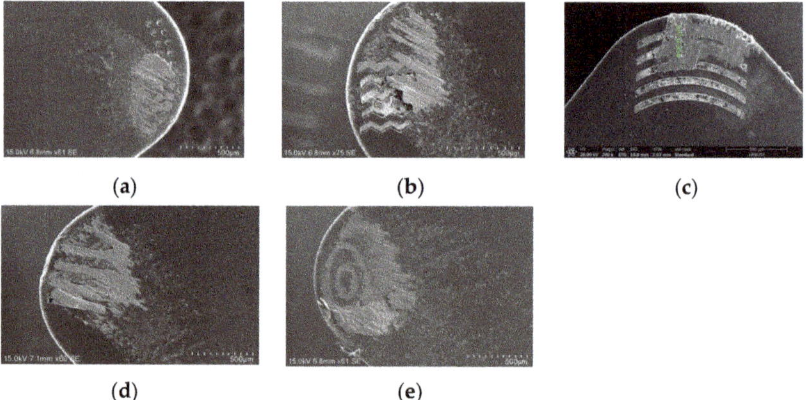

Figure 10. Tool wear of the rake face with different morphologies: (**a**) micro-pit; (**b**) micro-wave-groove; (**c**) micro-arc-groove; (**d**) micro-vertical-groove; (**e**) micro-elliptic-groove.

As shown in Figure 11, the wear area of the rake face is an irregular geometric shape, which enables us to compare the wear degree of the rake face of each set of tools in the experiment. This experiment uses the rectangular equivalent area method, and the wear area calculated using the infinitesimal method is equivalent to a rectangular wear zone of equal area to indicate the degree of wear on the rake face.

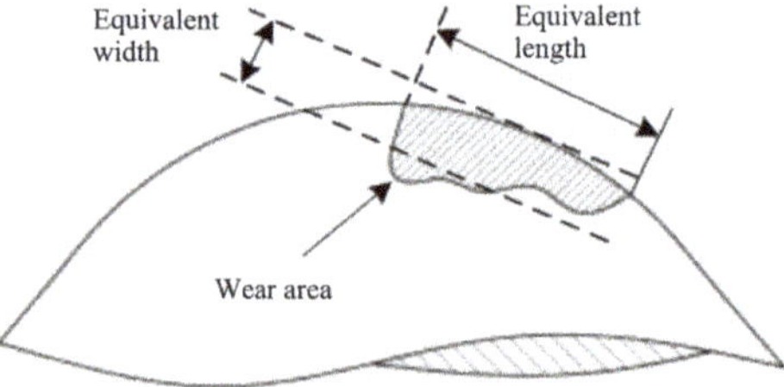

Figure 11. Wear area of the rake face.

In this experiment, the infinitesimal method is used to calculate the area of the wear area of the rake face. Divide the wear area into n small rectangles, the width of i is l_i, measuring the height of the rectangle separately; then, the whole area is:

$$S = l_1 \times h_1 + l_2 \times h_2 + \ldots + l_n \times h_n = \sum_{i=1}^{n} l_i \times h_i = w \times l \tag{5}$$

$$l = l_1 + l_2 + \ldots + l_n = \sum_{i=1}^{n} l_i \tag{6}$$

where S is the equivalent rectangular wear zone area; w is the equivalent width of the rectangular wear zone; and l is the equivalent rectangular wear zone length.

The calculation results of the wear area of the rake face of the five micro-textured tools using the infinitesimal method are shown in Table 3, and the histogram of the wear area of the rake face of the five micro-textured tools is shown in Figure 12.

Table 3. The wear area of rake face of five kinds of micro-textured tools.

Micro-Textured Morphology	Equivalent Band Width w/μm	Equivalent Belt Length l/μm	Wear Area S/μm²
micro-pit	37.8	337.1	12,762.5
micro-elliptic-groove	30.71	805.86	24,754.7
micro-arc-groove	23.4	659.3	15,395.6
micro-wave-groove	32.7	637.3	20,851.1
micro-vertical-groove	29.4	687.9	20,229.7

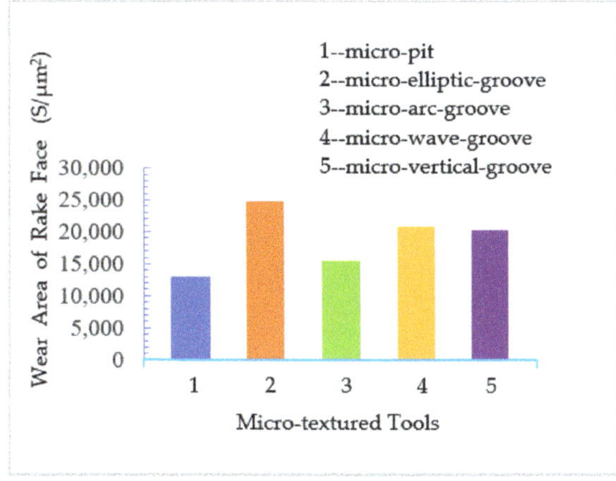

Figure 12. The wear area of rake face of micro-textured tools.

The experiment results showed that among the five designed micro-textured tools, the ones with micro-pits had the smallest wear area, and the wear area was relatively far away from the tool nose. The micro-textured tools with micro-pits achieved the best anti-wear ability. The experimental results verified the simulation results.

5. Conclusions

In this study, five micro-textured tools with different morphologies were used in cutting GH4169 during spray cooling. The research was conducted through both finite element simulations and experiments. The influence of micro-textures with different morphologies on the tool wear was revealed. The optimal morphology of the micro-texture was obtained. The conclusions are as follows:

(1)   A cutting simulation model was established in a spray cooling condition. The effect of the micro-texture morphologies on the wear of the rake face was obtained. A series of

experiments were conducted with the same spray cooling parameters. The results of the experiments on the tool wear conformed to those of the simulation. The reliability of the simulation was verified.

(2) The insertion of micro-textures can effectively improve the anti-wear ability of micro-textured tools on the tool–chip interface and reduce tool wear. The distribution of the micro-textures was set within the working area; it measured a distance of 60–120 μm from the top micro-texture to the edge of the tool nose, and the other micro-textures were set in the area 500 μm away from the tool nose.

(3) The simulation and experiment results showed that the micro-pits had a significantly better anti-wear ability than the other four morphologies. The wear depth of the rake face ranged between 0.354×10^{-3} and 3.07×10^{-3} mm, and the wear area of the rake face was 12,762.5 μm^2.

This article reveals the influence of micro-textured tools with different morphologies on tool wear. The conclusion of the article was drawn from a series of experiments and simulations, and it provides technical reference for both the machining of GH4169 and the application of micro-textured tools. However, many factors also contribute to tool wear in cutting, such as the size parameters of the micro-textures, the geometric parameters of the tools, the cutting parameters, etc. These factors were not considered in this article; therefore, it is necessary to conduct further related research.

Author Contributions: Methodology, J.H.; software, J.W.; validation, Z.L.; writing—original draft preparation, J.H.; writing—review and editing, X.F. All authors have read and agreed to the published version of the manuscript.

Funding: This study was supported by the National Natural Science Foundation of China (No. 51675144).

Data Availability Statement: Not applicable.

Conflicts of Interest: The authors declare no conflict of interest.

References

1. Zheng, J.J.; Guo, Y.; Liu, X.; Zhang, Z.H.; Zhang, T. Introduction on Research and Application of Nickel Base Superalloy GH4169. *IOP Conf. Ser. Earth Environ. Sci.* **2021**, *651*, 022081. [CrossRef]
2. Yu, W.W.; Ming, W.W.; An, Q.L.; Chen, M. Cutting performance and wear mechanism of honeycomb ceramic tools in interrupted cutting of nickel-based superalloys. *Ceram. Int.* **2021**, *47*, 18075–18083. [CrossRef]
3. Wu, M.Y.; Wu, S.J.; Chu, W.X.; Cheng, Y.N. Experimental Study on Tool Wear in Cutting Superalloy under High-Pressure Cooling. *Integr. Ferroelectr.* **2020**, *207*, 208–219. [CrossRef]
4. Liu, N.C.; Zheng, C.L.; Xiang, D.Y.; Huang, H.; Wang, J. Effect of cutting parameters on tool wear under minimum quantity cooling lubrication (MQCL) conditions. *Int. J. Adv. Manuf. Technol.* **2019**, *105*, 515–529. [CrossRef]
5. Sharma, S.; Das, P.P.; Ladakhi, T.Y.; Pradhan, B.B.; Phipon, R. Performance Evaluation and Parametric Optimization of Turning Operation of Ti6Al-4V Alloy Under Dry and Minimum Quantity Lubrication Cutting Environments. *J. Mater. Eng. Perform.* **2022**, *32*, 5353–5364. [CrossRef]
6. Ukamanal, M.; Mishra, P.C.; Sahoo, A.K. Effects of Spray Cooling Process Parameters on Machining Performance AISI 316 Steel: A Novel Experimental Technique. *Exp. Tech.* **2020**, *44*, 19–36. [CrossRef]
7. Zhang, H.P.; Wang, Z.J.; Liu, G.L. Tool Wear of High-speed Machining 300 M Steel Under CMQL Condition. *J. Harbin Univ. Sci. Technol.* **2020**, *25*, 75–82.
8. Ren, L.Q.; Han, Z.W.; Li, J.Q.; Tong, J. Experimental investigation of bionic rough curved soil cutting blade surface to reduce soil adhesion and friction. *Soil Tillage Res.* **2004**, *85*, 1–12. [CrossRef]
9. Zhang, N.; Yang, F.Z.; Liu, G.H. Cutting performance of micro-textured WC/Co tools in the dry cutting of Ti-6Al-4V alloy. *Int. J. Adv. Manuf. Technol.* **2020**, *107*, 3967–3979. [CrossRef]
10. Han, Z.L.; Wang, J.D.; Chen, D.R. Drag Reduction by Samples on Surfaces in Plane Contact Lubrication. *Tribology* **2009**, *29*, 10–16.
11. Han, Z.W.; Ren, L.Q.; Liu, Z.B. Investigation on Anti-wear Ability of Bionic Non-smooth Surfaces Made by Laser Texturing. *Tribology* **2004**, *4*, 289–293.
12. Kishawy, H.A.; Salem, A.; Hegab, H.; Hosseini, A.; Balazinski, M. Micro-textured cutting tools: Phenomenological analysis and design recommendations. *CIRP Ann.* **2021**, *70*, 65–68. [CrossRef]
13. Cao, T.K.; Li, Z.G.; Zhang, S.G.; Zhang, W.F. Cutting performance and lubrication mechanism of micro-texture tool with continuous lubrication on tool-chip interface. *Int. J. Adv. Manuf. Technol.* **2023**, *125*, 1815–1826. [CrossRef]

14. Liu, W.; Liu, S.; Liang, G.Q.; Yuan, H.C. Element analysis on cutting performance and friction reduction effect of micro-texture tools. *Surf. Technol.* **2022**, *51*, 338–346.
15. Yang, G.; Feng, W. Short communication: Experiment study on micro-textured tool with internal cooling. *Mech. Sci.* **2022**, *13*, 619–623. [CrossRef]
16. Wu, Z.; Deng, J.X.; Xin, Y.Q.; Cheng, J.; Zhao, J. Cutting performance of self-lubricating turning tools with elliptical micro-textures. *Trans. Chin. Soc. Agric. Mach.* **2012**, *43*, 228–234.
17. Huang, K.; Yang, F.Z.; Zheng, K.R.; Yi, B. Design and Manufacture of Micro Texture on Tool Surface and Its Influence on Friction Characteristics. *Tool Eng.* **2022**, *56*, 37–42.
18. Long, Y.Q.; Deng, J.X.; Zhou, H.M.; Zhao, W.; Wang, Z. Performance of Self-lubricating Micro-texture Tool on Dry Cutting 0Cr18Ni9 Austenitic Stainless Steel. *Mater. Mech. Eng.* **2015**, *39*, 75–79.
19. Qi, B.Y.; Li, L.; He, N.; Zhao, W.; Wang, Z. Experimental Study on Orthogonal Cutting of Ti6Al4V with Surface Micro-textured Cutting Tool. *Tribology* **2011**, *31*, 346–351. [CrossRef]
20. Tu, C.J.; Guo, X.H.; Guo, D.L.; Chen, Y.D.; Guan, J.J. Comparison of Machinability of Self-lubrication Ceramic Tools with Different Morphological Micro-textures. *Mater. Mech. Eng.* **2018**, *42*, 47–51, 57.
21. Arulkirubakaran, D.; Senthilkumar, V.; Kumawat, V. Effect of micro-textured tools on machining of Ti-6Al-4V alloy: An experimental and numerical approach. *Int. J. Refract. Met. Hard Mater.* **2016**, *54*, 165–177. [CrossRef]
22. Tatsuya, S.; Yuki, N.; Toshiyuki, E. Development of a novel cubic boron nitride cutting tool with a textured flank face for high-speed machining of Inconel 718. *Precis. Eng.* **2017**, *48*, 75–82.
23. Fang, Z.L.; Toshiyuki, O. Cooling performance of micro-texture at the tool flank face under high pressure jet coolant assistance. *Precis. Eng.* **2017**, *49*, 41–51. [CrossRef]
24. Zhang, Y.; Li, Q.H.; Yang, F.S. Simulation and Model of Micro-textured Tool Continuous Wear. *Tool Eng.* **2022**, *56*, 76–81.
25. Duan, Z.F.; Chen, L.; Li, B.B. Effect of micro-textured morphology with different wettabilities on tool cutting performance. *The Int. J. Adv. Manuf. Technol.* **2022**, *123*, 1745–1754. [CrossRef]
26. Yu, C. Research on Cutting Performance and Anti-wear Mechanism of Tools with Micro-texture under Spray Cooling. Master's Thesis, Harbin University of Science and Technology, Harbin, China, 2020.
27. Zhang, G.F.; Zhang, B.W. Influence of the micro-textured parametric variations on the cemented carbide fraction characteristics. *J. Mach. Des.* **2017**, *34*, 25–30.
28. Feng, X.M.; Dong, Q.S.; Hu, J.S.; Wang, B.H. Simulation and Experimental Analysis of Cutting Temperature in Cutting of Superalloy GH4169 under Spray. *Mech. Sci. Technol. Aerosp. Eng.* **2021**, *41*, 985–991.

Article

Measurement and Prediction of Sawing Characteristics Using Dental Reciprocating Saws: A Pilot Study on Fresh Bovine Scapula

Dedong Yu [1,†], Fan Zou [2,†], Wenran Zhang [1,†], Qinglong An [2,*] and Ping Nie [3,*]

[1] Department of Second Dental Center, National Clinical Research Center for Oral Diseases, National Center for Stomatology, Shanghai Ninth People's Hospital, College of Stomatology, Shanghai Jiao Tong University School of Medicine, Shanghai Jiao Tong University, Shanghai 201999, China
[2] School of Mechanical Engineering, Shanghai Jiao Tong University, Shanghai 200240, China
[3] Center of Cranio-Facial Orthodontics, Department of Oral and Cranio-Maxillofacial Surgery, National Clinical Research Center for Oral Diseases, National Center for Stomatology, Shanghai Ninth People's Hospital, College of Stomatology, Shanghai Jiao Tong University School of Medicine, Shanghai Jiao Tong University, Shanghai 201999, China
* Correspondence: qlan@sjtu.edu.cn (Q.A.); nieping@alumni.sjtu.edu.cn (P.N.)
† These authors contributed equally to this work.

Abstract: Bone sawing is one of the most common operations during traditional dental and cranio-maxillofacial surgery and training systems based on virtual reality technology. It is necessary to predict and update conditions (including the sawing force, temperature and tool wear) in real time during VR surgical training and surgical simulation. All the specimens used in this study were fresh bovine scapula. The forces and temperatures were measured during the sawing process. Additionally, the thermal conductivity was measured via a laser flash instrument. Response surface methodology (RSM) was adopted to analyze and model the sawing force and sawing temperature. Meanwhile, tool wear was observed using a scanning electron microscope. The regression models of the sawing force and temperature rise under different experimental conditions were acquired. To obtain the minimum force within the recommended parameter ranges of commonly used medical reference parameters for bone sawing, a higher rotational speed combined with a lower feed rate were recommended. When considering the sawing force and temperature rise comprehensively, the rotational speed should not be extremely high (about 13,000 rpm is recommended). Meanwhile, abrasive wear is the main wear mode of saw blades. In order to avoid surgery failure, it is necessary to replace the saw blade in time. The experimental data were confirmed to be scientific and accurate for the predicted models of sawing conditions. To minimize the main cutting force, a feed rate of 40 mm/min combined with a rotational speed of 13,700 rpm is recommended. High cutting temperatures have the potential to cause irreversible tissue damage, so surgeons using dental reciprocating saws need to avoid excessively high-speed gears.

Keywords: bone sawing; sawing force; tool wear; sawing temperature; surgical training; response surface methodology

Citation: Yu, D.; Zou, F.; Zhang, W.; An, Q.; Nie, P. Measurement and Prediction of Sawing Characteristics Using Dental Reciprocating Saws: A Pilot Study on Fresh Bovine Scapula. *Lubricants* **2023**, *11*, 441. https://doi.org/10.3390/lubricants11100441

Received: 29 August 2023
Revised: 23 September 2023
Accepted: 27 September 2023
Published: 12 October 2023

1. Introduction

Osteotomy is an essential segment during bone surgery for orthopedic and craino-maxillofacial surgery. In orthognathic surgery, including Le Fort I osteotomy, bilateral sagittal split ramus osteotomy, genioplasty, etc., the surgeon cuts maxilla or mandible as planned and moves to designed position in order to establish an ideal occlusal relationship and an aesthetic jaw shape. A micro reciprocating saw is widely applied in clinical practice. It is more precise and leads to less bone loss than traditional drills and saws. Meanwhile, maxilla and mandible are anatomically complex, containing abundant nerves and blood vessels compared to long bones. Therefore, the micro reciprocating saw is well suited for the osteotomy of the jaws.

Bone sawing is also a vital procedure in dental and craino-maxillofacial training systems based on virtual reality (VR) technology. The VR training system should be sufficiently realistic to perfectly imitate orthognathic surgery. A high-quality VR training system can train novice physicians to improve technical proficiency. The sawing procedure should minimize force and temperature rise to avoid or alleviate interference with the surgeon and the incidence of complications. Therefore, high-precision models of the sawing force and sawing temperature are becoming increasingly important for VR surgical training and surgical simulations [1–3].

Bone sawing is a process that causes tissue injury. To improve the safety of orthognathic surgery and avoid medical accidents, the reduction of damage during surgery is one of the main goals of related research [4]. The damage mainly arises due to two reasons. One is the stress damage caused by the cutting force [5]. A large mechanical stress will produce microcracks in the bone tissue and even lead to inflammation [6]. The other reason is the thermal damage caused by the sawing temperature [7]. Excessive cutting heat can accelerate the osteonecrosis of bone tissue cells, resulting in the loss of the regeneration function of the bone tissue [8]. Moreover, an inappropriate sawing force and temperature may destroy the surgical instruments and hinder patient recovery [9]. Therefore, accurate prediction and control of the sawing condition are critical to ensure the quality of surgical operation. The material removal mechanism is an important part of research in material machining [10]. The material removal mechanism of bone materials has also been widely studied.

To date, several researchers have made endeavors to develop sawing force prediction models, most of which have been based on material properties and mechanical models. Scholars have investigated methods for modeling cutting temperature and cutting force, including linear regression models, multiple regression models and neural network methods. James [11] et al. proposed a new sawing force model based on the concept of a single tool sawing at a depth of cut less than the cutting edge radius to predict bone sawing forces. Other studies have mainly focused on the establishment of a regression model of the sawing force [12]. Lin [13] et al. established a sawing force model for virtual surgery training based on virtual reality (VR) technology.

According to relevant studies, the overall rate of instrument breakage during surgery has been 0.35% [14]. It is dangerous for a saw blade to fail during surgery. The wear performance of common materials has been extensively studied [15–17]. The failure mechanism of the saw blade is also an essential research topic that is studied to avoid potential risks. Many studies have focused on the wear phenomenon and mechanism of bone drill [18–20]. However, to date, few studies have been reported that address the wear mechanism of a saw blade used to cut bone material. Additionally, the prediction of the sawing force is the main focus of researchers, and the existing sawing force models fail to take the anisotropy of bone materials into consideration. Moreover, the cutting temperature and tool wear are still not well understood.

In the current work, a precisely controlled experimental platform with various measurement functions was established to conduct sawing experiments. Stryker thin blades, which are commonly used in osteotomy, were utilized. Based on the experimental data, regression models of the sawing force and sawing temperature were established. Response surface methodology (RSM) was used to analyze the influence of sawing parameters on the sawing force and sawing temperature. In addition, the wear signatures of the sawing blades were observed and analyzed to reduce the risk of surgical tool failure. The models established in the current work can help to realize haptic feedback and surgical simulation, which contributes to orthognathic surgical training based on VR technology.

The sawing test was carried out on a specially designed experimental platform to determine the optimal parameters of the bone saw. Meanwhile, according to the experimental data, the model can be used for VR training.

2. Materials and Methods

2.1. Measurement of Thermal Diffusivity

The thermal diffusivity of cortical bone and cancellous bone was measured with a laser flash instrument (type NETZSCH LFA 467, Selb, Germany). In the measurement of the cortical bone thermal diffusivity, the orientations parallel to the osteons and perpendicular to the osteons were considered. The specimen was machined using a precision cutting machine with a size of 10 mm × 10 mm × 1 mm. The temperature of the specimen in the test was slowly increased from 0 °C to 100 °C, and the thermal diffusivity was measured at 10 °C intervals.

2.2. Sawing Experiments on Fresh Bovine Scapula

The specimens used in this study were provided by a slaughterhouse. Soft tissues were removed from the scapula, and the specimens were preserved in saline solution for experimentation. All the experiments were approved by the Ethics Committee of Shanghai Ninth People's Hospital.

The sawing experiments were conducted on a specially designed experimental platform, as shown in Figure 1. The experimental platform is fixed on a five-axis machining center (DMG, type DMU60, Stuttgart, Germany) with a position accuracy of 0.01 mm to realize precise motion control. The bone was firmly clamped by a specially designed fixture upon a Kistler dynamometer (KISTLER, type 9267, Winterthur, Switzerland), which can capture the in situ force signals generated during the sawing process. A temperature measuring system with an infrared thermal imager (FLIR, type A615, Wilsonville, OR, USA) was used to monitor and record the temperature signal in the sawing test. A fresh saw blade (STRYKER, type 5100-137-134, Portage, MI, USA), which is often used in orthognathic surgery, was utilized in the sawing experiments. The details of the saw blade are shown in Figure 2.

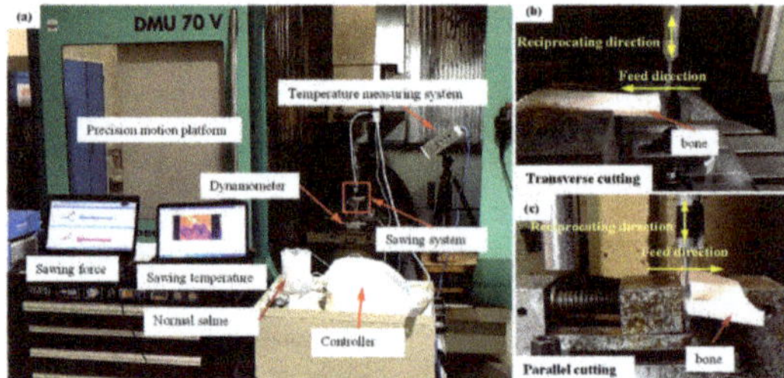

Figure 1. Experimental setup. (**a**) Sawing and measurement system, (**b**) Schematic diagram of transverse cutting, (**c**) Schematic diagram of parallel cutting.

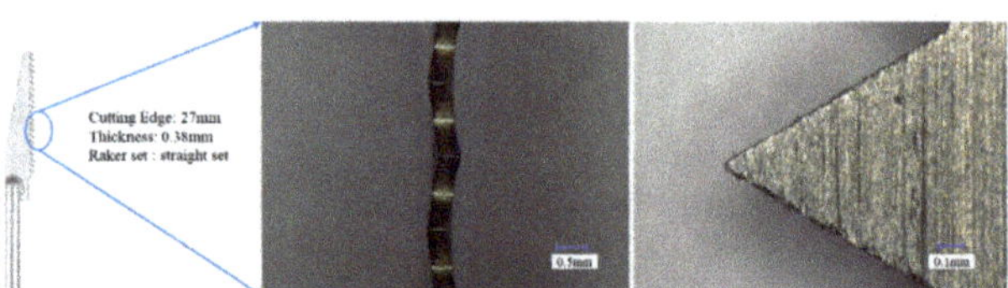

Figure 2. Saw blades used in the experiment.

In the current work, the feed rate and rotation speed were set as the main research variables. In addition, the angle between the saw blade and the osteons may influence the force and temperature. As summarized in Table 1, a full factorial experiment with 24 sawing conditions was designed; among the sawing conditions, "parallel" means that the sawing direction was parallel to the direction of the osteons, and "vertical" means that the sawing direction was perpendicular to the direction of the osteons.

Table 1. Sawing Test Parameters.

No.	Sawing Type	Feed Rate (mm/min)	Rotation Speed (r/min)
1	Parallel	40	9800
2	Parallel	60	9800
3	Parallel	80	9800
4	Parallel	40	11,200
5	Parallel	60	11,200
6	Parallel	80	11,200
7	Parallel	40	12,600
8	Parallel	60	12,600
9	Parallel	80	12,600
10	Parallel	40	14,000
11	Parallel	60	14,000
12	Parallel	80	14,000
13	Vertical	40	9800
14	Vertical	60	9800
15	Vertical	80	9800
16	Vertical	40	11,200
17	Vertical	60	11,200
18	Vertical	80	11,200
19	Vertical	40	12,600
20	Vertical	60	12,600
21	Vertical	80	12,600
22	Vertical	40	14,000
23	Vertical	60	14,000
24	Vertical	80	14,000

For reliability of the experimental results, three repeated tests were carried out under each cutting condition, and the result used in the analysis was the average of the results of the three tests.

2.3. Flow Chart of This Work

Figure 3 shows the flow chart of this work. The study was divided into three parts: sawing experiments, measurements and observations, analysis and optimization. The cutting force and cutting temperature were measured in situ in the bone sawing experiment, and the tool was observed after the test. The measured data were processed and analyzed, and a multiple regression model was used to fit the model of cutting force and temperature variation with parameters. The sawing parameters were also optimized.

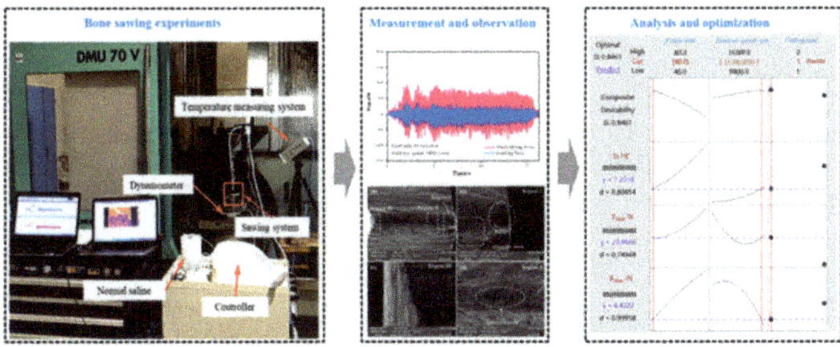

Figure 3. Flow Chart of This Work.

2.4. Modeling of Sawing Condition to Realize VR Surgical Training

Response surface methodology (RSM) was adopted to model the sawing force and temperature rise in the sawing procedure of fresh bovine scapula specimens. Response surface methodology is a collection of statistical and mathematical techniques used for the purpose of setting up a series of experiments for adequate predictions of a response and for fitting a hypothesized model to data obtained under the chosen design. Main effects and potential secondary effects can be identified by the RSM method. Meanwhile, the results of multi-objective optimization can also be obtained.

The general form of the multivariable regression formula can be expressed as Equation (1).

$$y = \beta_0 + \sum_{i=1}^{k} \beta_i x_i + \sum_{i=1}^{k} \beta_{ii} x_i^2 + \sum_i \sum_j \beta_{ij} x_i x_j + \varepsilon \tag{1}$$

where y is the response variable; β_0 is a constant; $\beta_1, \beta_2, \ldots, \beta_k$ are coefficients; $x_1, x_2, \ldots, x_k$ are the input variables and ε is an error term.

By inputting the sawing force model into the VR surgical training equipment, the magnitude and direction of the sawing force can be calculated in real time. Then, the real-time updated calculation results will be fed back to the operator through the force feedback handle until the operation is completed.

2.5. Observation of the Wear Characteristics of Saw Blades

The failure of the saw blade is detrimental to orthognathic surgery. An observation of the wear characteristics of saw blades is helpful to avoid such risks. After the tests, the saw blades were treated by ultrasonic cleaning for 5 min in acetone, and the wear characteristics of the cutting edges were rigorously investigated via a scanning electron microscope (TESCAN, type VEGA 3, Brno, Czech). TESCAN VEGA3 is a thermionic emission SEM system that comes either with a tungsten heated filament or lanthanum hexaboride (LaB$_6$) as the electron source.

3. Results

Before the sawing trials, the thermal diffusivity of the cortical bone and cancellous bone were measured, and the acquired results are shown in Figure 4. The thermal diffusivity of the cortical bone and cancellous bone fluctuated within the range of 0~100 °C, and the thermal diffusion coefficient parallel to the direction of the osteons was the greatest. It can be seen that the variation of thermal diffusivity of the bone materials fluctuated less in the range of 0~100 °C between 0.1–0.175. The thermal diffusivity of the cortical bone was significantly higher than that of the cancellous bone. The thermal diffusivity of the cortical bone parallel to the osteon direction was slightly higher than that perpendicular to the cortical bone direction. Figure 4 shows the morphology of the bone material and machined surface after the experiment. As shown in Figure 5a, the distance between each sawing trace was greater than 1 mm to reduce the interaction effect. The cutting traces could be observed on the machined surface of the bone material.

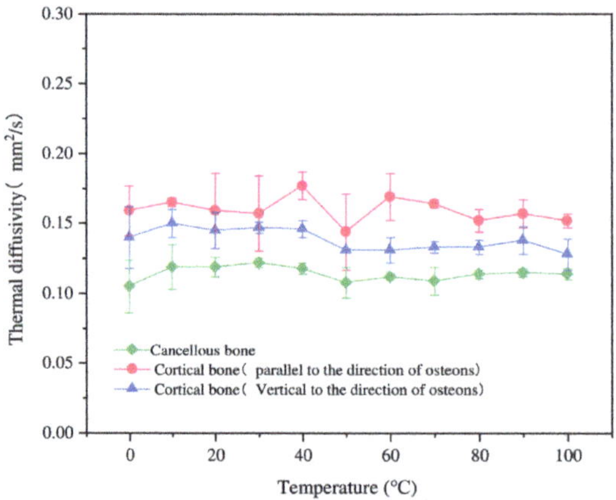

Figure 4. Measurement of the thermal diffusivity of the bone materials.

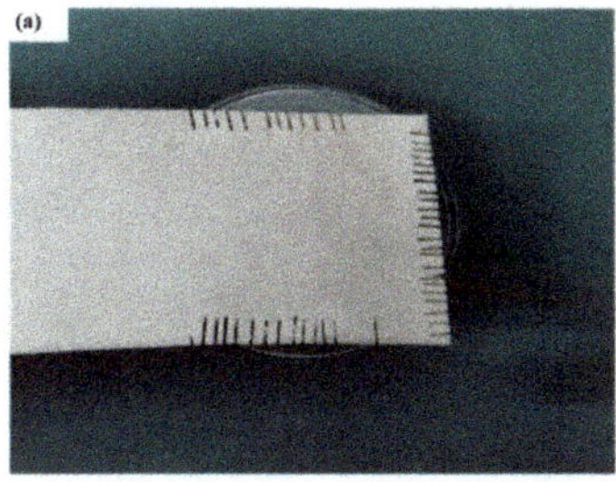
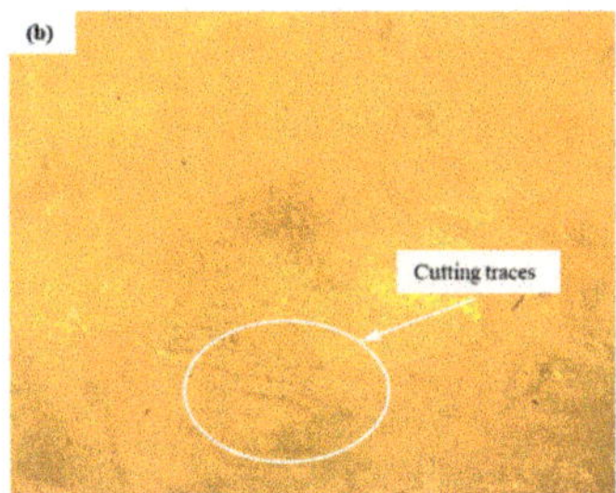

Figure 5. (**a**) Bone material after sawing and (**b**) a sawed surface.

The cutting force and temperature under different cutting parameters were obtained during the sawing process. Figure 6a is a schematic diagram of the sawing process, where the main cutting force is along the reciprocating direction of the saw blade, and the feed force is along the tool feed direction. The typical cutting force signal is shown in Figure 5b. The frequency of the cutting force signal was consistent with the reciprocating motion of the tool. The average values of the main sawing forces and feed forces under different cutting conditions are depicted in Figures 7 and 8. The main cutting force decreased with the increase in rotating speed and increased with the increase in the feed force. The variation trend of the feed force with sawing parameters was similar to that of the main cutting force. The evolution of the temperature rise with different cutting parameters is illustrated in Figure 9, and a scanning electron microscopy (SEM) image showing the wear signatures of the saw blade is shown in Figure 10. The experiment was carried out to observe the tool as a whole and focused on the wear in the cutting edge region. The worn blade was sawed 200 mm in length with the parameter of feed rate: 40 mm/min and rotation speed: 13,700 rpm.

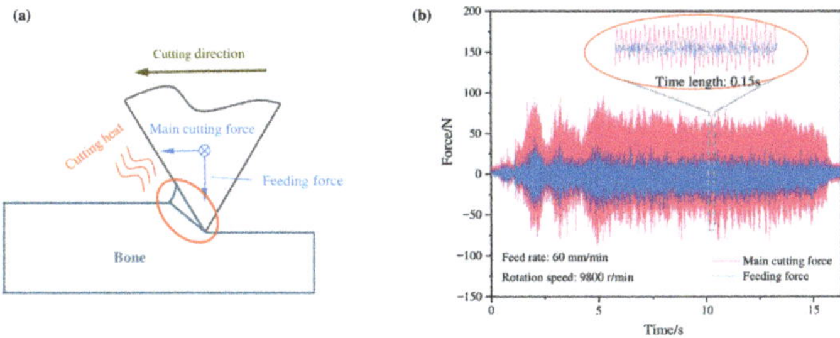

Figure 6. (**a**) Typical cutting force signal and (**b**) schematic diagram of the sawing process.

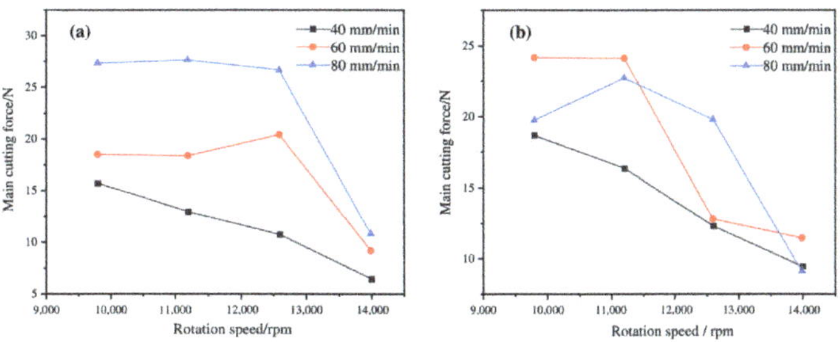

Figure 7. The main cutting force with different experimental parameters: (**a**) parallel sawing and (**b**) vertical sawing.

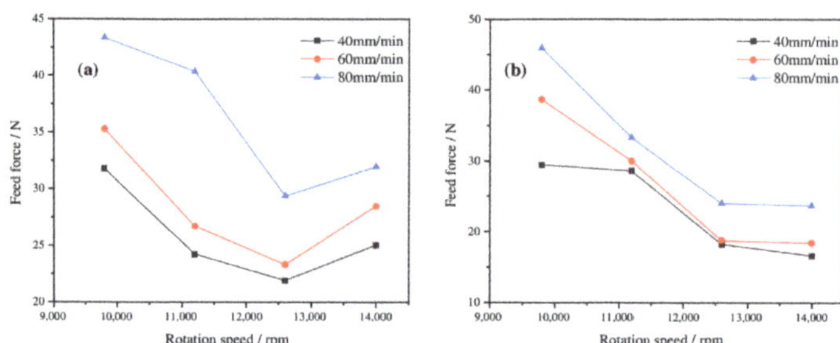

Figure 8. The feed sawing force with different experimental parameters: (**a**) parallel sawing and (**b**) vertical sawing.

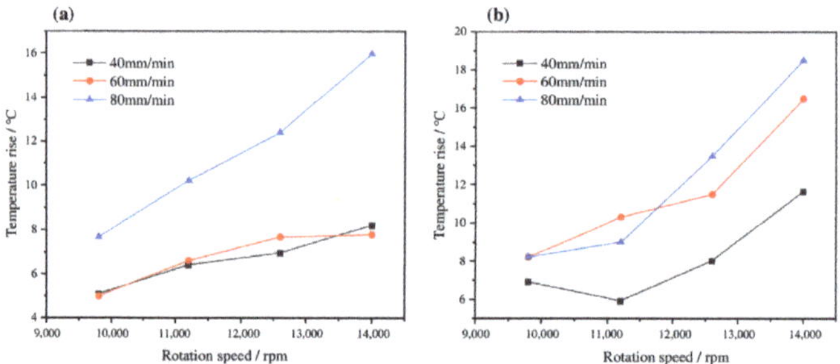

Figure 9. The temperature rise with different experimental parameters: (**a**) parallel sawing and (**b**) vertical sawing.

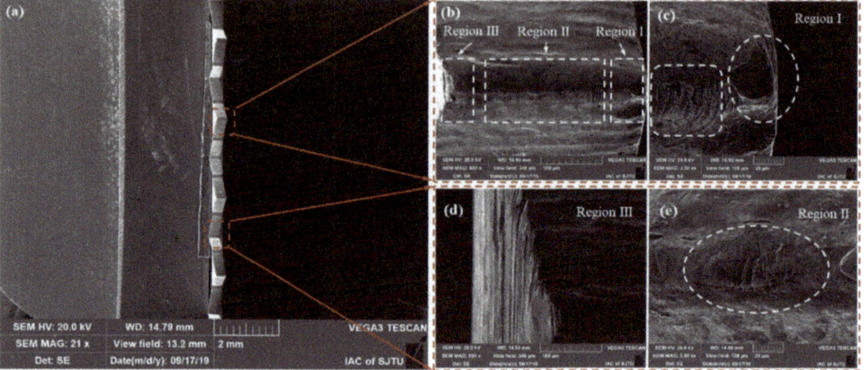

Figure 10. Observation of the wear characteristics of the saw blades. (**a**) Macro morphology of saw blades, (**b**) Wear characteristics of saw blade tooth tip, (**c**) Wear characteristics of the bottom part of the saw blade teeth. (**d**) Wear characteristics of the bottom part of the saw blade teeth, (**e**) Wear mechanism of the bottom part of the saw blade teeth.

According to Equation (2), cutting depths in this paper can be obtained. The cutting depth can be calculated as follows:

$$D_c = 500f/n \tag{2}$$

where D_c is the depth of the cut, f is the feed rate and n is the rotation speed.

According to Equation (2), the cutting depths under different cutting parameters are summarized in Table 2.

Table 2. Cutting depth under different cutting parameters.

Parameters	Value								
f (mm/min)	40	40	40	60	60	60	80	80	80
n (r/min)	9800	11,200	12,600	9800	11,200	12,600	9800	11,200	12,600
D_c (μm)	2.04	1.78	1.59	3.06	2.68	2.38	4.08	3.57	3.17

Multiple regression models of the sawing force and sawing temperature were established based on the acquired force and temperature signals. Full quadratic formulas were introduced to express the variation in the main cutting force, feed force and temperature rise, as shown in Equations (3)–(5), respectively. The multiple regression model is widely used for modeling cutting temperatures and cutting forces [21], including those of bone materials [22]. It has the advantages of fast computation and high applicability. The model is commonly used for force feedback in virtual surgical systems [23].

$$\begin{cases} F_{Main} = -122.4 + 0.763f + 0.02112v - 0.00183f^2 - 0.000001v^2 - 0.000021f \times v \quad (P) \\ F_{Main} = -180.0 + 0.562f + 0.02090v - 0.00183f^2 - 0.000001v^2 - 0.000021f \times v \quad (P) \end{cases} \quad (3)$$

$$\begin{cases} F_{Feed} = 171.0 + 0.261f - 0.02477v + 0.00444f^2 + 0.000001v^2 - 0.000044f \times v \quad (P) \\ F_{Feed} = 201.0 + 0.210f - 0.02728v + 0.00444f^2 + 0.000001v^2 - 0.000021f \times v \quad (P) \end{cases} \quad (4)$$

$$\begin{cases} T_{Rise} = 31.9 - 0.407f - 0.00377v + 0.00167f^2 + 0.000001v^2 + 0.000028f \times v \quad (P) \\ T_{Rise} = 29.5 - 0.439f - 0.00324v + 0.00167f^2 + 0.000001v^2 + 0.000028f \times v \quad (P) \end{cases} \quad (5)$$

where "P" indicates that the sawing direction is parallel to the direction of the osteons, and "V" signifies that the sawing direction is perpendicular to the direction of the osteons.

Within the scope of the current experiment, the accuracy of the developed models is acceptable. The analysis of the residual plot and the normal probability plot of the residuals was another effective method used to estimate the accuracy of the established models [24]. Figure 11 shows that the normal probability plots of the residuals reflect approximately straight lines, which verifies that the residuals are normally distributed, and that the regression results are fitted well. The effects of the experimental parameters (feed rate and rotation speed) on the sawing force and temperature are discussed for sawing bone materials. Figures 12–14 show contours corresponding to the minimax response surface, which was obtained using the RSM method. The contours depict the behavior of the force and temperature with the interaction effects from the input variables based on the full quadratic formulas. To obtain a lower cutting force and temperature, higher rotational speeds and lower feed rates are preferred.

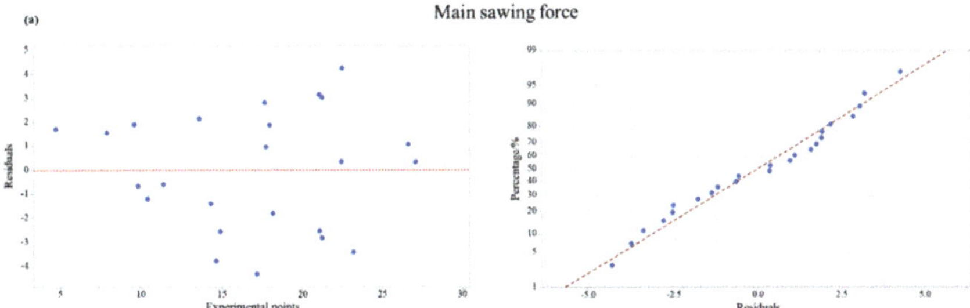

Figure 11. *Cont.*

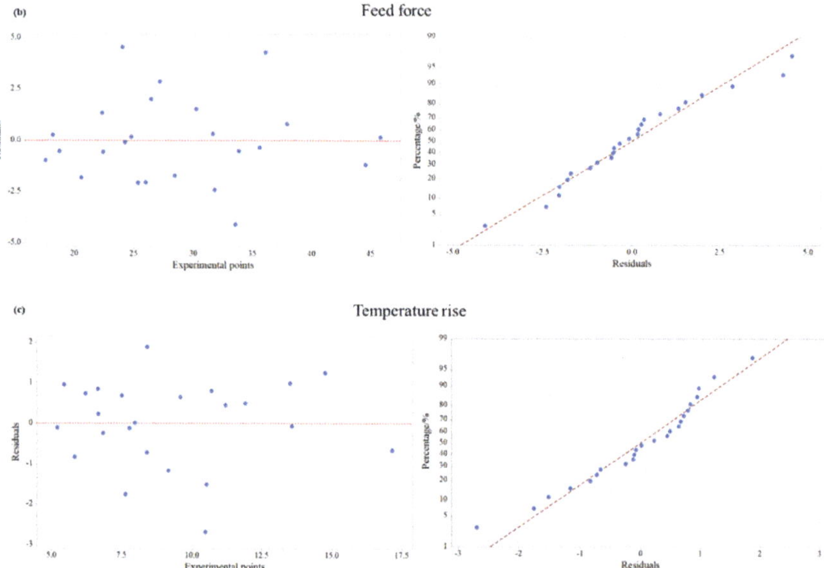

Figure 11. Residual fitted values for the models and experimental results, and their expectations assuming a normal distribution: (**a**) main cutting force; (**b**) feed force; (**c**) temperature rise.

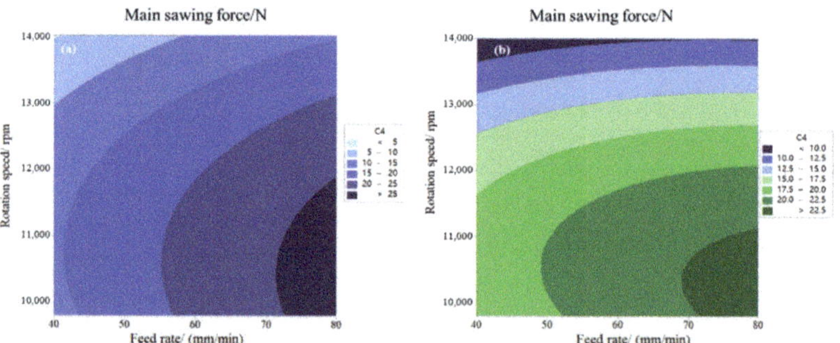

Figure 12. The effects of the sawing parameters on the main cutting force (**a**) parallel sawing and (**b**) vertical sawing.

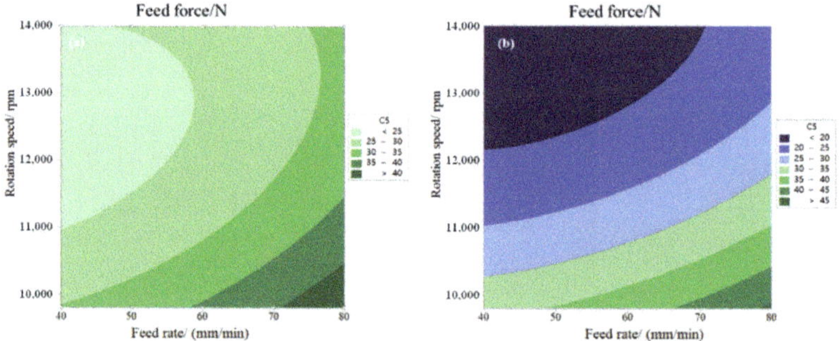

Figure 13. The effects of the sawing parameters on the feed force. (**a**) parallel sawing and (**b**) vertical sawing.

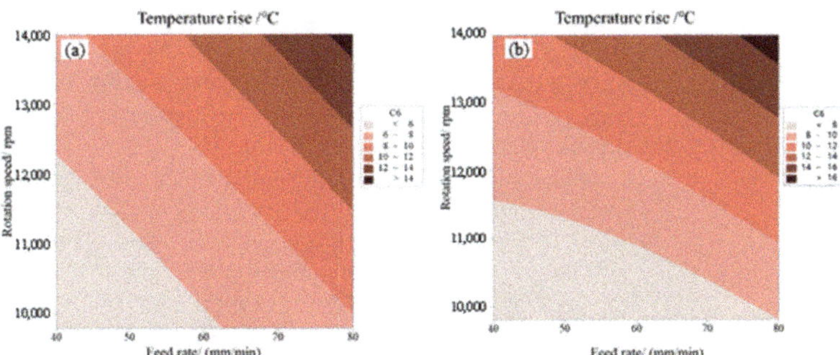

Figure 14. The effects of the sawing parameters on the temperature rise. (**a**) parallel sawing and (**b**) vertical sawing.

4. Discussion

Potential intraoperative adverse events in orthognathic surgery include abnormal hemorrhage due to infection or injury, fracture and split, osteonecrosis, etc. [25]. The mandible is identified as adverse-event-prone, probably due to its unique anatomical conditions, such as its masticatory muscle attachment, articulatory function, difficulties in visualizing the inferior alveolar neurovascular bundle and strict demands for osteotomy and fixation [26]. Therefore, it is necessary to optimize the procedure comprehensively in order to minimize the incidence of adverse events. In this study, we analyzed the sawing force, sawing temperature and the wear signatures of the sawing blades, targeting a set of sawing parameters standard for both safety and efficiency.

In most studies, the temperature measured in the experiment is used as the index of sawing temperature. However, the temperature rise is more suitable as an evaluation index because there is a difference between the experimental temperature (usually at room temperature) and the operating temperature. Meanwhile, it needs to be determined that the thermal diffusivity of the bone material changes with temperature. Information about the thermal diffusivity of bone materials at different temperatures cannot be found in the existing literature. Bone material is a typical heterogeneous anisotropic material. Directionality should be considered. The thermal diffusivity of the cortical bone and cancellous bone change slightly, within the range of 0~100 °C. This indicates that the temperature rise (Tr) can be used to evaluate the temperature during bone material sawing.

During surgery, the cutting force is one of the most important indexes used to evaluate the sawing process, and it affects the control accuracy and structural tissue damage. There is a negative correlation between the main cutting force and rotation speed. Meanwhile, the main cutting force has a positive correlation with the feed rate. The variation in the feed force with the different sawing parameters is similar to that of the main cutting force. However, when the speed reaches 14,000 rpm, the feed force exhibits an accelerating trend.

Previous studies have shown that excess heat in the bone could lead to irreversible damage or even necrosis when reaching 47 °C [27]. In many procedures, an irrigation channel is added to offset the heat accumulation of the sawing. However, on the one hand, water flow may interfere with the operation view. On the other hand, digital surgical guide plates, which are increasingly clinically popular, could block the water flow from heat transfer with bone tissue, making it difficult to achieve the goal of cooling [28]. Therefore, reducing the heat generated by the saw itself is the fundamental way to reduce the damage. As shown in Figure 9, the sawing temperature shows a positive correlation with the feed rate and rotation speed. During the sawing process, the heat generated by the friction between the saw blade and the bone material is the main contribution to the increase in the cutting temperature. As the rotating speed increases, the heat generated by the friction increases proportionally, which leads to a sharp rise in the sawing temperature. In comparison with the rotating speed, the feed rate has less positive impact on the sawing temperature. In general, the higher the feed rate is, the larger the cutting force and thus the greater the work of friction between the saw blade and the bone material.

According to the related literature [29], the material removal mechanism of the bone is shear cutting at a relatively low cutting depth (less than 20 μm). When the cutting depth exceeds 80 μm, bone chips form from the bone material mainly due to fracturing. Since the cutting mode and crack growth often lead to many very small cracks forming in the bone material, promoting inflammation during healing [30], the cutting mode of shear cutting is recommended in surgery. Therefore, the feed speed must be controlled within a certain range. When cutting at a depth greater than 20 μm, the cutting mode and crack growth often lead to many very small cracks forming in the bone material, promoting inflammation during healing. During surgery, it is necessary to appropriately reduce the feed (moving speed) when the rotational speed decreases. Table 2 shows that the cutting depths are less than 10 μm regardless of the experimental parameters used and can meet the actual requirements of orthognathic surgery. With the cutting mode of shear cutting, the friction coefficient is approximately 0.82 when the cutting direction is parallel to the osteons, and the shear stress (τ_s) is 85 MPa [31]. When the cutting direction is perpendicular to the osteons, the friction coefficient is approximately 0.85 and the resulting shear stress is 191 MPa. This explains why the cutting force and cutting temperature tend to be higher when the cutting direction is perpendicular to the osteons. We consider "perpendicular" and "parallel" to build a more accurate model. This technology can improve the reality of VR training.

With regard to tool wear, abrasive wear is the main wear form of saw blades when cutting such materials. The tool wear is mainly affected by the friction between the high-density minerals in the bone material and the saw blade during the sawing process. Since the sawing temperature and force are low in the bone material, it is difficult to generate other forms of tool wear, such as adhesive wear and diffusion wear.

A higher rotational speed in bone sawing reduces cutting forces and increases cutting temperatures. In surgery, lower cutting forces can reduce the labor intensity of the surgeon to perform the operation more accurately. However, excessive temperatures will lead to irreversible thermal damage to tissues. Therefore, choosing the right speed (taking into account both cutting force and cutting temperature) can help in surgical procedures. With the aim of achieving a minimum sawing force and temperature, the experimental parameters are optimized. Considering the range of commonly used parameters for orthognathic in Table 1, the optimized curve is shown in Figure 15. The minimum main cutting force, feed force and temperature rise are 6.4322 N, 23.9666 N and 7.2 °C with the following optimized parameters: a feed rate of 40 mm/min, a rotational speed of 13,700 rpm and a sawing direction parallel to the osteons. Moreover, in future experiments or surgeries, attention should be paid to replacing saw blades at appropriate intervals and fixing the specimen firmly during sawing.

In addition, the study had several limitations. Firstly, it was an in vitro study using bovine scapulae. The mechanical properties of bovine and human bones are not identical. And heat dissipation occurs in living bodies due to mechanisms such as blood flow, which could not be simulated in our experiments. Secondly, the infrared thermal imager measured temperature changes on the surface, which deviated somewhat from the temperature rise inside the bone. Finally, the parameters derived from our study still need abundant, emulation and probative tests before being applied in clinical practice.

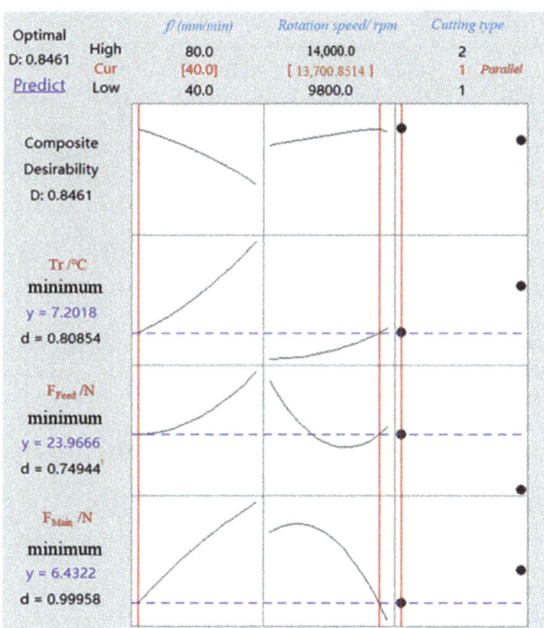

Figure 15. Optimization of the parameters to achieve a minimum sawing force and temperature.

5. Conclusions

This work carried out sawing experiments on bone materials. The changing law of cutting force and cutting temperature was studied, and the mechanism of tool wear was analyzed. Meanwhile, a multivariate regression model capable of being used for virtual surgery training was obtained. The conclusions obtained are as follows.

To minimize the main cutting force, a feed rate of 40 mm/min combined with a rotational speed of 13,700 rpm is recommended. When considering the sawing force and temperature rise comprehensively, the rotational speed should not be extremely high (about 13,000 rpm is recommended). When surgeons use dental reciprocating saws, they need to avoid high-speed gears for reducing thermal damage to tissues. The experimental data were confirmed to be scientific and accurate for the predicted models of sawing condition, which can be used in VR training systems.

Author Contributions: Statement: (1) D.Y., F.Z. and W.Z. were responsible for the design, conducting of the experiments, data collection and drafting of the article, and accountable for all aspects of the work. F.Z., D.Y. and W.Z. contributed equally. (2) P.N. and Q.A. were responsible for the experimental design, data interpretation, critical revision of the article and approval of article. P.N. and Q.A. contributed equally. All authors have read and agreed to the published version of the manuscript.

Funding: This work was funded by the National Natural Science Foundation of China (Grant No. 52175422 and 32101094); Clinical Research Project of Health Industry of Shanghai Health Commission (Grant No. 202240194) and Cross-disciplinary Research Fund of Shanghai Ninth People's Hospital, Shanghai Jiao Tong University School of Medicine (Grant No. JYJC202114).

Institutional Review Board Statement: Not applicable.

Informed Consent Statement: Not applicable.

Data Availability Statement: Not applicable.

Conflicts of Interest: The authors declare no conflict of interest.

References

1. Wu, Y.; Wang, F.; Fan, S.; Chow, J.K.-F. Robotics in Dental Implantology. *Oral Maxillofac. Surg. Clin.* **2019**, *31*, 513–518. [CrossRef] [PubMed]
2. Srivastava, R.; Jyoti, B.; Priyadarshi, P.K. Computer Aided Navigation for Predictable Dental Implantology: A Review. *Natl. J. Integr. Res. Med.* **2019**, *10*, 3.
3. Yan, Y.; Li, Q.; Wang, Q.; Peng, Y. Real-Time Bone Sawing Interaction in Orthopedic Surgical Simulation Based on the Volumetric Object. *J. Vis.* **2018**, *21*, 239–252. [CrossRef]
4. Yu, D.; Liu, C.; Wu, Y.; An, Q. Measurement and Prediction of Drilling Force in Fresh Human Cadaver Mandibles: A Pilot Study. *Clin. Implant Dent. Relat. Res.* **2020**, *22*, 4–12. [CrossRef] [PubMed]
5. Tawy, G.F.; Rowe, P.J.; Riches, P.E. Thermal Damage Done to Bone by Burring and Sawing with and without Irrigation in Knee Arthroplasty. *J. Arthroplast.* **2015**, *31*, 1102–1108. [CrossRef] [PubMed]
6. He, H.; Wang, C.; Zhang, Y.; Zheng, Y.; Xu, L.; Xie, G.; Zhao, D.; Chen, B.; Chen, H. Investigating Bone Chip Formation in Craniotomy. *Proc. Inst. Mech. Eng. Part H J. Eng. Med.* **2017**, *231*, 959–974.
7. Lee, J.; Chavez, C.L.; Park, J. Parameters Affecting Mechanical and Thermal Responses in Bone Drilling: A Review. *J. Biomech.* **2018**, *71*, 4–21. [CrossRef] [PubMed]
8. Reingewirtz, Y.; Szmuklermoncler, S.; Senger, B. Influence of Different Parameters on Bone Heating and Drilling Time in Implantology. *Clin. Oral Implant. Res.* **2010**, *8*, 189–197. [CrossRef]
9. Donath, K. A Method for the Study of Undecalcified Bones and Teeth with Attached Soft Tissues. *J. Oral Pathol. Med.* **1982**, *11*, 318–326. [CrossRef]
10. Liu, H.; Zhao, P.; Guo, Y.; Li, D.; Wang, Y.; Sun, S.; Wu, J. Material removal behaviors of FCC metals in nanoscale and microscale scratching: Theoretical model and experiments. *J. Mater. Process. Technol.* **2023**, *312*, 117855. [CrossRef]
11. James, T.P.; Pearlman, J.J.; Saigal, A. Rounded Cutting Edge Model for the Prediction of Bone Sawing Forces. *J. Biomech. Eng.* **2012**, *134*, 071001. [CrossRef] [PubMed]
12. Hsieh, M.S.; Tsai, M.I.G.-D.A.; Yeh, Y.-D. An Amputation Simulator with Bone Sawing Haptic Interaction. *Biomed. Eng. Appl. Basis Commun.* **2006**, *18*, 229–236. [CrossRef]
13. Conward, M.; Samuel, J. A Microstructure-Based Mechanistic Model for Bone Sawing: Part 1—Cutting Force Predictions. *J. Manuf. Sci. Eng.* **2021**, *143*, 121009. [CrossRef]
14. Pichler, W.; Mazzurana, P.; Clement, H.; Grechenig, S.; Grechenig, W. Frequency of Instrument Breakage During Orthopedic Procedures and Its Effects on Patients. *J. Bone Jt. Surg.* **2009**, *90*, 2652–2654. [CrossRef] [PubMed]
15. Kumaraswamy, J.; Anil, K.C.; Shetty, V.; Shashishekar, C. Wear behaviour of the Ni-Cu alloy hybrid composites processed by sand mould casting. *Adv. Mater. Process. Technol.* **2023**, *9*, 351–367. [CrossRef]
16. Kumaraswamy, J.; Shetty, V.; Sanman, S.; Naik, P.; Shedthi, S.; Gnanavel, C.; Mogose, I.N. Thermal Analysis of Ni-Cu Alloy Nanocomposites Processed by Sand Mold Casting. *Adv. Mater. Sci. Eng.* **2022**, *2022*, 2530707. [CrossRef]
17. Kumaraswamy, J.; Anil, K.C.; Shetty, V. Development of Ni-Cu based alloy hybrid composites through induction furnace casting. *Mater. Today Proc.* **2023**, *72*, 2268–2274. [CrossRef]
18. Allsobrook, O.F.; Leichter, J.; Holborow, D.; Swain, M. Descriptive Study of the Longevity of Dental Implant Surgery Drills. *Clin. Implant Dent. Relat. Res.* **2011**, *13*, 244–254. [CrossRef]
19. Koopaie, M.; Kolahdouz, S.; Kolahdouz, E.M. Comparison of Wear and Temperature of Zirconia and Tungsten Carbide Tools in Drilling Bone: In Vitro and Finite Element Analysis. *Br. J. Oral Maxillofac. Surg.* **2019**, *57*, 557–565. [CrossRef]
20. Queiroz, T.P.; Souza, F.Á.; Okamoto, R.; Margonar, R.; Pereira-Filho, V.A.; Garcia, I.R., Jr.; Vieira, E.H. Evaluation of Immediate Bone-Cell Viability and of Drill Wear after Implant Osteotomies: Immunohistochemistry and Scanning Electron Microscopy Analysis. *J. Oral Maxillofac. Surg.* **2008**, *66*, 1233–1240. [CrossRef]
21. Tahmasbi, V.; Qasemi, M.; Ghasemi, R.; Gholami, R. Experimental study and sensitivity analysis of force behavior in cortical bone milling. *Med. Eng. Phys.* **2022**, *105*, 103821. [CrossRef] [PubMed]
22. Lin, Y.; Yu, D.; Chen, X.; Wang, X.; Shen, G.; Wang, C. Simulation and evaluation of a bone sawing procedure for orthognathic surgery based on an experimental force model. *J. Biomech. Eng.* **2014**, *136*, 034501.
23. James, T.P.; Pearlman, J.J.; Saigal, A. Predictive force model for haptic feedback in bone sawing. *Med. Eng. Phys.* **2013**, *35*, 1638–1644. [CrossRef] [PubMed]
24. Heydari, H.; Kazerooni, N.C.; Zolfaghari, M.; Ghoreishi, M.; Tahmasbi, V. Analytical and Experimental Study of Effective Parameters on Process Temperature During Cortical Bone Drilling. *Proc. Inst. Mech. Eng. Part H J. Eng. Med.* **2018**, *232*, 871–883. [CrossRef] [PubMed]
25. Knoedler, S.; Baecher, H.; Hoch, C.C.; Obed, D.; Matar, D.Y.; Rendenbach, C.; Kim, B.-S.; Harhaus, L.; Kauke-Navarro, M.; Hundeshagen, G.; et al. Early Outcomes and Risk Factors in Orthognathic Surgery for Mandibular and Maxillary Hypo- and Hyperplasia: A 13-Year Analysis of a Multi-Institutional Database. *J. Clin. Med.* **2023**, *12*, 1444. [CrossRef] [PubMed]
26. Zaroni, F.M.; Cavalcante, R.C.; da Costa, D.J.; Kluppel, L.E.; Scariot, R.; Rebellato, N.L.B. Complications Associated with Orthognathic Surgery: A Retrospective Study of 485 Cases. *J. Cranio-Maxillofac. Surg.* **2019**, *47*, 1855–1860. [CrossRef] [PubMed]
27. Eriksson, A.R.; Albrektsson, T. Temperature Threshold Levels for Heat-Induced Bone Tissue Injury: A Vital-Microscopic Study in the Rabbit. *J. Prosthet. Dent.* **1983**, *50*, 101–107. [CrossRef] [PubMed]

28. Liu, Y.-F.; Wu, J.-L.; Zhang, J.-X.; Peng, W.; Liao, W.-Q. Numerical and Experimental Analyses on the Temperature Distribution in the Dental Implant Preparation Area When Using a Surgical Guide. *J. Prosthodont.* **2018**, *27*, 42–51. [CrossRef]
29. Yeager, C.; Nazari, A.; Arola, D. Machining of Cortical Bone: Surface Texture, Surface Integrity and Cutting Forces. *Mach. Sci. Technol.* **2008**, *12*, 100–118. [CrossRef]
30. Augustin, G.; Davila, S.; Mihoci, K.; Udiljak, T.; Vedrina, D.S.; Antabak, A. Thermal Osteonecrosis and Bone Drilling Parameters Revisited. *Arch. Orthop. Trauma Surg.* **2008**, *128*, 71–77. [CrossRef]
31. Liao, Z.; Axinte, D.A. On Chip Formation Mechanism in Orthogonal Cutting of Bone. *Int. J. Mach. Tools Manuf.* **2016**, *102*, 41–55. [CrossRef]

 *lubricants*

Article

An Experimental Study on Ultrasonic Vibration-Assisted Turning of Aluminum Alloy 6061 with Vegetable Oil-Based Nanofluid Minimum Quantity Lubrication

Guoliang Liu [1,2,*], Jin Wang [2], Jintao Zheng [2], Min Ji [2] and Xiangyu Wang [3]

[1] State Key Laboratory of Mechanical Transmission for Advanced Equipment, Chongqing University, Chongqing 400044, China
[2] School of Mechanical and Automotive Engineering, Qingdao University of Technology, Qingdao 266520, China; wangjinss1018@163.com (J.W.); zhengjintao_2024@163.com (J.Z.); jimin@qut.edu.cn (M.J.)
[3] School of Mechanical Engineering, University of Jinan, Jinan 250022, China; me_wangxy@ujn.edu.cn
* Correspondence: liuguoliang@qut.edu.cn

Abstract: Minimum quantity lubrication (MQL) is a potential technology for reducing the consumption of cutting fluids in machining processes. However, there is a need for further improvement in its lubrication and cooling properties. Nanofluid MQL (NMQL) and ultrasonic vibration-assisted machining are both effective methods of enhancing MQL. To achieve an optimal result, this work presents a new method of combining nanofluid MQL with ultrasonic vibration assistance in a turning process. Comparative experimental studies were conducted for two types of turning processes of aluminum alloy 6061, including conventional turning (CT) and ultrasonic vibration-assisted turning (UVAT). For each turning process, five types of lubricating methods were applied, including dry, MQL, nanofluid MQL with graphene nanosheets (GN-MQL), nanofluid MQL with diamond nanoparticles (DN-MQL), and nanofluid MQL with a diamond/graphene hybrid (GN+DN-MQL). A specific cutting energy and areal surface roughness were adopted to evaluate the machinability. The results show that the new method can further improve the machining performance by reducing the specific cutting energy and areal surface roughness, compared with the NMQL turning process and UVAT process. The diamond nanoparticles are easy to embed on the workpiece surface under the UVAT process, which can increase the specific cutting energy and Sa as compared to the MQL method. The graphene nanosheets can produce the interlayer shear effect and be squeezed into the workpiece, thus reducing the specific cutting energy. The results provide a new way for the development of eco-friendly machining.

Keywords: MQL; nanofluid; ultrasonic vibration-assisted turning; specific cutting energy; areal surface roughness; aluminum alloy 6061

Citation: Liu, G.; Wang, J.; Zheng, J.; Ji, M.; Wang, X. An Experimental Study on Ultrasonic Vibration-Assisted Turning of Aluminum Alloy 6061 with Vegetable Oil-Based Nanofluid Minimum Quantity Lubrication. *Lubricants* **2023**, *11*, 470. https://doi.org/10.3390/lubricants11110470

Received: 4 October 2023
Revised: 19 October 2023
Accepted: 31 October 2023
Published: 2 November 2023

1. Introduction

Cutting fluid has been frequently used to improve machinability, including reducing cutting forces and temperature, extending tool life, and improving machined surface integrity, since the early 20th century [1]. However, several drawbacks of the traditional flooded cutting fluids have been confirmed. For example, the cost of cutting fluids accounts for 7–17% of the total manufacturing cost, while its effective availability is relatively low [2,3]. It is difficult to provide the fluid into the cutting area. In addition, most of the cutting fluids used in the processing industry are mineral-based oils, which have huge risk to human health and environmental ecology [4]. Therefore, finding a new reliable machining method with green cutting lubrication technologies has been an extremely urgent task.

In recent years, several sustainable cutting methods have been proposed, including dry cutting, minimum quantity lubrication (MQL), cryogenic cutting, and so on [5–7]. Dry cutting avoids the usage of cutting fluid during the machining process, but often leads to severe tool wear and damaged machined surface integrity because of the bad lubricating and cooling conditions [8]. Cryogenic cutting employs liquid nitrogen, liquid carbon dioxide, or cryogenic gas as the coolant in the process, providing efficient cooling performance without generating pollutants [9]. However, these cryogenic fluids have limited lubricating properties, and their cost is relatively high. In comparison, MQL mixes compressed gas with a tiny amount of cutting fluid to form a droplet mist and generates a high spraying speed, which can obtain a good lubrification effect with a significantly reduced cutting fluid consumption [10]. Furthermore, researchers proved that the adverse effects of mineral-based cutting fluid on the environment and human health can be greatly eliminated by using vegetable oils as the lubricant media [11]. Vegetable oil-based MQL technology has been adopted by many researchers and proved that it can obtain good cutting performance during the machining of different kinds of materials [12].

However, there is a need for further enhancement of the lubrication and cooling properties of the vegetable oil-based MQL droplet mist. The amount of fluid permeating into the cutting zone is minor, although the compressed gas can generate a high spraying speed of droplet mist. According to the existing research, two primary methods are used to enhance the lubrication and cooling properties of MQL. The first involves improving the heat-conducting and lubricating properties of the lubricant media to achieve an effective lubricating and cooling effect with the limited lubricant media. The second approach involves increasing the effective flow rate of droplet mist into the cutting zone to lead to more lubricant media being produced [13,14].

Adding a proper amount of nanoparticles into the vegetable oils to form the nanofluid minimum quantity lubrication (NMQL) process has been proved to be an effective way for further improving the heat-conducting and lubricating properties of the MQL droplets [15]. Researchers have extensively studied the cutting performance of the NMQL process with various kinds of nanoparticles/nanosheets, such as Al_2O_3, graphite, CNTs, TiO_2, MoS_2, hBN, Ag, ZnO, Fe_2O_3, SiO_2, and diamond, in different machining methods, including turning, milling, drilling, and grinding [15–20]. It is found that these nanoparticles/nanosheets have different lubricating and cooling properties based on their shape and structural characteristics. The Al_2O_3 and diamond nanoparticles can reduce the tool-chip/workpiece friction coefficient by transforming the sliding friction into rolling friction [15,16]. MoS_2 and graphite have a layered structure with outstanding deposition and film formation properties, which can reduce the friction coefficient by transforming the two sliding metal surfaces to relative slippage of molecular layers [19]. Moreover, the graphite possesses exceptionally high thermal conductivity, ensuring excellent cooling performance during the machining processes [20]. In order to obtain better lubricating and cooling properties, hybrid nanofluid with two or more types of nanoparticles was also designed and tested by some researchers. Sharma et al. [21] performed NMQL turning of AISI 304 steel with Al_2O_3 and graphite as the hybrid nanofluid, and excellent results were observed by considering surface roughness, cutting force, cutting temperature, and tool wear. Jamil et al. [22] combined Al_2O_3 and CNT to form a hybrid nanofluid, and then added cryogenic media to enhance the cooling effect, during the machining of TC4 titanium alloy. A good machining performance was observed.

However, cutting tools are in close contact with the workpieces and chips during the conventional continuous cutting processes. Thus, the nanofluids are blocked from entering the tool-chip/tool-workpiece interfaces, making the lubricating and cooling effect unsatisfying, which is a restriction that has to be urgently settled for NMQL machining processes. In recent years, ultrasonic vibration-assisted machining (UVAM) has been a widely and successfully applied technology in the field of turning, grinding, milling, and drilling [23,24]. The high-frequency vibration of cutting tools can transform the continuous cutting process to an intermittent machining process, thus reducing the cutting force, cutting temperature, tool wear, and improving the machined surface integrity. In addition, the interrupted cutting action of ultrasonic vibration can break the close contact between the cutting tool and the workpiece/chip, and produce a pumping phenomenon, which can greatly increase the effective flow rate of the cutting fluid. Therefore, combining UVAM with NMQL may produce satisfactory results. Some researchers have attempted to combining different UVAM technologies with the NMQL method. Hoang et al. [25] combined graphene NMQL with an ultrasonic vibration-assisted deep hole drilling process (UVAD) for AISI SUS 304 stainless steel, and found a higher production rate, longer tool life, and better machining performance in terms of cutting forces and processing ability. Gao et al. [26,27] used different types of hybrid nanofluids as an MQL lubricant in ultrasonic vibration-assisted grinding (UVAG) processes. They all proved that the combination of NMQL and UVAG can further decrease the temperature, grinding forces, and friction coefficient.

The above literature review shows the obvious processing advantages of simultaneously using UVAG/UVAD and NMQL methods; however, the experimental confirmation of processing advantages of combining UVAT and NMQL techniques is missing. Thus, this paper aims to investigate the machining performance of UVAT and NMQL methods on aluminum alloy 6061 to experimentally confirm the feasibility of this new machining method. The experimental tests were conducted for different cooling techniques, including dry, MQL using palm oil, and NMQL using diamond nanoparticles, graphene nanosheets, and a hybrid of diamond/graphene mixed in palm oil. Two modes of turning processes, including conventional turning (CT) and ultrasonic vibration-assisted turning (UVAT), were combined with the above cooling techniques. The machining performance of various methods was fully evaluated and compared in terms of their specific cutting energy and areal surface roughness.

2. Experimental Setup and Scheme

The experimental study design aligns with the scheme presented in Figure 1. All cutting experiments are carried out on a CAK3665 CNC lathe (Shenyang Machine Tool Co., Ltd., Shenyang, Liaoning, China)with a self-developed ultrasonic vibration excitation system, as shown in Figure 2a. The Kennametal CCMT 060204 carbide insert (Kennametal, USA) with a tool nose radius of 0.4 mm is used in this work. After being mounted on the ultrasonic horn, the rake angle $\gamma 0$, clearance angle $\alpha 0$, and side cutting edge angle Kr of the tool are $0°$, $7°$, and $5°$, respectively, as shown in Figure 2b. Aluminum alloy 6061 is used as the workpiece, and its mechanical properties are shown in Table 1, which were tested by our co-author in a previous study [28]. The radial turning mode is adopted in this work, and ten experiments designed by the single factor experimental design method are shown in Table 2. The cutting speed is selected to guarantee the separation between the tool and chip, while the feed rate and depth of each cut are selected according to the recommendations within the tool manual. During the UVAT processes, an ultrasonic elliptic vibration is applied on the tool nose, and the ultrasonic amplitude in cutting speed direction and depth of cut direction are 3 and 1 μm, respectively. The ultrasonic frequency is 40.8 kHz and the phase difference is $90°$. For both MQL and NMQL conditions, the KS-2106 MQL system from Jinzhao Energy Saving Technology Co., Ltd. (Shanghai, China) is adopted. The flow rate for the lubricant is 60 mL/h with a gas pressure of 0.7 MPa. For the NMQL condition, the mass fraction of nanofluids is 0.5%, and the mass ratio of diamond nanoparticles and graphene nanosheets in GN+DN-MQL is 1:1. The size of graphene nanosheets is about 1–3 μm in length and 1–5 nm in thickness. The diameter of diamond nanoparticles is 50 nm. The nanofluid was prepared by dispersing the nanoparticles/nanosheets in palm oil via a two-step method with the Span-80 as the dispersant (mass fraction of 20% of the nanoparticles/nanosheets), i.e., adding a certain amount of nanoparticles/nanosheets and dispersant to base oils followed by sonification (40 kHz, 100 W) for 1 h. According to previous works, the nanoparticles/nanosheets can be dispersed evenly in the palm oil after sonification [29,30].

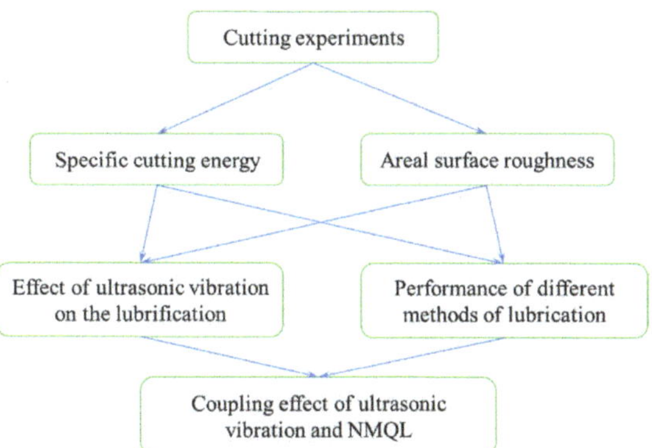

Figure 1. Flow chart of the experimental study.

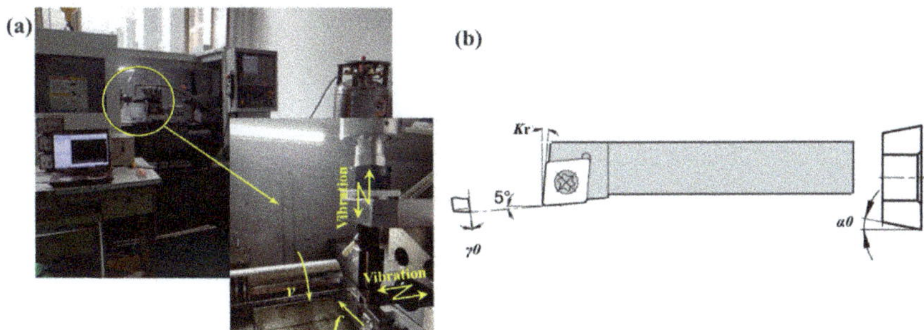

Figure 2. (**a**) Experimental equipment of the UVAT and NMQL method and (**b**) the cutting tool angles.

Table 1. The mechanical properties of AA 6061 [28].

Density (g/cm³)	Tensile Strength (MPa)	Elongation (%)	Hardness (Hv$_{9.8}$/MPa)
2.75	255	8.17	107.3

Table 2. Cutting test parameters.

Cutting Method	Lubrication Condition	Cutting Speed v (m/min)	Feed Rate f (mm/rev)	Depth of Cut a_p (mm)
CT	Dry MQL GN-MQL DN-MQL GN+DN-MQL	20	0.1	0.1
UVAT	Dry MQL GN-MQL DN-MQL GN+DN-MQL			

CT: conventional turning; UVAT: ultrasonic vibration-assisted turning; MQL: minimum quantity lubrication with palm oil as lubricant; GN-MQL: MQL using graphene nanosheets mixed in palm oil; DN-MQL: MQL using diamond nanoparticles mixed in palm oil; GN+DN-MQL: MQL using diamond/graphene hybrid mixed in palm oil.

During the cutting process, the cutting force in the cutting velocity direction for each cutting condition is measured with a YDC-III89B piezoelectric dynamometer (Dalian University of Technology, Dalian, Liaoning, China), and the specific cutting energy is then calculated. After the turning process, the KEYENCE VK-X1000 laser microscope

(Tokyo, Japan) is used to observe the surface morphology, and then used to measure the areal surface roughness *Sa* and *Sz*. The scanning electron microscope (SEM) and energy dispersive spectrometer (EDS) (ZEISS, Tokyo, Japan) are also used to analyze the machined surface morphology and components. A new insert is used in each cutting process to eliminate the errors induced by tool wear and the mounting error, and three tests are conducted for each machining condition to minimize the experimental error.

3. Results and Discussion

3.1. Specific Cutting Energy

Specific cutting energy refers to the energy consumed for removing per unit volume of a workpiece material [31]. The expression for this is $e_s = F_v/(a_p \times f)$, where e_s is the specific cutting energy, F_v is the cutting force in tangential force, a_p is the depth of cut, and f is the feed rate. A lower specific energy indicates an improved lubrication condition [32]. The specific cutting energy of different cutting conditions is shown in Figure 3, and the error bars show the standard deviation of the results among three repetitions. It can be seen from Figure 3 that the ultrasonic vibration reduces the specific cutting energy for all lubrication conditions. It is well-acknowledged that UVAT can reduce the cutting forces by achieving separation between the cutting tool and the chip/workpiece, and thus reduces the specific cutting energy.

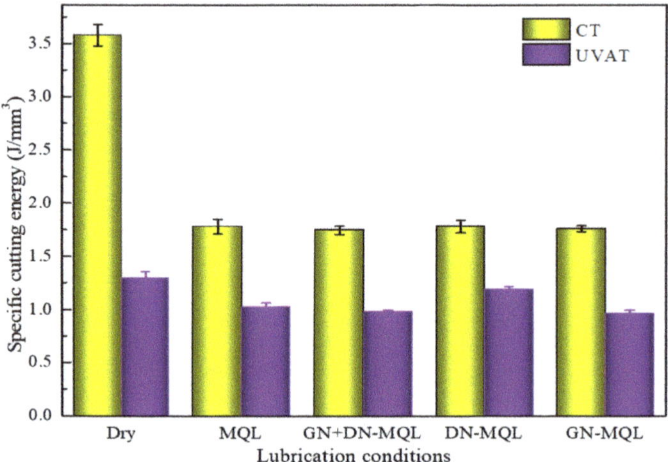

Figure 3. Specific cutting energy of different cutting conditions.

It can also be noticed that, during the CT process, the MQL and NMQL methods with different lubricants all generate significantly lower specific cutting energy in comparison to dry cutting. However, there is no obvious difference between the different MQL and NMQL methods. During the CT process, the cutting tool is in close contact with the workpiece and chip during the whole cutting process, and only a small amount of lubricant can permeate into the cutting zone. In this work, the mass fraction of nanoparticles or nanosheets in nanofluids is very low (only 0.5%). In this case, the nanoparticles or nanosheets that can permeate into the tool/chip or tool/workpiece interfaces to exert a lubrication effect are negligible. Thus, the MQL and NMQL methods fail to produce significant differences among each other regarding specific cutting energy. Certainly, the small quantity of lubricant permeating into the cutting zone can still produce remarkable differences in comparison to dry cutting conditions.

On the contrary, during the UVAT process, not only all the MQL and NMQL methods with different lubricants generate lower specific cutting energy than dry cutting, but the NMQL methods with graphene nanosheets, including GN-MQL and GN+DN-MQL, also generate lower specific cutting energy than the MQL method which uses pure palm oil. This means that adding graphene nanosheets into the palm oil can further increase the lubrication effect. This was also confirmed by Jia et al. during their investigation of UVAG processes [33]. The results in Figure 3 also prove that the diamond nanoparticles mixed in palm oil have a negative effect on the lubrication effect. This occurs because the DN-MQL method increases the specific cutting energy more in comparison to the MQL method which uses pure palm oil. The ultrasonic vibration can separate the cutting tool and chip/workpiece, and can thus permeate more of the lubricant into the cutting zone. Therefore, the nanoparticles or nanosheets that permeate into the tool/chip or tool/workpiece interfaces can exert a more obvious lubrication effect, resulting in affecting the specific cutting energy differently, as shown in Figure 3.

As for the different effects of graphene nanosheets and diamond nanoparticles on the specific cutting energy, their effects can be attributed to their different shapes and lubrication mechanisms. It has been proved by molecular dynamics simulations, in our previous work, that the tribological behavior of the nano-diamond is dominated by rolling, while the behavior of graphene nanosheets is governed by the interlayer shear effect [34]. During the UVAT processes, the cutting tool constantly bombards the workpiece surface, which can push the nano-diamond and graphene nanosheets into the workpiece. Furthermore, the workpiece material in this work (aluminum alloy 6061) exhibits a low hardness and high plasticity. Therefore, the near-spherical nano-diamond is easy to be squeezed into the workpiece. Once this happens, the nano-diamond will be tightly wrapped by the high-plastic aluminum alloy, and the rolling bearing effect will be severely weakened. Moreover, the nano-diamonds that are embedded on the workpiece surface can work as hard spots and interact with the cutting tool to increase the friction coefficient. This phenomenon is confirmed by the experimental surface morphology shown in Figure 4. It must be noted that the workpiece is cleaned ultrasonically before being observed. Hence, it is reasonably to believe that the large bulge is embedded on the workpiece surface. The results of elemental analysis (Figure 4) also show that the bulge is a mixture of numerous nano-diamonds and aluminum alloy.

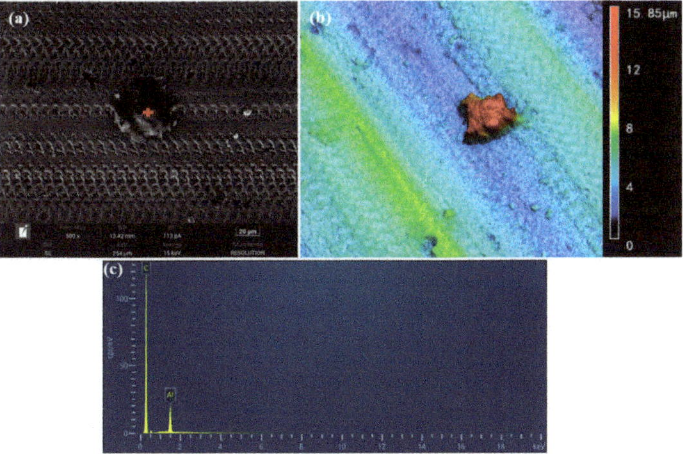

Figure 4. (**a**) SEM and (**b**) 3D images (×1000) of the surface morphology of the UVAT and DN-MQL method, and (**c**) EDS results of the bulge.

While for the graphene nanosheets, the interlayer shear effect occurs inside the graphene nanosheets. In addition, the graphene nanosheets are relatively large, soft, and more likely to attach to the workpiece and cutting tool surfaces [35]. In this condition, the interlayer shear effect can still work to reduce the cutting forces, even if the graphene nanosheets are squeezed into the workpiece. Figure 5 shows the experimental surface morphology of the UVAT and GN-MQL method. It can be seen that the graphene nanosheets are squeezed into the workpiece and attached onto the surface. However, the 3D images show that there is no definite upward bulge around the graphene nanosheets. In this case, shear sliding occupies the main position and reduces the friction coefficient.

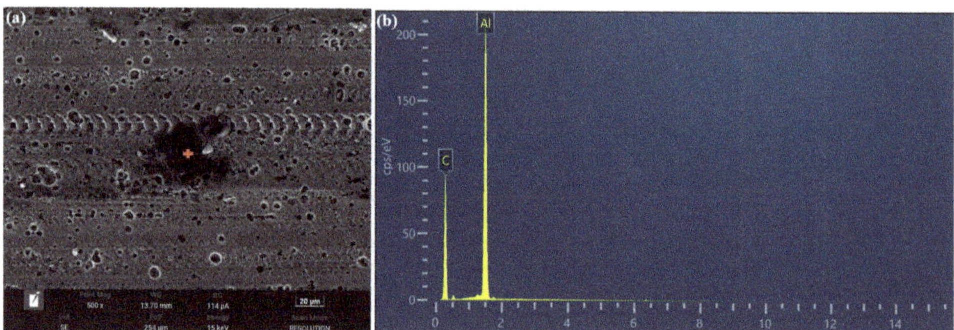

Figure 5. (**a**) SEM and (**b**) EDS images of the surface morphology of the UVAT and GN-MQL method.

3.2. Areal Surface Roughness

Figure 6 shows the measured areal surface roughness results under different cutting conditions, and the error bars show the standard deviation of measurement results among three repetitions. Sa is the arithmetic average surface roughness and Sz is the maximum peak-to-valley height of the whole measuring area. It can be seen from Figure 6a that the ultrasonic vibration increases the surface roughness during the dry cutting processes, but generates smoother surfaces under MQL and NMQL conditions. In addition, the lubrication methods show different effects on surface roughness Sa under CT and UVAT processes. During the CT process, MQL with palm oil produces a better surface finish quality than the dry cutting condition, and all the NMQL methods further reduce the surface roughness compared to the MQL condition. While in the UVAT process, the NMQL method with nano-diamond, i.e., DN-MQL, generates a rougher surface in comparison to the MQL condition with palm oil.

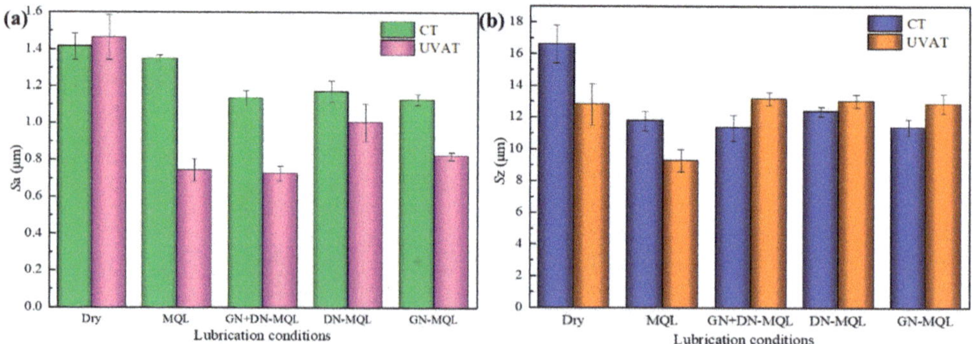

Figure 6. Areal surface roughness, (**a**) Sa and (**b**) Sz, of different cutting conditions.

However, when comparing the results of Sz in Figure 6b, the conclusion is different than the one from above. The ultrasonic vibration reduces Sz during dry and MQL conditions in comparison to the CT processes, but generates larger Sz during all NMQL conditions. Among all different lubrication methods, the MQL condition generates the lowest *Sz* during both the CT and UVAT processes, and the NMQL methods all reduce *Sz* in comparison to the dry cutting condition.

Figure 7 shows the surface morphology of the CT and UVAT processes under dry cutting conditions. It can be seen that the ultrasonic vibration generates regular dimples on the machined surface because of the intermittent separation of the cutting tool and workpiece, which increases the average roughness value *Sa*. However, the ultrasonic vibration amplitude is minor, i.e., 1 μm in the depth of cut direction, and thus the depth of the dimples is very small. Therefore, its effect on surface roughness is minor; thus, the increase rate of *Sa* is small. On the contrary, the ultrasonic vibration weakened the adhesion between the cutting tool and chip/workpiece, thus reducing the chippings adhered on the machined surfaces. This was also observed by researchers during the UVAG processes [24]. Figure 7a and the EDS results show that the chippings adhered on the machined surfaces are mainly workpiece materials, which are formed by the chips that were squeezed into the surface. Therefore, the *Sz* value of the UVAT process is lower than that of the CT process.

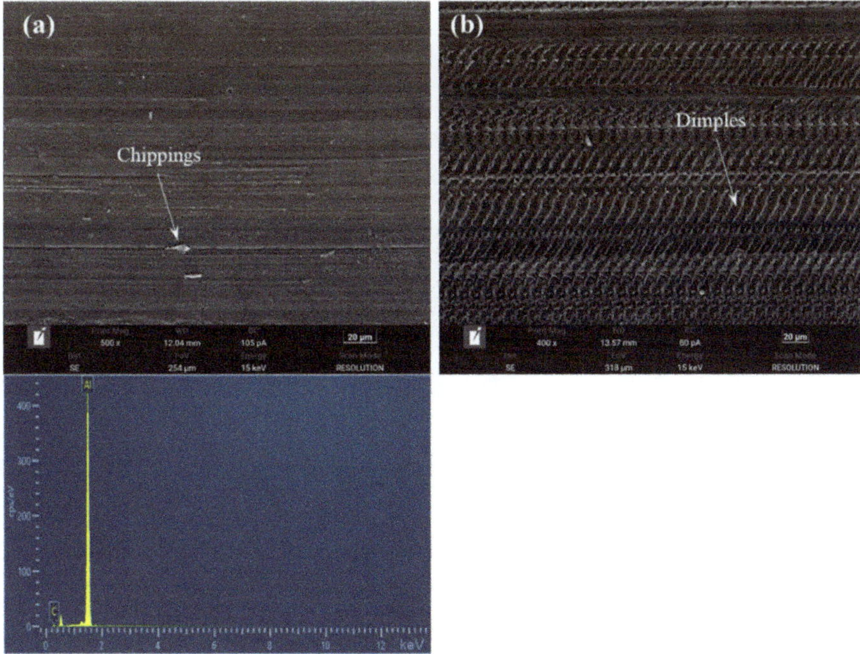

Figure 7. SEM images of the surface morphology of (**a**) CT and (**b**) UVAT processes under dry cutting conditions and the EDS results of the chipping in (**a**).

During the MQL cutting condition, the lubrication effect of palm oil drops can be significantly reinforced via ultrasonic vibration. This can occur as the ultrasonic vibration leads more palm oil drops into the cutting zone. The better lubrication condition can reduce the tool-chip/workpiece adhesion and the side flow of workpiece materials, thus reducing the scallop height between tool paths. This phenomenon can be proved by the 3D images of the surface morphology (Figure 8), and is consistent with the results of Ni et al. [23]. It can be seen from the 3D image, in Figure 8a, that there are obvious plowing marks on the surfaces of the CT process, and the depths of valleys are significantly different. The profile curve fluctuates in a large range from −6.3 μm to 5.7 μm. On the contrary, the surface of the UVAT process, in Figure 8b, is much smoother. The good lubrication condition eliminates the obvious plowing marks. The profile curve in Figure 8b fluctuates in a stable range from

−2.7 μm to 4.2 μm. In this case, the surface roughness of the UVAT and MQL process, including the *Sa* and *Sz*, is obviously reduced in comparison to the CT and MQL process.

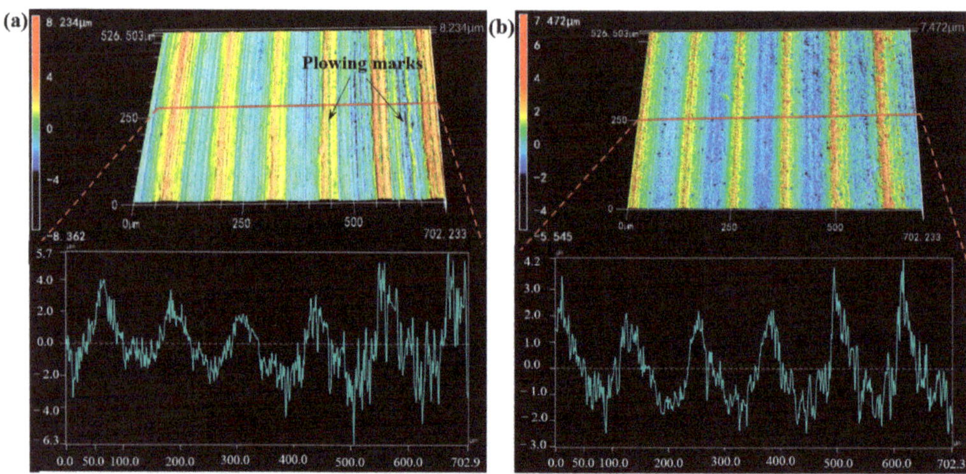

Figure 8. Three-dimensional images (×400) of the surface morphology of (**a**) CT and (**b**) UVAT processes under MQL condition.

When adding nano-diamond or graphene nanosheets into the oil, the lubrication effect of the UVAT and NMQL process can be further improved, thus reducing the average surface roughness *Sa* in comparison to the CT processes. On the other hand, the nanoparticles/nanosheets can be squeezed into the workpiece using the vibrating cutting tool, thus generating lots of pits after being washed away, as shown in Figure 9. As a result, the *Sz* value of the UVAT and NMQL process is greater than that of the CT and NMQL process.

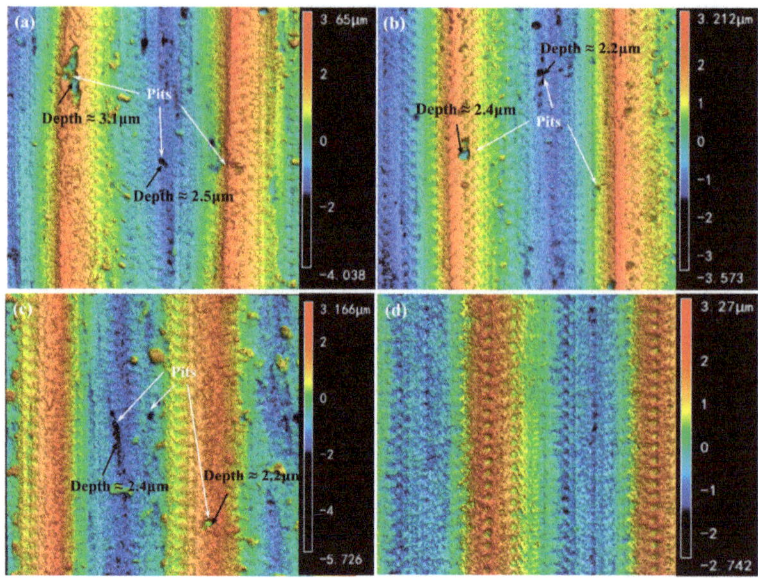

Figure 9. Three-dimensional images (×1000) of the surface morphology of UVAT processes under (**a**) DN-MQL, (**b**) GN-MQL, (**c**) GN+DN-MQL, and (**d**) MQL conditions.

In addition, the larger value of Sz in the UVAT and NMQL process than the UVAT and MQL process can also be attributed to the pits generated with the embedded nanoparticles/nanosheets. It is obvious that the surface of the UVAT and MQL process (as shown in Figure 9d) is uniform, and the microstructure produced with ultrasonic vibration is regular, without obvious random defects. On the contrary, the pits resulting from the embedded nanoparticles/nanosheets can be observed on all surfaces of the UVAT and NMQL process. Furthermore, a comparison of the depth of pits in Figure 9 reveals that the UVAT and GN-NMQL process generates relatively shallow pits (the largest depth of pits is 2.4 μm) in comparison to the NMQL conditions which use a nano-diamond material (the largest depth of pits is 3.1 μm).

4. Conclusions

(1) Combining nanofluid MQL with ultrasonic vibration assistance in turning processes has great potential in further improving the machinability in comparison to the NMQL turning process and UVAT process, and has the ability to reduce the specific cutting energy and areal surface roughness.

(2) Due to the low hardness of aluminum alloy 6061, the diamond nanoparticles are easy to embed onto the workpiece surface under the UVAT process, which can significantly weaken their rolling bearing effect and increase the specific cutting energy compared to the MQL method which uses pure palm oil.

(3) The graphene nanosheets can produce the interlayer shear effect by being squeezed into the workpiece. Therefore, the NMQL method which uses a mixture of graphene nanosheets and nano-diamond/graphene nanosheets can further reduce the specific cutting energy compared to the MQL method which uses pure palm oil.

(4) The ultrasonic vibration can significantly reinforce the lubrication effect of the MQL and NMQL methods. The superior lubrication condition can reduce the tool-chip/workpiece adhesion, cutting tool plowing, and workpiece material side flow, thus reducing the average areal surface roughness Sa compared to the CT and MQL as well as CT and NMQL processes.

(5) During the UVAT processes, the nanoparticles/nanosheets can be squeezed into the workpiece with the vibrating cutting tool, thus generating lots of pits on the machined surfaces. Therefore, the surface roughness of the UVAT and NMQL process is increased more in comparison to the UVAT and MQL process.

(6) Future research should be performed to reveal the cutting performance of the UVAT and NMQL process for different kinds of workpiece materials, such as materials with high strength (titanium alloy, for example) and high hardness (quenched steel, for example). Furthermore, more kinds of nanoparticles/nanosheets should be used during the UVAT and NMQL process and the optimal mass fraction of the nanoparticles/nanosheets should be tested.

Author Contributions: Project administration and writing—original draft preparation, G.L.; methodology, J.W.; data curation, J.Z.; writing—review and editing, M.J.; resources, X.W. All authors have read and agreed to the published version of the manuscript.

Funding: This work is financially supported by the Open Research Fund of State Key Laboratory of Mechanical Transmission for Advanced Equipment, Chongqing University (SKLMT-MSKFKT-202105), National Natural Science Foundation of China (52375446 and 52005281), and Natural Science Foundation of Shandong Province (ZR2020QE181).

Data Availability Statement: Data will be made available on request.

Conflicts of Interest: The authors declare no conflict of interest.

References

1. Said, Z.; Gupta, M.; Hegab, H.; Arora, N.; Khan, A.M.; Jamil, M.; Bellos, E. A comprehensive review on minimum quantity lubrication (MQL) in machining processes using nano-cutting fluids. *Int. J. Adv. Manuf. Technol.* **2019**, *105*, 2057–2086. [CrossRef]

2.   Amiril, S.A.S.; Rahim, E.A.; Syahrullail, S. A review on ionic liquids as sustainable lubricants in manufacturing and engineering: Recent research, performance, and applications. *J. Clean. Prod.* **2017**, *168*, 1571–1589. [CrossRef]

3.   Yıldırım, Ç.V.; Sarıkaya, M.; Kıvak, T.; Şirin, Ş. The effect of addition of hBN nanoparticles to nanofluid-MQL on tool wear patterns, tool life, roughness and temperature in turning of Ni-based Inconel 625. *Tribol. Int.* **2019**, *134*, 443–456. [CrossRef]

4.   Abd Rahim, E.; Dorairaju, H. Evaluation of mist flow characteristic and performance in minimum quantity lubrication (MQL) machining. *Measurement* **2018**, *123*, 213–225. [CrossRef]

5.   Liu, G.L.; Zheng, J.T.; Huang, C.Z.; Sun, S.F.; Liu, X.F.; Dai, L.J.; Wang, X.Y. Coupling effect of micro-textured tools and cooling conditions on the turning performance of aluminum alloy 6061. *Adv. Manuf.* **2023**, *11*, 663–681. [CrossRef]

6.   Guolong, Z.H.A.O.; Lianjia, X.I.N.; Liang, L.I.; Zhang, Y.; Ning, H.E.; Hansen, H.N. Cutting force model and damage formation mechanism in milling of 70wt% Si/Al composite. *Chin. J. Aeronaut.* **2023**, *36*, 114–128.

7.   Liu, K.; Zhang, J.; Li, J.; Bao, R.; Zuo, Y.; Liu, H.; Wang, Y. Experiment study of surface formation mechanism during cryogenic turning of PEEK. *J. Manuf. Process.* **2023**, *104*, 322–333. [CrossRef]

8.   Şirin, Ş.; Sarıkaya, M.; Yıldırım, Ç.V.; Kıvak, T. Machinability performance of nickel alloy X-750 with SiAlON ceramic cutting tool under dry, MQL and hBN mixed nanofluid-MQL. *Tribol. Int.* **2021**, *153*, 106673. [CrossRef]

9.   Albertelli, P.; Strano, M.; Monno, M. Simulation of the effects of cryogenic liquid nitrogen jets in Ti6Al4V milling. *J. Manuf. Process.* **2023**, *85*, 323–344. [CrossRef]

10.   He, T.; Liu, N.; Xia, H.; Wu, L.; Zhang, Y.; Li, D.; Chen, Y. Progress and trend of minimum quantity lubrication (MQL): A comprehensive review. *J. Clean. Prod.* **2022**, *386*, 135809. [CrossRef]

11.   Hybská, H.; Mitterpach, J.; Sameš ová, D.; Schwarz, M.; Fialová, J.; Veverková, D. Assessment of ecotoxicological properties of oils in water. *Arch. Environ. Prot.* **2018**, *44*, 31–37.

12.   Sankaranarayanan, R.; Krolczyk, G.M. A comprehensive review on research developments of vegetable-oil based cutting fluids for sustainable machining challenges. *J. Manuf. Process.* **2021**, *67*, 286–313.

13.   Wang, X.; Li, C.; Zhang, Y.; Ding, W.; Yang, M.; Gao, T.; Ali, H.M. Vegetable oil-based nanofluid minimum quantity lubrication turning: Academic review and perspectives. *J. Manuf. Process.* **2020**, *59*, 76–97. [CrossRef]

14.   Kumar, A.; Sharma, A.K.; Katiyar, J.K. State-of-the-art in sustainable machining of different materials using nano minimum quality lubrication (NMQL). *Lubricants* **2023**, *11*, 64. [CrossRef]

15.   Gupta, M.K.; Jamil, M.; Wang, X.; Song, Q.; Liu, Z.; Mia, M.; Imran, G.S. Performance evaluation of vegetable oil-based nano-cutting fluids in environmentally friendly machining of inconel-800 alloy. *Materials* **2019**, *12*, 2792. [CrossRef]

16.   Yıldırım, Ç.V. Experimental comparison of the performance of nanofluids, cryogenic and hybrid cooling in turning of Inconel 625. *Tribol. Int.* **2019**, *137*, 366–378. [CrossRef]

17.   Darshan, C.; Jain, S.; Dogra, M.; Gupta, M.K.; Mia, M. Machinability improvement in Inconel-718 by enhanced tribological and thermal environment using textured tool. *J. Therm. Anal. Calorim.* **2019**, *138*, 273–285. [CrossRef]

18.   Musavi, S.H.; Davoodi, B.; Niknam, S.A. Effects of reinforced nanoparticles with surfactant on surface quality and chip formation morphology in MQL-turning of superalloys. *J. Manuf. Process.* **2019**, *40*, 128–139. [CrossRef]

19.   Das, A.; Pradhan, O.; Patel, S.K.; Das, S.R.; Biswal, B.B. Performance appraisal of various nanofluids during hard machining of AISI 4340 steel. *J. Manuf. Process.* **2019**, *46*, 248–270. [CrossRef]

20.   Lim, S.K.; Azmi, W.H.; Jamaludin, A.S.; Yusoff, A.R. Characteristics of Hybrid Nanolubricants for MQL Cooling Lubrication Machining Application. *Lubricants* **2022**, *10*, 350. [CrossRef]

21.   Sharma, A.K.; Tiwari, A.K.; Dixit, A.R.; Singh, R.K.; Singh, M. Novel uses of alumina/graphene hybrid nanoparticle additives for improved tribological properties of lubricant in turning operation. *Tribol. Int.* **2018**, *119*, 99–111. [CrossRef]

22.   Jamil, M.; Khan, A.M.; Hegab, H.; Gong, L.; Mia, M.; Gupta, M.K.; He, N. Effects of hybrid Al 2 O 3-CNT nanofluids and cryogenic cooling on machining of Ti–6Al–4V. *Int. J. Adv. Manuf. Technol.* **2019**, *102*, 3895–3909. [CrossRef]

23.   Ni, C.; Zhu, L. Investigation on machining characteristics of TC4 alloy by simultaneous application of ultrasonic vibration assisted milling (UVAM) and economical-environmental MQL technology. *J. Mater. Process. Technol.* **2020**, *278*, 116518. [CrossRef]

24.   Kumar, M.N.; Subbu, S.K.; Krishna, P.V.; Venugopal, A. Vibration assisted conventional and advanced machining: A review. *Procedia Eng.* **2014**, *97*, 1577–1586. [CrossRef]

25.   Hoang, T.D.; Ngo, Q.H.; Chu, N.H.; Mai, T.H.; Nguyen, T.; Ho, K.T.; Nguyen, D. Ultrasonic assisted nano-fluid MQL in deep drilling of hard-to-cut materials. *Mater. Manuf. Process.* **2022**, *37*, 712–721. [CrossRef]

26.   Gao, T.; Zhang, X.; Li, C.; Zhang, Y.; Yang, M.; Jia, D.; Zhu, L. Surface morphology evaluation of multi-angle 2D ultrasonic vibration integrated with nanofluid minimum quantity lubrication grinding. *J. Manuf. Process.* **2020**, *51*, 44–61. [CrossRef]

27.   Rabiei, F.; Rahimi, A.R.; Hadad, M.J. Performance improvement of eco-friendly MQL technique by using hybrid nanofluid and ultrasonic-assisted grinding. *Int. J. Adv. Manuf. Technol.* **2017**, *93*, 1001–1015. [CrossRef]

28.   Wang, J.; Duan, P.; Wang, T.; Wang, X.; Qiao, Y. Effect of Cryogenic Cooling on Mechanical Properties and Cutting Force of 6061 Aluminum Alloy. *J. Phys. Conf. Ser.* **2023**, *2459*, 012032. [CrossRef]

29.   Zhang, Y.; Li, C.; Jia, D.; Zhang, D.; Zhang, X. Experimental evaluation of the lubrication performance of MoS2/CNT nanofluid for minimal quantity lubrication in Ni-based alloy grinding. *Int. J. Adv. Manuf. Technol.* **2015**, *99*, 19–33. [CrossRef]

30.   Su, Y.; Gong, L.; Li, B.; Liu, Z.; Chen, D. Performance evaluation of nanofluid MQL with vegetable-based oil and ester oil as base fluids in turning. *Int. J. Adv. Manuf. Technol.* **2016**, *83*, 2083–2089. [CrossRef]

31.   Shaw, M.C.; Cookson, J.O. *Metal Cutting Principles*; Oxford University Press: New York, NY, USA, 2005.

32. Liu, G.; Özel, T.; Li, J.; Wang, D.; Sun, S. Optimization and fabrication of curvilinear micro-grooved cutting tools for sustainable machining based on finite element modelling of the cutting process. *Int. J. Adv. Manuf. Technol.* **2020**, *110*, 1327–1338. [CrossRef]
33. Jia, D.; Li, C.; Zhang, Y.; Yang, M.; Zhang, X.; Li, R.; Ji, H. Experimental evaluation of surface topographies of NMQL grinding ZrO2 ceramics combining multiangle ultrasonic vibration. *Int. J. Adv. Manuf. Technol.* **2019**, *100*, 457–473. [CrossRef]
34. Wang, D.; Zhang, Y.; Zhao, Q.; Jiang, J.; Liu, G.; Li, C. Tribological mechanism of carbon group nanofluids on grinding interface under minimum quantity lubrication based on molecular dynamic simulation. *Front. Mech. Eng.* **2023**, *18*, 17. [CrossRef]
35. Hoang, T.D.; Mai, T.H.; Nguyen, V.D. Enhancement of Deep Drilling for Stainless Steels by Nano-Lubricant through Twist Drill Bits. *Lubricants* **2022**, *10*, 173. [CrossRef]

Article

The Study of Tool Wear Mechanism Considering the Tool–Chip Interface Temperature during Milling of Aluminum Alloy

Xinxin Meng [1], Youxi Lin [2,*], Shaowei Mi [2] and Pengyu Zhang [2]

[1] College of Engineering, Fujian Jiangxia University, Fuzhou 350108, China; mxx@fjjxu.edu.cn
[2] School of Mechanical Engineering and Automation, Fuzhou University, Fuzhou 350108, China; mswfzu123@163.com (S.M.); zpy6239@163.com (P.Z.)
* Correspondence: lyx@fzu.edu.cn

Abstract: ADC12 aluminum alloy has been widely used in the aerospace, ship, and automotive fields because of its high specific strength, excellent die-casting performance, and wear resistance. Adhesion wear is the main wear mechanism of high-speed milling ADC12 aluminum alloy. The most important factor affecting adhesion wear is the tool–chip interface friction, which is directly manifested in the tool–chip interface temperature. Therefore, the temperature variation during the milling of aluminum alloy is analyzed using a temperature field model and infrared temperature measurement technology. Then, the tool wear morphology and the tool wear land width are observed using a scanning electron microscope. Finally, the tool wear mechanism considering the tool–chip interface temperature is discussed. The tool–chip interface temperature is related to the friction angle, tool–chip contact length, and friction force at the rake face, which increases first and then decreases as the cutting speed and feed rate increase. During the formation of the adhesive layer, the tool–chip interface temperature increases, the change rate of the cutting force and the tool wear rate increase, and adhesion, oxidation, and abrasive and delamination wear are generated on the tool surface. With the increase in temperature, the tool wear rate increases, the molten adhesive layer on the tool surface is accompanied by crack propagation, and adhesion wear, oxidation wear, and abrasive wear occur on the tool surface.

Keywords: ADC12 aluminum alloy; high-speed milling; tool–chip interface temperature; adhesion wear

Citation: Meng, X.; Lin, Y.; Mi, S.; Zhang, P. The Study of Tool Wear Mechanism Considering the Tool–Chip Interface Temperature during Milling of Aluminum Alloy. *Lubricants* **2023**, *11*, 471. https://doi.org/10.3390/lubricants11110471

Received: 25 August 2023
Revised: 12 September 2023
Accepted: 21 October 2023
Published: 2 November 2023

1. Introduction

Recently, ADC12 aluminum alloy (YZAlSi11Cu3) has been widely used in engine cylinder bodies and heads because of its low density, good casting performance, wear resistance, and small thermal-expansion coefficient [1]. Tool adhesion wear in the cutting process of aluminum alloy has always been a bottleneck restricting the improvement of machining efficiency and surface quality. The milling process has the characteristics of changeable tool paths and interrupted tool–workpiece engagement, which causes the tool to experience periodically varying cutting forces and local high temperatures at the cutting zone, aggravating tool wear [2]. In order to analyze the tool wear mechanism and monitor the wear process, scholars have tried to establish the tool wear model considering the milling force, spindle box vibration, and cutting torque during milling [3]. In addition, cutting processes with different cutting depths and widths result in different polar plots, so the polar plot of the bending moment during cutting can be used to investigate cutting behavior [4].

During the milling of aluminum–silicon alloys, the adhesion wear of the tool rake face is the main wear mechanism of uncoated carbide cutting tools [5]. Adhesion wear is related to the tool–chip interface temperature, which is determined by the tool materials, coating, tool geometry, and cutting conditions. For aluminum–silicon alloys with different silicon contents, it has been found that high-silicon (30 wt%) aluminum alloy is prone to silicon particle spalling and abrasive wear, while low-silicon (21 wt%) aluminum alloy presents higher hardness and better wear resistance [6]. Rapid tool wear results from a high cutting force and hard abrasion, and damaged machine surfaces are the main problems in machining 70 wt% Si/Al composite. The cutting edge radius has significant effects on the cutting force, surface roughness, and damage formation [7]. Milling–induced surface damage mainly consists of cracks, pits, scratches, matrix coating, and burrs. The wear mechanism has been analyzed during high-speed cutting of aluminum alloy using titanium–substrate ceramic cutting tools with different carbon contents [8]. It has been found that the tool with a lower carbon content has no obvious wear, but the chip easily adheres to the tool–chip interface, and the tool particles peel off. With the increase in carbon content, the tool breakage is more serious. Therefore, abrasive wear and adhesion wear are the main wear mechanisms of high-speed cutting of aluminum alloy with cermet tools.

The coating of a tool can significantly reduce wear, but when the critical temperature is reached, the tool wear rate increases with the increase in cutting temperature [9]. The tool wear mechanism of diamond-coated tools for cutting Al–Si alloy has been studied [10]. The main wear forms of single-layer diamond-coated tools are chipping and cracking. The diamond layer peels off from the base layer, affected by subsurface deformation and crack propagation [11]. Diamond film, composed of multilayer and single-layer structure grains, reduces chip collapse and cracks and has excellent wear characteristics. In addition, diamond-like coated tools and CVD (Chemical Vapour Deposition)–coated tools can also reduce the adhesive aluminum and chip buildup during the high-speed cutting of aluminum–silicon alloys [12]. For titanium-based multilayer MoS_2-coated tools, the MoS_2 transfer layer is formed between the coating and the aluminum alloy as the experimental temperature increases, which reduces the friction coefficient and tool wear rate [13].

Tool damage and chipping are the main wear forms of uncoated tools, while crater wear is the main wear form of TiN–coated tools [14]. In order to ensure surface quality, tools with a smaller rake angle and larger helical angle should be selected to reduce cutting force and tool wear [15]. During the milling of 7075 aluminum matrix composites, it was found that supercritical CO_2–oil–water-based lubricants had outstanding performance in inhibiting adhesion wear and reducing abrasive wear [16]. In order to control the cutting zone temperature, Venugopal et al. [17] studied the tool wear and tool life during machining alloys at different lubricated conditions. It was found that the chip material was attached to the tool at all lubricated conditions. Especially at higher cutting speeds, tool wear is accelerated and tool life is reduced. The local high temperature near the cutting edge and the chemical interaction of tool–workpiece materials are the main causes of tool wear.

When milling aluminum alloy with uncoated carbide tools, the feed per tooth has the greatest influence on tool wear, followed by the cutting speed, cutting width, and cutting depth. Adhesion wear is the main tool-wear mechanism. To ensure machining accuracy, high-speed and small feed rates should be adopted to effectively reduce tool wear and improve tool life [18].

The cutting temperature during milling can be measured by embedding a thermocouple under the workpiece surface or near the tool face [19,20]. The data acquisition system and wireless transmission unit are integrated into the special tool rest, or the data collector is installed on the tool handle to achieve synchronous rotation with the blade [20,21]. Because the method must ensure that the thermocouple wire is accurately embedded in a certain position on the blade, the machining vibration must be considered when the tool rotates. Infrared thermal imaging technology has become the most common method of temperature measurement [22]. With increasing cutting speed, the workpiece temperature gradually increases until the cutting speed reaches the critical speed, then the temperature begins to decrease.

The chip-forming mechanism and tool wear are influenced by the tool–chip interface friction effects. With increasing cutting speed, the tribological behavior of the tool–chip interface changes from sliding to adhesion [23]. Once the adhesion is formed, thermoplastic shear occurs in the second shear zone, which leads to local high temperature at the tool–chip interface. Owing to the diffusion wear mechanism, the tool dissolves with the chip, which accelerates the crater wear. Based on the theory of Shaw [23], the tool–chip interface temperature, considering the effect of crater wear, has been calculated [24]. The maximum depth of the crater wear is far from the tool cutting edge. At high cutting speed, fracture chips are produced because of thermoplastic shear at the primary shear zone, which increases the temperature near the cutting edge and accelerates tool wear.

Cui et al. [25] studied the influence of cutting temperature on the chip morphology and the cutting force during milling. It was found that the higher the chip temperature, the stronger the chip plasticity, then long spiral chips are formed. Considering the flank wear, complex tool geometry, dynamic heat flow, and heat distribution, a new tool temperature prediction model was proposed based on the heat source method [26], and the theoretical model was verified by measuring the temperature of the new tool and the worn tool. It was found that tool temperature and cutting force firstly depend on the feed rate of each tooth, and secondly on the cutting speed. With the increasing feed per tooth, the heat distribution of the tool–chip interface increases. With the increasing cutting speed, the heat distribution of the tool–chip interface decreases.

Most researchers have only analyzed the tool wear mechanism from the macroscopic tool wear morphology. However, few scholars have analyzed the adhesion wear mechanism from the wear morphology changes of microscopic adhesives. In this paper, the change law of tool–chip interface temperature with the cutting parameters is analyzed. The tool wear morphology and the tool wear land width was observed using a scanning electron microscope with the increasing milling length. The tool wear mechanism affected by the tool–chip interface temperature is discussed.

2. The Mathematical Model of Temperature Field

2.1. The Temperature Field Model of Milling

Figure 1 shows the simplification of the milling process. Along the cross-section U perpendicular to the direction z, the milling process can be viewed as an oblique cutting, as shown in Figure 1c. The most important planes for oblique cutting are shear plane, tool rake plane, cutting plane x–y, normal plane x–z, and velocity plane P_v. The cutting mechanism of the normal plane in oblique cutting is equivalent to the orthogonal cutting [27]. The heat source distribution of two-dimensional orthogonal cutting is shown in Figure 2. There are three heat sources, which are shear effect at the first shear zone, friction effect of the tool–chip interface at the second deformation zone, and heat production of the tool–work interface at the third deformation zone.

The instantaneous chip thickness changes periodically with the cutting time during milling, as shown in Figure 3. Based on Altintas theory [28], tangential force $F_t\ (\varphi)$, radial force $F_r\ (\varphi)$, and axial force $F_a\ (\varphi)$ can be regarded as functions of instantaneous uncut chip area (ah) and edge contact length (a), as per Formula (1).

$$
\begin{aligned}
F_t(\varphi) &= K_{tc}ah(\varphi) + K_{te}a \\
F_r(\varphi) &= K_{rc}ah(\varphi) + K_{re}a \\
F_a(\varphi) &= K_{ac}ah(\varphi) + K_{ae}a
\end{aligned}
\tag{1}
$$

where K_{tc}, K_{rc}, and K_{ac} are the cutting force coefficients contributed by the shearing action in tangential, radial, and axial directions, respectively. K_{te}, K_{re}, and K_{ae} are the edge constants. According to Figure 3, the horizontal direction F_x, normal force F_y, and axial cutting force F_z can be calculated, as per Formula (2).

$$
\begin{aligned}
F_x(\varphi) &= -F_t \cos \varphi - F_r \sin \varphi \\
F_y(\varphi) &= +F_t \sin \varphi - F_r \cos \varphi \\
F_z(\varphi) &= +Fa
\end{aligned}
\tag{2}
$$

where c is the feed per tooth and φ is the instantaneous immersion angle.

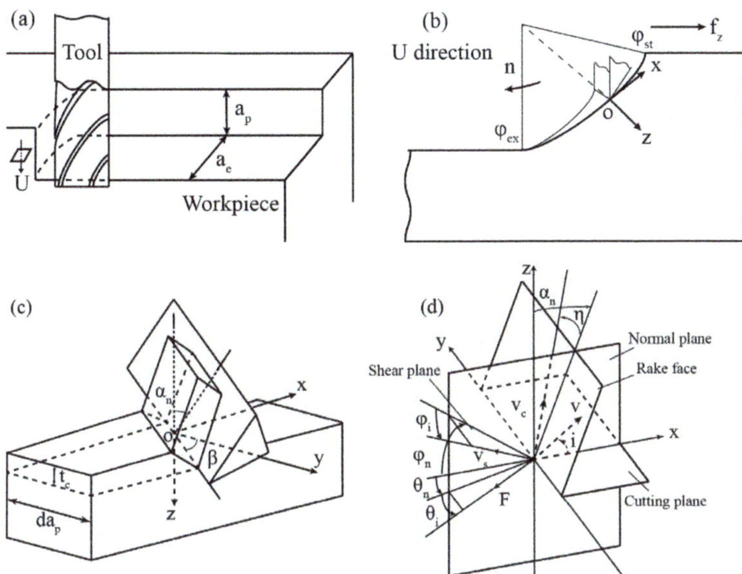

Figure 1. The simplified diagram of the milling process. (**a**) The milling process; (**b**) the geometrical relationship of the selected cross-section; (**c**) the oblique cutting model; (**d**) the geometrical relationship of oblique cutting.

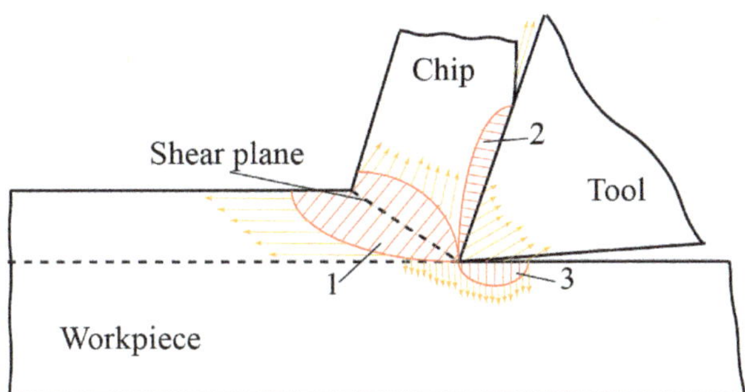

Figure 2. The heat distribution of two-dimensional orthogonal cutting.

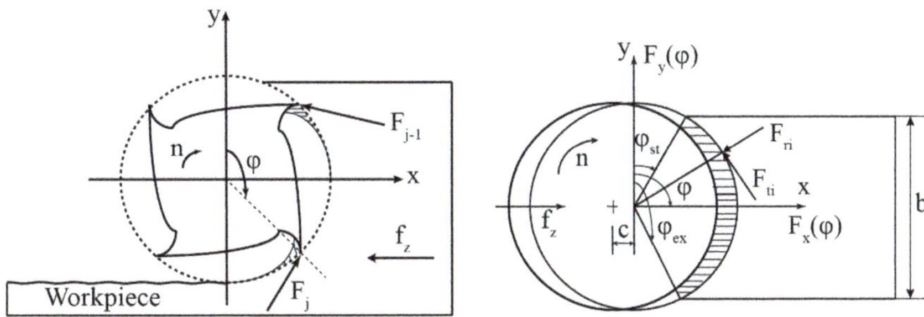

Figure 3. The milling force of the milling process.

Based on the heat source method, the milling heat source can be regarded as a continuous motion spiral linear heat source. The temperature field of continuous motion and heating by finite length helix heat source, namely, the milling temperature field, can be obtained; see Formula (3) [29].

$$\theta = \frac{Rq_s}{c\rho(4\pi at)^{3/2}\sin\beta}\int_0^\tau\int_0^\psi e^{-\frac{(x-vt_i-R\cos\psi)^2+(y-R\sin\psi)^2+(z-R\psi\cot\beta)^2}{4at}}\,d\psi dt \tag{3}$$

2.2. The Tool–Chip Interface Temperature

According to the milling heat transfer model, the temperature rise at any point at any time can be obtained by the workpiece heat source intensity q_w, and the workpiece temperature field can be obtained by adding the initial temperature. Only the shear heat source in the primary shear zone and the tool–chip interface friction heat source in the second deformation zone are considered. Assuming the plastic deformation of the primary shear zone is concentrated on the shear plane, the heat q_t generated by the shear plane per unit time can be obtained. In addition, the heat q_{ts1} generated per unit area per unit time of the shear face is shown in Formula (4), where a_c is the cutting thickness and a_w is the cutting width. According to metal-cutting theory, the heat generated by the shear plane is mainly transmitted to chip and workpiece, and the ratio of chip is set as R_1. Assuming the heat source of the second deformation zone is not considered and the chip temperature leaving the shear plane is uniformly distributed, the chip temperature rise by the first shear zone heat can be obtained. Adding the initial workpiece temperature θ_0, the mean shear plane temperature can be obtained, as per Formula (5). In terms of the workpiece, the workpiece is simplified as a semi-infinite object, and the mean temperature of the shear plane can be calculated according to the formula of the semi-infinite object moving with heat source, as shown in Formula (6). Simultaneously, R_1 can be obtained. By substituting it into Formula (5), the formula of average shear plane temperature can be obtained, as per Formula (8).

$$q_{ts1} = \frac{F_s v_s}{a_c a_w \csc\varphi} \tag{4}$$

$$\bar{\theta}_s = \Delta\theta_s + \theta_0 = \frac{R_1 q_{ts1}}{c_1\rho_1 v \sin\varphi} + \theta_0 \tag{5}$$

$$\bar{\theta}_s = 0.752\frac{(1-R_1)q_{ts1}}{\lambda_1}\sqrt{\frac{a_1 a_c}{v_s \sin\varphi}} + \theta_0 \tag{6}$$

$$R_1 = \frac{1}{1 + 1.33\sqrt{\frac{a_1\varepsilon}{va_c}}} \tag{7}$$

$$\bar{\theta}_s = \Delta\theta_s + \theta_0 = \frac{R_1\tau a_c a_w \cos\alpha_r}{c_1\rho_1 \sin^2\varphi_c \cos(\varphi_c - \alpha_r)} + \theta_0 \tag{8}$$

The second deformation zone can also be regarded as a plane heat source, and the heat generated per unit time q_f can be obtained; thus, the heat production q_{ts2} per unit time per unit area and the tool–chip contact length l_t can be calculated. The heat generated by friction heat source of the second deformation area is mainly transferred into the chip and tool, and the proportion of friction heat source into the chip is set as R_2. The contact area heat source of the tool rake face is connected with the tool and is fixed, but the chip moving speed is v_c. The cutting width is much larger than the cutting thickness during milling. The temperature along the cutting edge changes little. Therefore, the milling temperature field can be regarded as a two-dimensional issue, and the heat source on the tool rake face can be regarded as a moving heat source of a semi-infinite object surface. Then, the average chip temperature rise caused by the heat source on the rake face is shown in Formula (10).

In summary, the average tool–chip interface temperature can be obtained, as shown in Formula (11).

$$q_{ts2} = \frac{F_f v_c}{l_t a_w} \tag{9}$$

$$\overline{\theta}_f = \frac{0.752 R_2 q_{ts2}}{\lambda_2} \sqrt{\frac{a_2 l_f}{v_c}} \tag{10}$$

$$\overline{\theta}_t = \overline{\theta}_f + \overline{\theta}_s = \frac{0.752 R_2 q_{ts2}}{\lambda_2} \sqrt{\frac{a_2 l_f}{v_c}} + \overline{\theta}_s \tag{11}$$

According to the heat source method [30], the heat source of the tool rake face can be regarded as a surface heat source which is stationary and continuously acting. Therefore, from the aspect of the cutting tool, the average tool–chip interface temperature is calculated according to the face heat source temperature rise formula, as shown in Formula (12). By combining with Formula (11), the heat transfer ratio R_2 in the second deformation zone can be calculated, and the average tool–chip interface temperature can be obtained.

$$\overline{\theta}_t = \frac{(1 - R_2) q_{ts2} l_f}{\lambda_t} \overline{A} + \theta_0' \tag{12}$$

$$R_2 = \frac{\frac{F_f v_c \overline{A}}{a_w \lambda_t} - \overline{\theta}_s + \theta_0'}{\frac{F_f v_c \overline{A}}{a_w \lambda_t} + \frac{0.752 F_f}{a_w c_2 \rho_2} \sqrt{\frac{v}{a_2 l_f \xi}}} \tag{13}$$

2.3. Experimental Setup

2.3.1. The Workpiece and Tool Materials

The VMC–850E vertical machining center (SHENYANG SHENYI LATHE MANU-FACTURING CO., LTD., Shenyang, China) was used as the machine tool for the milling experiment, and the spindle speed ranged from 50~8000 r/min. The experimental settings are shown in Figure 4. The size of the workpiece material used in the experiment was 200 mm × 100 mm × 10 mm. The workpiece surface had six through holes connected with the dynamometer, and its chemical composition is shown in Table 1. The tool shank was BT40 series (Xiamen Golden Egret Special Alloy Co., Ltd., Xiamen, China), the tool head was 400R–63–22–4T (Xiamen Golden Egret Special Alloy Co., Ltd., Xiamen, China), the diameter of the tool shank was 63 mm, and it can carry 4 inserts. The inserts used in the milling experiment were APKT1604PDFR–MA H01(Xiamen Golden Egret Special Alloy Co., Ltd., Xiamen, China) series CNC milling tools for aluminum. The insert material was uncoated carbide with a rake angle of 15° and a clearance angle of 11°. Table 2 lists the mechanical properties of the ADC12 aluminum alloy and carbide cutting tools [31]. All experiments were carried out by dry cutting.

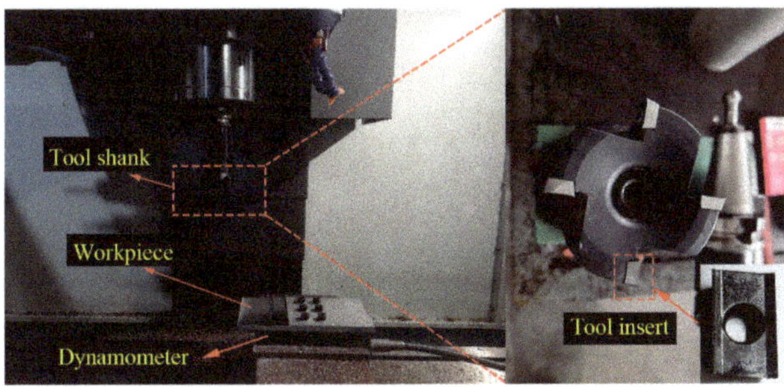

Figure 4. Milling experimental equipment.

Table 1. The chemical composition of ADC12 aluminum alloy (wt%).

Si	Fe	Cu	Mg	Mn	Zn	Ni	Sn	Al
9.6–12	<1.3	1.5–3.5	<0.3	<0.5	<1.0	<0.5	$\leq$0.3	others

Table 2. The mechanical properties of the ADC12 aluminum alloy and cutting tool.

Material Parameter	Workpiece	Tool
Density/kg·m^{-3}	2.67×10^3	15×10^3
Young's modulus/GPa	76	800
Poisson's ratio	0.33	0.2
Specific heat/J·kg^{-1}·K^{-1}	962	200
Thermal conductivity/W·m^{-1}·K^{-1}	92.6	46
Expansion coefficient/K^{-1}	2.06×10^{-5}	4.7×10^{-6}

2.3.2. The Experimental Scheme Design

Since the high-speed cutting of aluminum alloy is defined as a cutting speed greater than 1000 m/min [32], considering the maximum spindle rotation speed limit (8000 r/min), the maximum cutting speed was set at 1200 m/min, and the spindle rotate speed was 6066 r/min. Studying the effect of cutting speed on tool wear behavior, the milling experiments were carried out at the low speed of 300 m/min, medium speed of 600 m/min and 900 m/min, and the high speed of 1200 m/min.

According to recommended cutting parameters of the carbide blade, the feed rate was 0.05 mm/rev. The previous experiment found that the larger vibration and poor workpiece machining quality were generated when the cutting depth was 2 mm. To reduce the machine vibration and improve the surface quality of the workpiece, the cutting depth here was set to 0.5 mm. Since the uncoated carbide blade width was 9.525 mm, the recommended cutting width did not exceed 4 mm, so the cutting width was set to 3 mm. According to the recommended cutting parameters of cemented carbide blades, when analyzing the effect of the feed rate on tool wear behavior, the milling experiments were carried out under the conditions of 0.03 mm/rev, 0.07 mm/rev, 0.09 mm/rev, and 0.11 mm/rev, respectively. The cutting speed was fixed at 900 m/min, the cutting depth was 0.5 mm, and the cutting width was 3 mm. The experimental parameters are shown in Table 3.

Table 3. Milling experiment parameters of ADC12 aluminum alloy.

Cutting Parameters	Value
Cutting speed/(m/min)	300~1200
Feed/(mm/rev)	0.03~0.11
Cutting width/mm	3
Cutting depth/mm	0.5
Milling length/m	0~40

The cutting force measurement system adopted in this milling experiment was the Kistler 9257B piezoelectric three–way dynamometer (Kistler, Switzerland), which was connected to a multi-channel charge amplifier 5070A (Kistler, Switzerland). The cutting force data of the milling process were obtained through DynoWare software (Kistler, Switzerland), and the sampling frequency was set to 1000 Hz. The tool temperature during milling was measured with the FLIR T440 infrared thermal imager, and the emissivity of the uncoated carbide tool was set to 0.35 [33]. Repeated experiments were carried out under each cutting parameter to ensure the measured data accuracy and reduce the experimental error.

3. Results and Discussion

3.1. Tool–Chip Interface Temperature

As the tool cuts in and cuts out the workpiece, the cutting force changes periodically with the instantaneous chip thickness during milling, and the tool is subjected to periodic stress and temperature cycles.

Based on Formula (1), the cutting force can be expressed as a function of the feed per tooth. Then, cutting force coefficients and edge coefficients can be fitted using the milling experiment. Combined with Formula (2), the horizontal, normal, and axial theoretical cutting forces acting on the tool can be calculated. To ensure that the instantaneous uncut chip thickness is equal to the feed per tooth, the immersion angle φ is assumed to be 90°. The theoretical calculated cutting forces (F_{cal}) and the experimental cutting forces (F_{exp}) under different cutting parameters are shown in Figure 5.

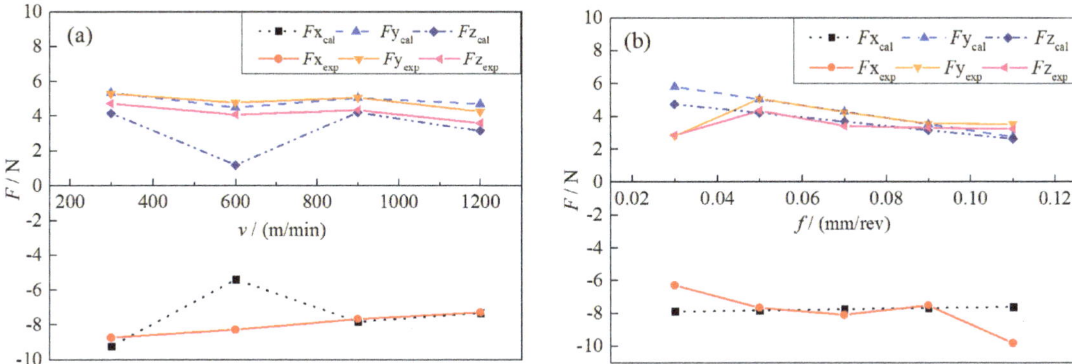

Figure 5. (**a**) The cutting force at different cutting speeds; (**b**) the cutting force at different feed rate.

By comparing the value of three directional cutting forces, it was found that the F_x cutting force was dominant. With increasing cutting speed, the cutting forces F_x, F_y, and F_z showed a decreasing trend. With increasing feed rate, the cutting forces F_y and F_z increased first and then decreased, and reached the maximum value at 0.05 mm/rev. By comparison, it was found that there was a deviation between the experimental value and the theoretical value at the cutting speed of 600 m/min, the feed rate of 0.03 mm/rev, and 0.11 mm/rev, because the cutting force was also affected by cutting temperature, tool spindle vibration, tool wear, and other factors during milling. This section mainly analyzes the changing law of F_x cutting force under different cutting parameters.

According to the experimental cutting force, the heat distribution ratio R_1 and R_2 can be calculated, as shown in Figure 6. According to the formula in Section 2.2, the chip temperature rise and tool temperature rise caused by the heat source in the primary shear zone and the secondary shear zone can be calculated. The average tool–chip interface temperature can be obtained by reverse method according to the measured tool temperature. With increasing feed rate, R_1 gradually increased. R_2 increased first and then decreased, reaching the maximum value at 0.09 mm/rev. With increasing cutting speed, both R_1 and R_2 gradually increased, that is, the proportion of heat generated by the primary shear zone and tool–chip contact zone transferred to the chip gradually increased.

Figure 7a shows that the tool–chip interface temperature first decreases and then rises with increasing cutting speed, and reaches the lowest at 900 m/min. With increasing feed rate, the tool–chip interface temperature also decreases first and then rises, and reaches the lowest at 0.09 mm/rev. The previous study found that the adhesion of tool rake face was not obvious at 900 m/min under different cutting speeds. Under different feed rate, 0.09 mm/rev had the least adhesion on the tool rake face. This indicates that the more serious the adhesion of tool rake face, the higher the temperature at the tool–chip interface.

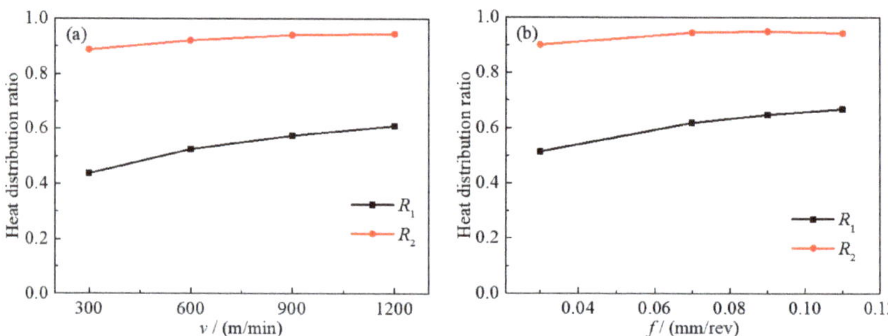

Figure 6. The variation of heat partition coefficient at different cutting parameters: (**a**) cutting speed; (**b**) feed rate.

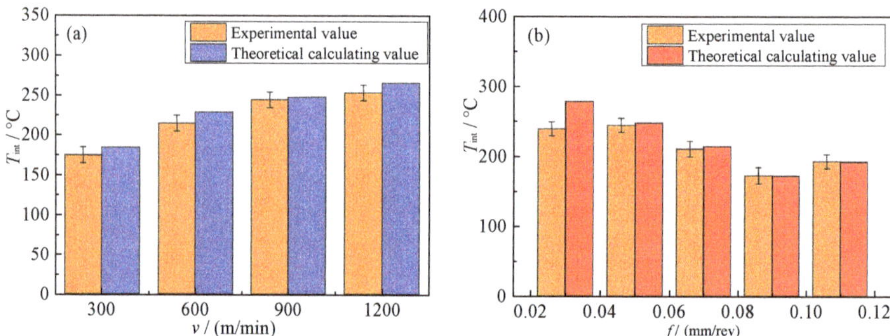

Figure 7. The variation of tool–chip interface temperature by (**a**) cutting speed; (**b**) feed rate.

Figure 8a,b shows the variation of the tool temperature measured by the thermal infrared imager with increasing milling length at different cutting speeds and different feed rates, respectively. Figure 9 shows the variation of the tool–chip interface temperature by milling length. Under different cutting speeds, with increasing milling length, the tool–chip interface temperature increases gradually, and reaches the highest value when the milling length reaches 30 m. As the milling length continues increasing, the tool–chip interface temperature decreases obviously. Under different feed rates, the tool–chip interface temperature increases first and then decreases with increasing milling length, and reaches the highest temperature when the milling length reaches 20 m. As milling length continues increasing, the tool–chip interface temperature decreases gradually.

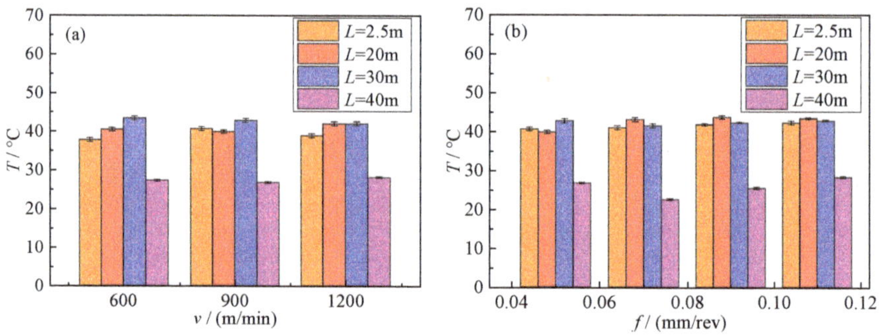

Figure 8. The variation of tool temperature by (**a**) cutting speed; (**b**) feed rate.

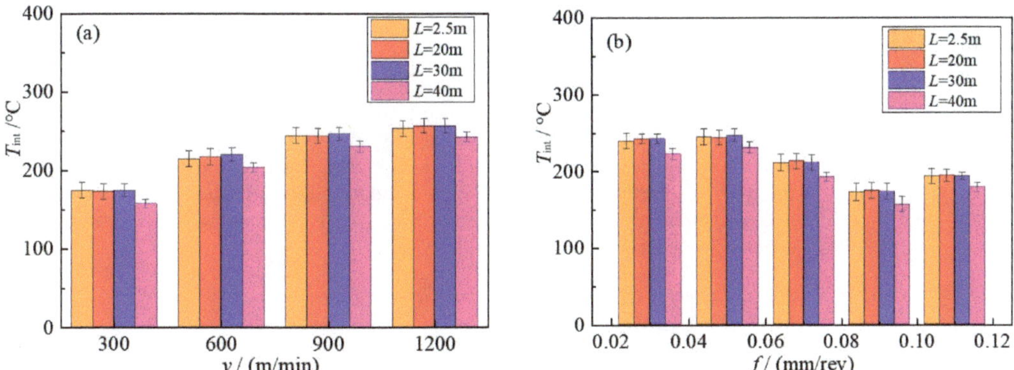

Figure 9. The variation of tool–chip interface temperature at different cutting length by (**a**) cutting speed; (**b**) feed rate.

To analyze the effect of tool–chip interface temperature on the cutting force, the change curve of F_x cutting force with milling length is discussed at different cutting speeds, as shown in Figure 10a–d. In this case, the feed rate is 0.05 mm/rev.

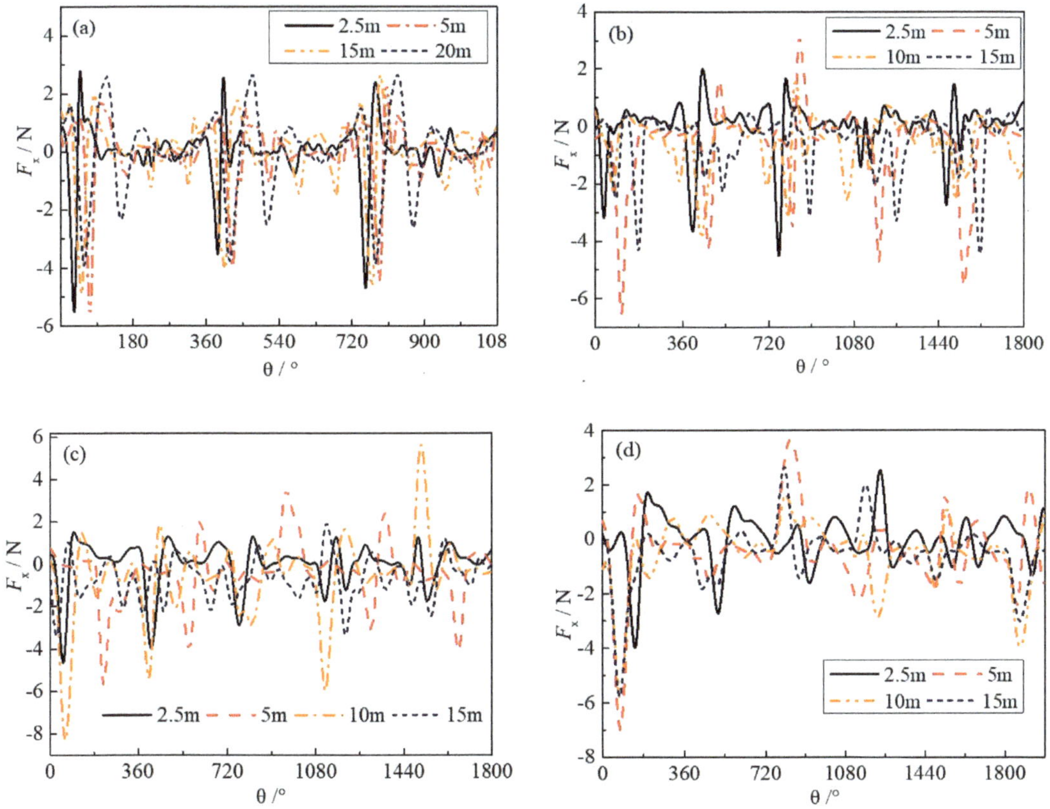

Figure 10. The cutting force F_x at different milling lengths: (**a**) 300 m/min; (**b**) 600 m/min; (**c**) 900 m/min; (**d**) 1200 m/min.

It can be seen that the F_x cutting force increases first and then decreases with increasing milling length under different cutting speeds. Under different cutting speeds, the processing time is longer under low-speed conditions with the same milling length, so the F_x cutting force reaches the maximum value when milling length reaches 2.5 m under low speed ($v = 300$ m/min). With increasing cutting speed, the milling length at which the F_x cutting force reaches its maximum value gradually increases to 5 m and 10 m. However, under high speed ($v = 1200$ m/min), the F_x cutting force reaches its maximum value when the milling length reaches 5 m, which is inconsistent with the above conclusions, so it is necessary to analyze the cutting force and the tool–chip interface temperature considering the tool wear state.

3.2. The Effect of the Tool–Chip Interface Temperature

3.2.1. The Tool Wear Morphology

Figures 11–13 show the tool wear morphology of the rake face with increasing milling length at 600 m/min, 900 m/min, and 1200 m/min, respectively.

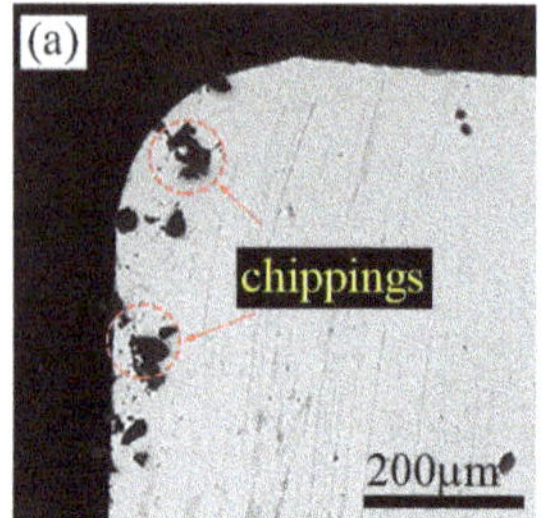

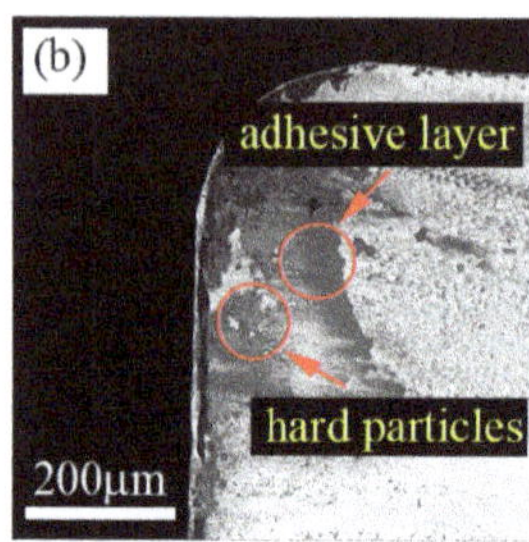

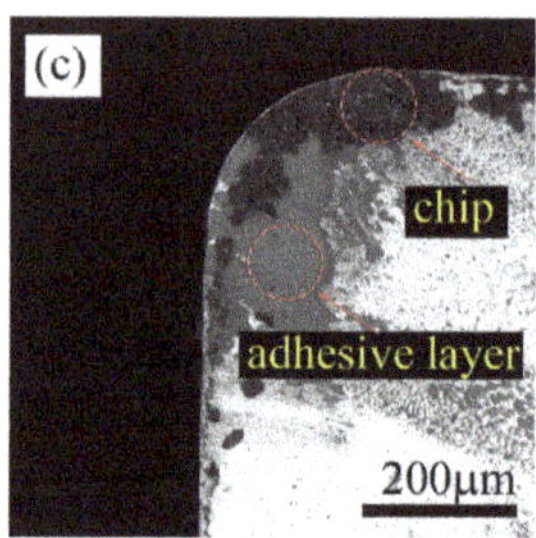

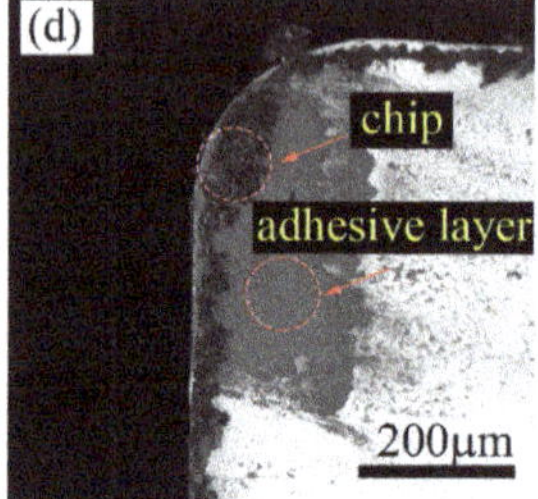

Figure 11. The tool wear morphology of rake face at 600 m/min at (**a**) 2.5 m; (**b**) 20 m; (**c**) 30 m; (**d**) 40 m.

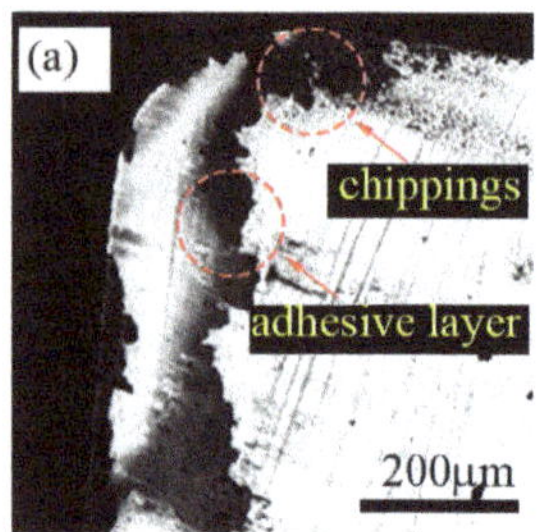

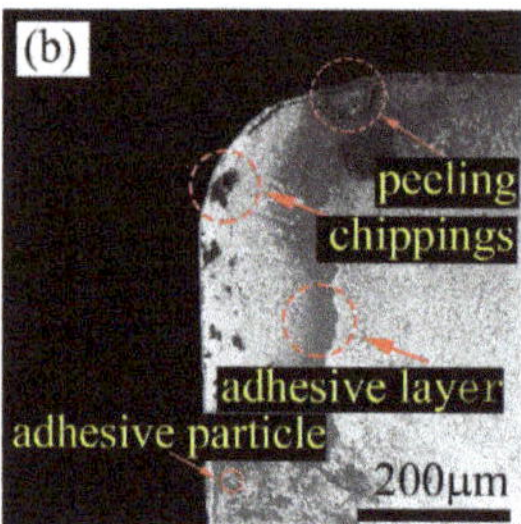

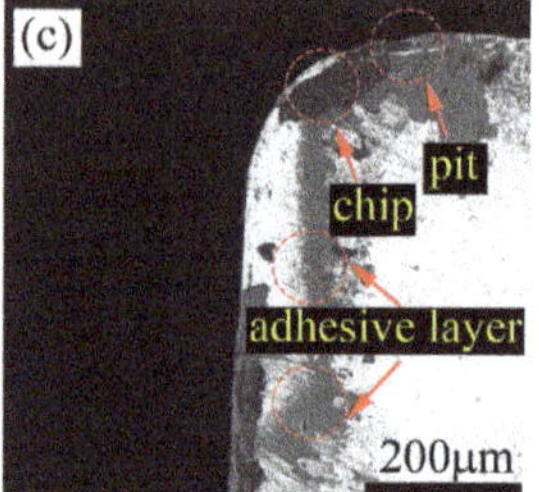

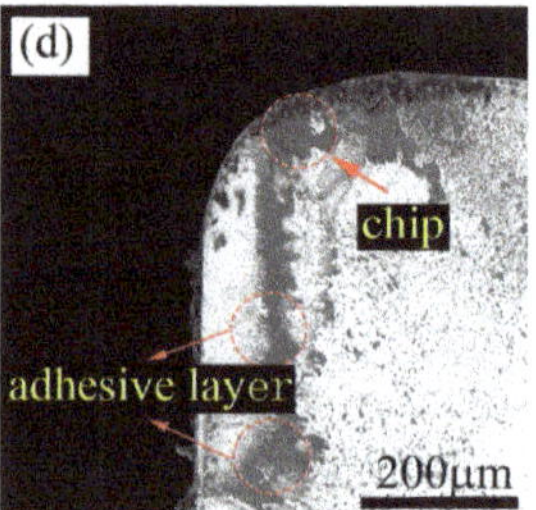

Figure 12. The tool wear morphology of rake face at 900 m/min at (**a**) 2.5 m; (**b**) 20 m; (**c**) 30 m; (**d**) 40 m.

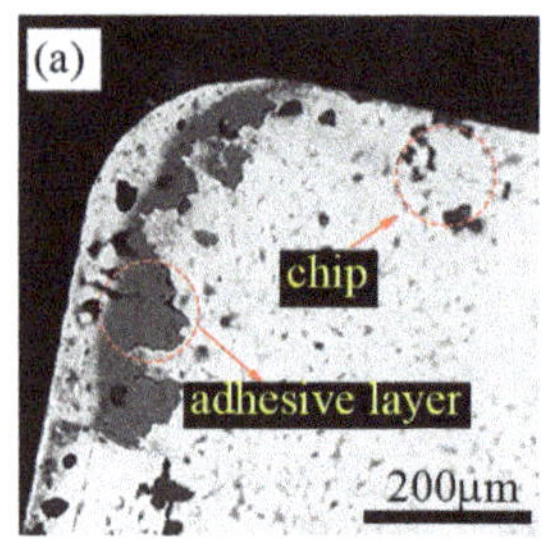

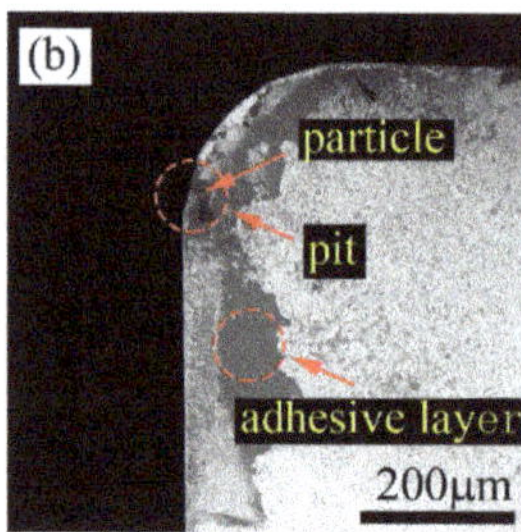

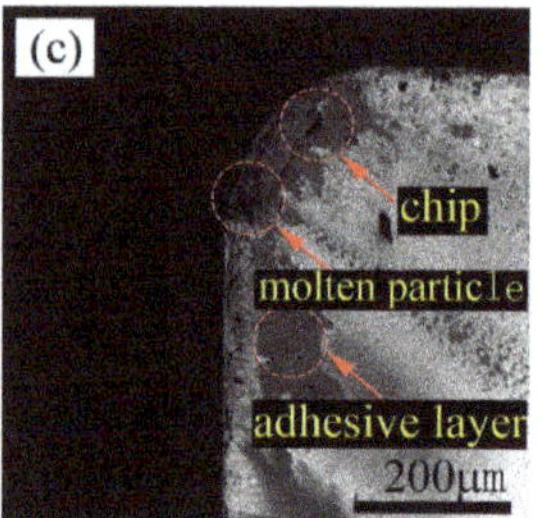

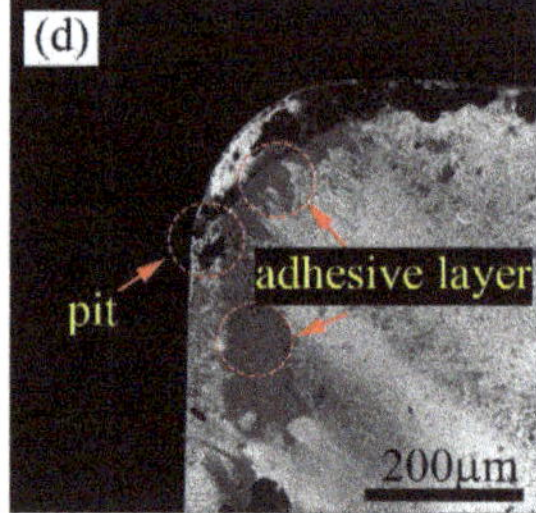

Figure 13. The tool wear morphology of rake face at 1200 m/min at (**a**) 2.5 m; (**b**) 20 m; (**c**) 30 m; (**d**) 40 m.

At 600 m/min, an adhesive layer and hard particles are generated on the tool rake face when the milling length reaches 20 m. The adhesive layer gradually accumulates and the chips are adhesive on the cutting edge when the milling length reaches 30 m. At 900 m/min, the delamination is also generated on the tool rake face in addition to hard particles and adhesive layer when the milling length reaches 20 m. The pit is formed after the tool material falls off when the milling length reaches 30 m. Meanwhile, the adhesive particles transform into adhesive layers. At 1200 m/min, the adhesion layer and the pit wear are generated on the rake face when the milling length reaches 20 m. Adhesion particles appear on the pit. The pit wear of the rake face expands and the adhesion particles at the pits are transformed into the adhesion layer when the milling length reaches 30 m.

To sum up, it is proven that the adhesive morphology change of the tool rake face is a process where the hard particles first transform into the adhesive layer and the adhesion particles coexist, and finally the adhesive layer is stably formed. The adhesion particles and adhesion layer coexist on the tool rake face at different cutting speeds when the milling length reaches 20 m. The adhesion particles basically change into the adhesion layer when the milling length reaches 30 m. Figure 9a shows that the tool–chip interface temperature increases gradually with increasing milling length when the milling length is less than 30 m. This indicates that the tool–chip interface temperature increases gradually during the formation of the adhesive layer. The adhesive layer is formed after the milling length reaches 30 m. After the adhesive layer is stabilized, the tool–chip interface temperature decreases slightly.

Figures 14–16 show the tool wear morphology of the rake face with increasing milling length at 0.07 mm/rev, 0.09 mm/rev, and 0.11 mm/rev, respectively.

At 0.07 mm/rev, the adhesive layer on the tool rake face steadily forms when the milling length reaches 20 m. Considering the change of cutting temperature, the tool–chip interface temperature gradually increases with increasing milling length when the adhesives changes from adhesive particle to adhesive layer. The tool–chip interface temperature gradually decreases with increasing milling length after the adhesive layer is stably formed.

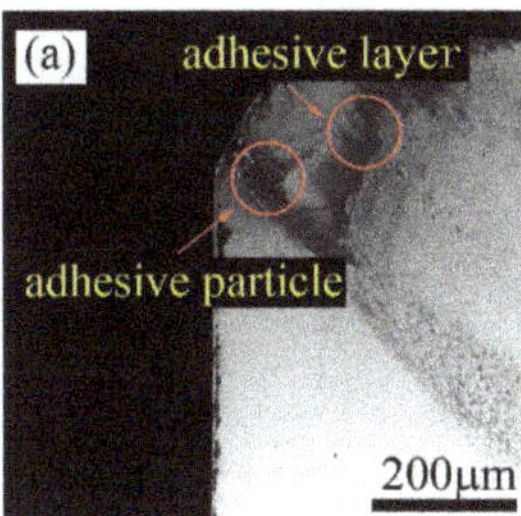

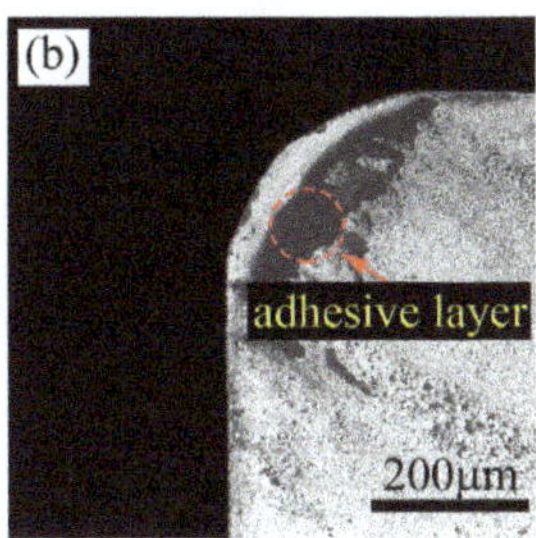

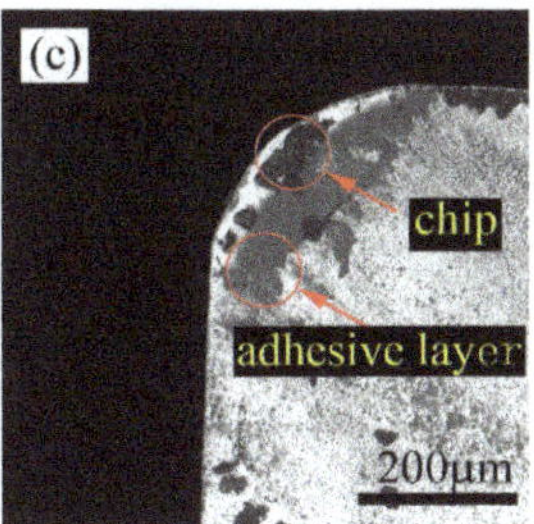

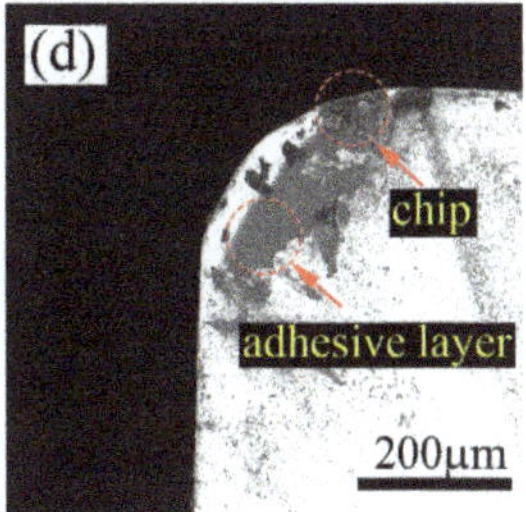

Figure 14. The wear morphology of rake face at 0.07 mm/rev at (**a**) 2.5 m; (**b**) 20 m; (**c**) 30 m; (**d**) 40 m.

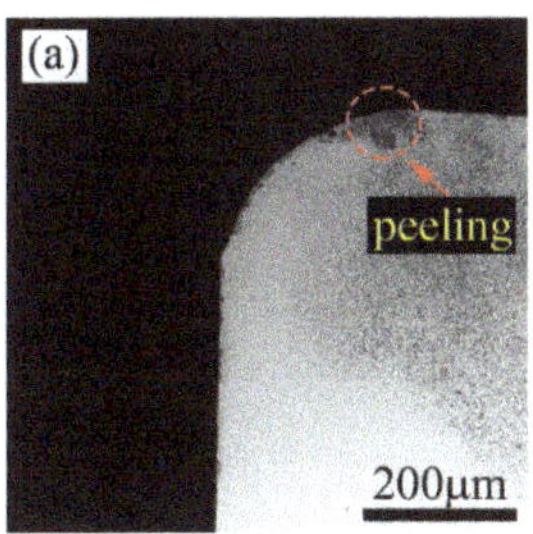

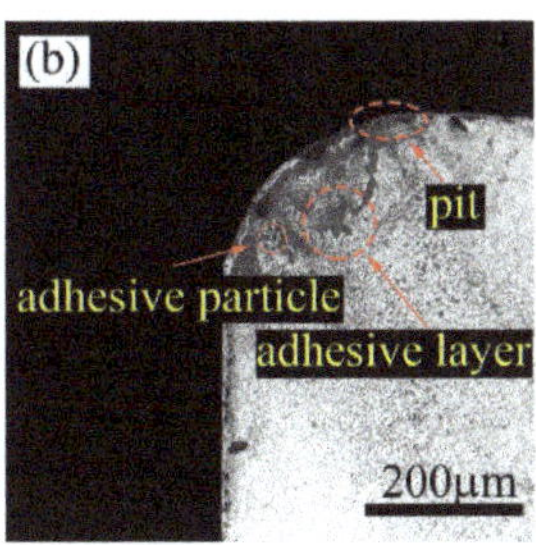

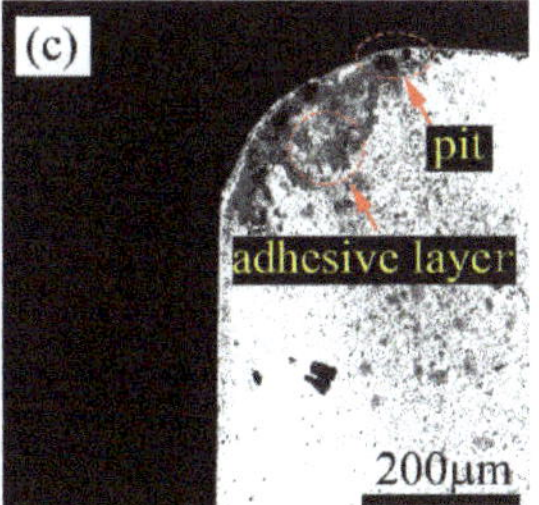

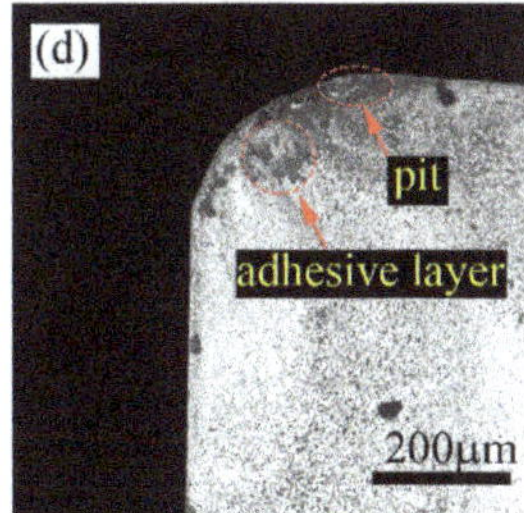

Figure 15. The wear morphology of rake face at 0.09 mm/rev at (**a**) 2.5 m; (**b**) 20 m; (**c**) 30 m; (**d**) 40 m.

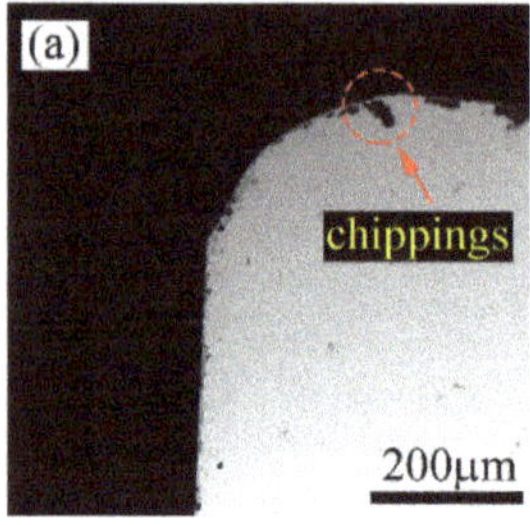
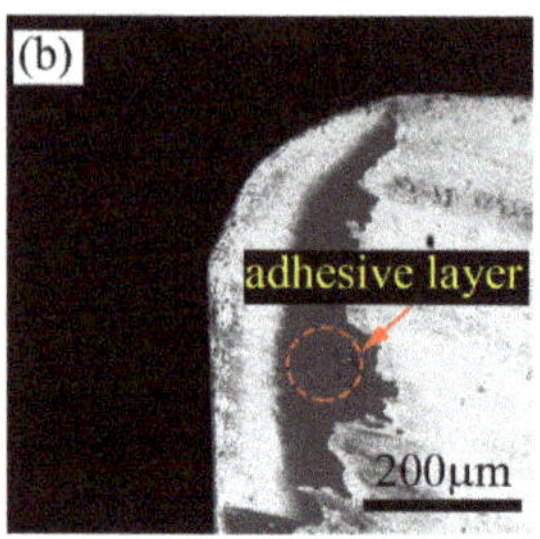
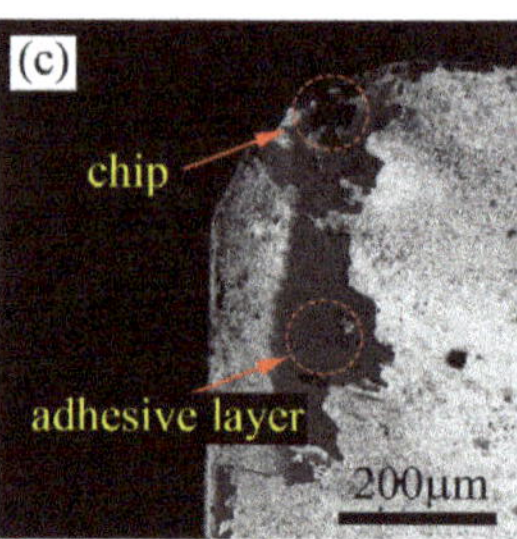
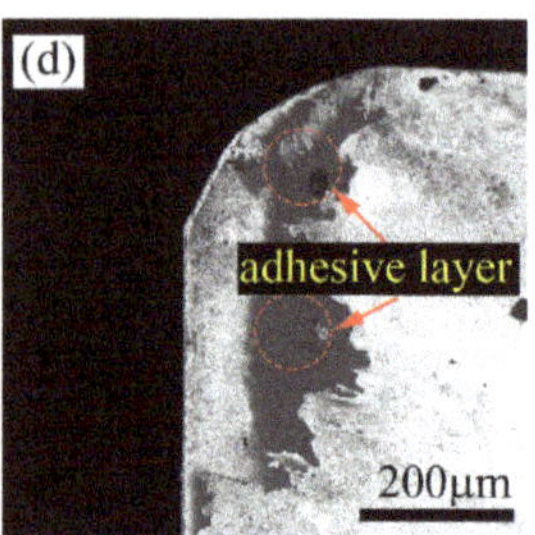

Figure 16. The wear morphology of rake face at 0.11 mm/rev at (**a**) 2.5 m; (**b**) 20 m; (**c**) 30 m; (**d**) 40 m.

At 0.11 mm/rev, the tool wear morphology of the rake face is similar to that of 0.07 mm/rev. An obvious adhesive layer forms on the tool rake face when the milling length reaches 20 m. Then, with increasing milling length, the wear morphology of adhesion layer does not change significantly. Therefore, the tool–chip interface temperature also rises first, and reaches the highest temperature at 20 m, and then decreases gradually. At 0.09 mm/rev, a small pit wear appears on the tool rake face at the beginning of milling. With increasing milling length, the pits gradually expand, the adhesion of the tool rake face is mainly hard particles, and no obvious adhesion layer is formed. It is found that the adhesion is the least obvious at 0.09 mm/rev, and the tool–chip interface temperature is the lowest.

3.2.2. The Effect of the Tool–Chip Interface Temperature on the *VB*

Figure 17a–c shows the relationship between the tool–chip interface temperature and wear land width at different cutting speeds. With increasing milling length, the tool–chip interface temperature firstly increases and then decreases, reaching the maximum value at the milling length of 30 m. With increasing milling length, the tool wear land width increases, but it decreases when the milling length reaches 30 m. The above shows that when the tool–chip interface temperature is high, the tool wear rate is higher. When the tool–chip interface temperature decreases, the tool wear rate decreases and the tool wear tends to be stable.

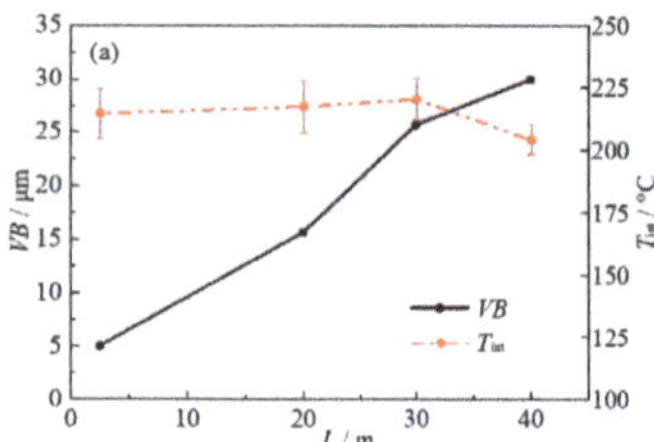
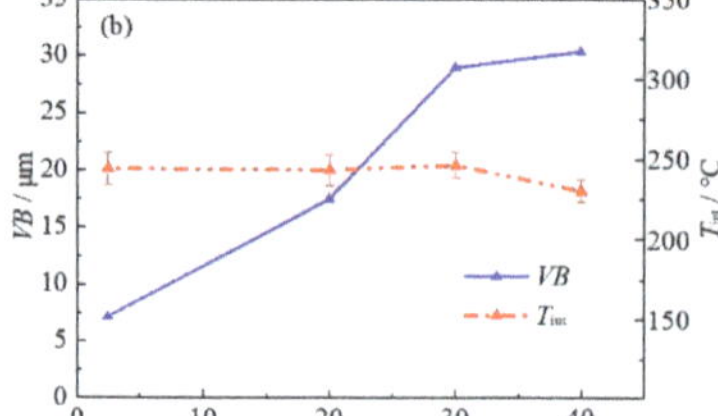
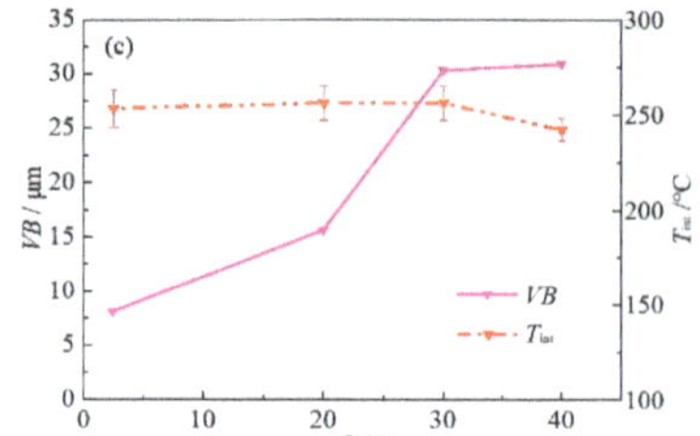

Figure 17. The tool–chip interface temperature and *VB*: (**a**) v = 600 m/min; (**b**) v = 900 m/min; (**c**) v = 1200 m/min.

Combined with the tool wear morphology of the rake face, it can be seen that under different cutting speeds, when the milling length is less than 30 m, with increasing milling length, the tool wear morphology of the rake face is transformed from hard particles to molten particles, the adhesive layer coexists, and is finally completely converted into a stable adhesive layer. During the formation of the adhesive layer, the tool–chip interface temperature gradually increases, and the tool wear rate also gradually increases. When the milling length is larger than 30 m, the adhesion layer is stably generated on the rake face. With increasing milling length, the tool–chip interface temperature decreases slightly, the tool wear rate also decreases slightly, and the tool wear tends to be stable.

Figure 18a–c shows the relationship between the tool–chip interface temperature and tool wear land width at different feed rates. With increasing milling length, the tool–chip interface temperature firstly increases and then decreases, reaching the maximum value at the milling length of 20 m. When the milling length is less than 20 m, the tool wear rate is higher.

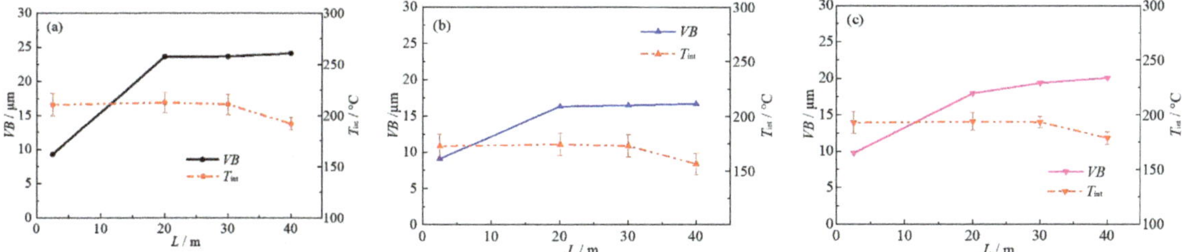

Figure 18. The tool–chip interface temperature and *VB*: (**a**) $f = 0.07$ mm/rev; (**b**) $f = 0.09$ mm/rev; (**c**) $f = 0.11$ mm/rev.

When the milling length reaches 20 m, the tool wear rate decreases and the tool wear increases slowly. This indicates that the tool wear rate is high when the tool–chip interface temperature increases gradually, which decreases when the tool–chip interface temperature tends to be stable or decreases slightly, and the tool wear increases more steadily.

Combined with the tool wear morphology of the rake face, it can be seen that under different feed rates, when the milling length is less than 20 m, with increasing milling length, the tool–chip interface temperature increases. The tool wear morphology of the rake face changes from hard particles to molten adhesive layer and hard particles coexisting, and finally stable formation of the adhesive layer. The tool wear rate is high in this process. When the milling length reaches 20 m, a stable adhesive layer forms on the rake face. With increasing milling length, the tool–chip interface temperature decreases slightly, the tool wear rate decreases, and the tool wear increases slowly.

In summary, during high-speed milling of ADC12 aluminum alloy, due to the friction effect of the tool–chip interface, the hard particles of chips rub against the tool rake face, resulting in the tool–chip interface temperature gradually increasing and the tool wear rate increasing. With increasing milling length, the hard particles melt and gradually form the stable adhesion layer. After that, the tool–chip interface temperature decreases slightly, the tool wear rate decreases, and the tool wear increases steadily.

3.2.3. The Wear Mechanism

The tool wear morphology of the rake face was observed at different milling lengths using a tungsten filament scanning electron microscope. The tool wear mechanism of the rake face can be explained by EDS (Energy Disperse Spectroscopy) analysis. It was found that the tool wear morphology of adhesives at the rake face changed with increasing milling length. Based on this, it is considered that the change in adhesive shape is related to the tool–chip interface temperature during the formation of adhesion wear. At the beginning of milling, the hard particles of the chip adhere to the tool rake face, and the tool–chip interface temperature rises gradually with increasing milling length. The adhesive layer gradually forms after the particles melt. Adhesion wear is formed by the accumulation or shedding of adhesive layers. The details are shown in Figure 19.

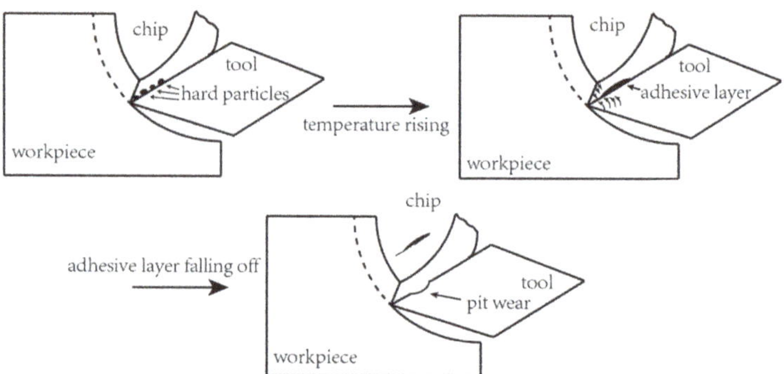

Figure 19. The formation of adhesion wear.

Based on the above analysis, the cutting speed of 1200 m/min and feed rate of 0.11 mm/rev are taken as examples to analyze the formation mechanism of adhesion wear.

At 1200 m/min, the tool wear morphology of the rake face at milling lengths of 20 m and 40 m are shown in Figure 20a,b. It can be seen that the rake face wear morphology consists of a pit wear of the cutting-edge area, A, and a stable adhesion layer on the tool surface, B, at 20 m. The molten hard particles are also attached to pit A, and the nonadhesion area of the pit exposes the tool matrix with pore structure. The formation of the adhesion layer is covered layer by layer. At 40 m, the pit is further enlarged and cracks are generated on the rake face. The thickness of the adhesion layer on the tool surface increases.

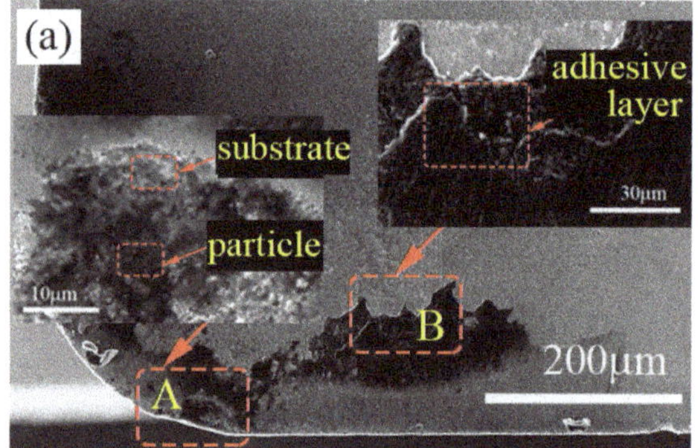

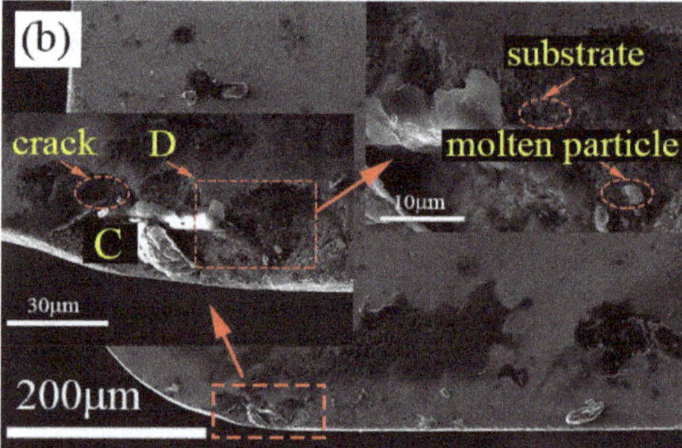

Figure 20. The tool wear morphology of rake face at 1200 m/min: (**a**) L = 20 m; (**b**) L = 40 m.

The EDS energy spectrum results for each point of the rake face are shown in Figure 21a–d. When milling length reaches 20 m, the pit A point is mainly distributed with aluminum and tungsten. The point B of the adhesion layer mainly distributes aluminum, a large amount of oxygen and magnesium, and iron and silicon appear. This indicates that there are mainly peeling wear and adhesion wear at the pit. During the formation of the adhesion layer, there are adhesion, oxidation, and abrasive wear on the rake face. When the milling length reaches 40 m, the line energy spectrum at pit C indicates that there is mainly aluminum. The pit D of the adhesive particles has a large amount of aluminum, oxygen, and magnesium. Therefore, during the expansion of the pit, the hard particles are gradually transformed into the adhesion layer, which then peels off with the milling process. At this time, the rake face wear mechanism is adhesion-peeling, abrasive, and oxidation wear.

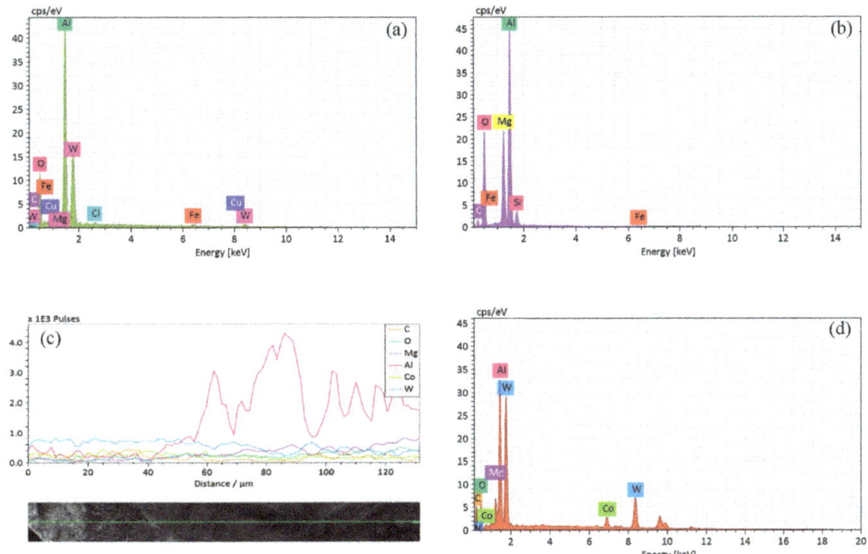

Figure 21. The EDS results of (**a**) point A; (**b**) point B; (**c**) point C; (**d**) point D at 1200 m/min

In summary, under different cutting speeds, when milling length is less than 20 m, the tool–chip interface temperature gradually increases with increasing milling length, the rake surface adhesion particles and the adhesion layer coexist, and the rake face wear mechanism is adhesion, oxidation, and abrasive wear. When milling length is more than 20 m, the tool–chip interface temperature decreases slightly with increasing milling length, and the wear morphology of the rake face is basically stable, mainly as the pit at the cutting edge and stable adhesion layer at the surface, and the rake face wear mechanism is adhesion and peeling wear.

At 0.11 mm/rev, Figure 22a,b shows the wear morphology of the rake face with milling lengths of 20 m and 40 m, respectively. It can be seen that a stable adhesive layer and surface cracks form at the rake face when the milling length reaches 20 m. The thickness of the adhesion layer becomes thicker and the adhesion particles peel off at 40 m. Figure 23a–d explains the EDS spectroscopy results for each point. This shows that aluminum, magnesium, and oxygen are mainly distributed at points A and B, indicating that the rake face wear mechanism is adhesion, oxidation, and abrasive wear at 20 m. There are large amounts of aluminum, magnesium, and oxygen distributed at points C and D, but there are no adhesion particles on the rake face, which indicates that the rake face wear mechanism is adhesion and oxidation wear at 40 m, and no abrasive wear.

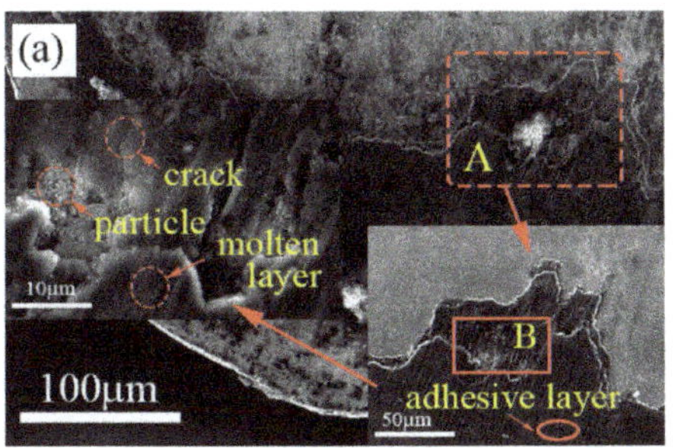

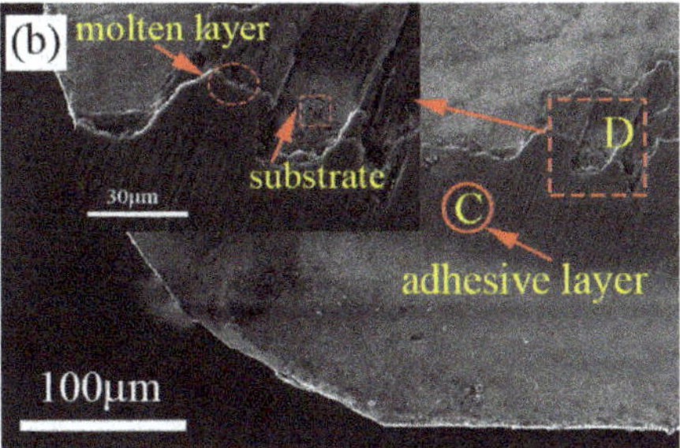

Figure 22. The wear morphology of rake face at 0.11 mm/rev: (**a**) *L* = 20 m; (**b**) *L* = 40 m.

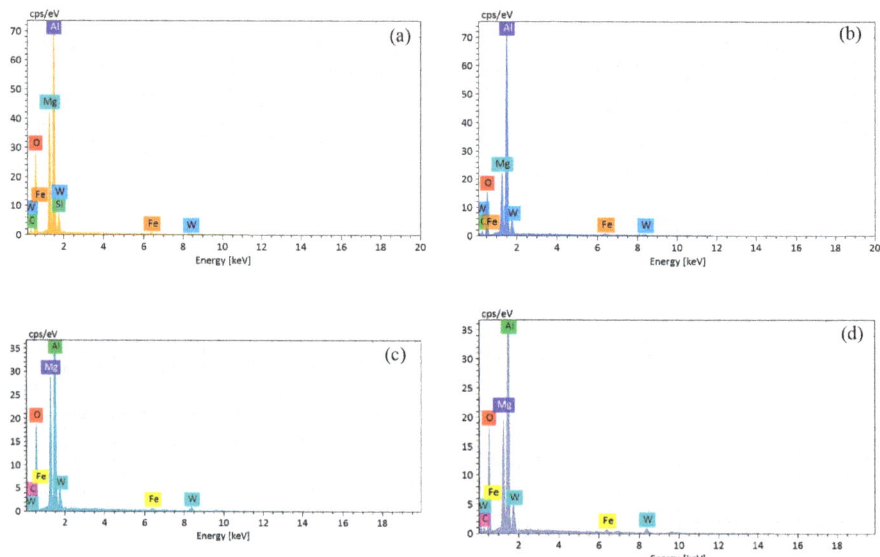

Figure 23. The EDS results of (**a**) point A; (**b**) point B; (**c**) point C; (**d**) point D at 0.11 mm/rev.

In summary, under different feed rates, when the milling length reaches 20 m, the tool–chip interface temperature reaches the maximum value, and the rake face forms a stable adhesion layer and attaches hard particles, resulting in scratches on the rake face, and the wear mechanism of the rake face is adhesion, oxidation, and abrasive wear. When the milling length reaches 40 m, the tool–chip interface temperature decreases, the hard particles of the rake face peel off, and the rake face wear mechanism is adhesion and oxidation wear.

4. Conclusions

Considering the tool–chip interface temperature, the tool wear mechanism of the rake face was analyzed during high-speed milling of ADC12 aluminum alloy in this paper. The cutting temperature was measured under different cutting parameters, and the variation of the tool wear morphology of the rake face and the EDS results were analyzed with increasing milling length. Some conclusions are drawn as follows.

(1) It was found that the tool–chip interface temperature first decreases and then rises with increasing cutting speed, and reaches the lowest at 900 m/min. With increasing feed rate, the tool–chip interface temperature also decreases first and then rises, and reaches the lowest at 0.09 mm/rev. Meanwhile, the tool wear morphology explains that the more serious the adhesion of the tool rake face, the higher the tool–chip interface temperature.

(2) With increasing milling length, the forms of adhesives at the rake face change from hard particles to the coexistence of hard particles and the adhesion layer. The tool–chip interface temperature increases gradually during the formation of the adhesion layer. After that, with increasing milling length, the adhesives gradually stably transform into the adhesion layer. The tool–chip interface temperature decreases gradually with increasing milling length after the adhesion layer forms steadily.

(3) Therefore, the formation mechanism of adhesion wear was discussed. First of all, the hard particles of the chip material adhere to the tool rake face at the beginning of milling. Due to the tool–chip interface frictional effect, the cutting temperature increases gradually. The hard particles melt and transform into the adhesion layer. Owing to the tool–chip interface frictional effect, the delamination appears on the tool rake face, and pits are formed.

Author Contributions: Conceptualization, X.M. and Y.L.; methodology, Y.L.; software, X.M.; validation, X.M.; formal analysis, X.M.; investigation, X.M.; resources, X.M.; data curation, X.M., S.M., and P.Z.; writing—original draft preparation, X.M.; writing—review and editing, X.M.; visualization, X.M.; supervision, Y.L.; project administration, Y.L.; funding acquisition, Y.L. and X.M. All authors have read and agreed to the published version of the manuscript.

Funding: This research was funded by the National Natural Science Foundation of China, grant number 51975123; the funder is Youxi Lin. This research was also funded by the School Research Project of Fujian Jiangxia University, grant number JXZ2022003; the funder is Xinxin Meng.

Data Availability Statement: The data presented in this study are available in the article.

Conflicts of Interest: The authors declare no conflict of interest.

Glossary

Symbol	Meaning	Unit
v	Cutting speed	m/min
f	Feed rate	mm/rev
L	Milling length	m
a_p	Cutting depth	mm
a_e	Cutting width	mm
VB	Wear land width	μm
T_{int}	Tool–chip interface temperature	°C
F	Cutting force	N
φ_c	Shear angle	°
L_t	Tool–chip contact length	mm
v_s	Velocity at shear plane	m/min
v_c	Velocity at rake face	m/min
α_r	Rake angle	°
h	Instantaneous chip thickness	mm
q_t	Heat production at shear face	J
q_f	Heat production at rake face	J
θ	Tool rotation angle	°

References

1. Luo, A.A.; Sachdev, A.K.; Apelian, D. Alloy development and process innovations for light metals casting. *J. Mater. Process. Technol.* **2022**, *306*, 117606. [CrossRef]
2. Xu, J.; Li, L.; Lin, T.; Gupta, M.K.; Chen, M. Machinability analysis in high-speed milling of AlSi7Mg alloys under EMQL conditions: An approach toward sustainable manufacturing. *J. Manuf. Process.* **2022**, *81*, 1005–1017. [CrossRef]
3. Zhang, X.; Gao, Y.; Guo, Z.; Zhang, W.; Yin, J.; Zhao, W. Physical model-based tool wear and breakage monitoring in milling process. *Mech. Syst. Signal Process.* **2023**, *184*, 109641. [CrossRef]
4. Tseng, H.C.; Tsai, M.S.; Yeh, B.C.; Li, K.M. Analysis of Tool Wear by Using a Cutting Bending Moment Model for Milling Processes. *Int. J. Precis. Eng. Manuf.* **2022**, *23*, 943–955. [CrossRef]
5. Meng, X.; Lin, Y.; Mi, S. The Research of Tool Wear Mechanism for High-Speed Milling ADC12 Aluminum Alloy Considering the Cutting Force Effect. *Materials* **2021**, *14*, 1054. [CrossRef] [PubMed]
6. Alshmri, F.; Atkinson, H.V.; Hainsworth, S.V.; Haidon, C.; Lawes, S.D.A. Dry sliding wear of aluminium-high silicon hypereutectic alloys. *Wear* **2014**, *313*, 106–116. [CrossRef]
7. Zhao, G.; Xin, L.; Li, L.; Zhang, Y.; Ning, H.E.; Hansen, H.N. Cutting force model and damage formation mechanism in milling of 70wt% Si/Al composite. *Chin. J. Aeronaut.* **2023**, *36*, 114–128. [CrossRef]
8. Zheng, Z.P.; Lin, N.; Zhao, L.B.; Li, X.; He, Y.H. Fabrication and wear mechanism of Ti(C,N)-based cermets tools with designed microstructures used for machining aluminum alloy. *Vacuum* **2018**, *156*, 30–38. [CrossRef]
9. Oliver, H.; Pete, C.; Martin, J. On the mechanism of tool crater wear during titanium alloy machining. *Wear* **2017**, *374–375*, 15–20.
10. Yoshikawa, H.; Nishiyama, A. CVD diamond coated insert for machining high silicon aluminum alloys. *Diam. Relat. Mater.* **1999**, *8*, 1527–1530. [CrossRef]
11. Rahaman, M.L.; Liangchi, Z. An investigation into the friction and wear mechanisms of aluminium high silicon alloy under contact sliding. *Wear* **2017**, *376–377*, 940–946. [CrossRef]
12. Bhowmick, S.; Banerji, A.; Alpas, A.T. Tribological behavior of Al-6.5%, -12%, -18.5% Si alloys during machining using CVD diamond and DLC coated tools. *Surf. Coat. Technol.* **2015**, *284*, 353–364. [CrossRef]

13. Banerji, A.; Bhowmick, S.; Alpas, A.T. Role of temperature on tribological behaviour of Ti containing MoS_2 coating against aluminum alloys. *Surf. Coat. Technol.* **2017**, *314*, 2–12. [CrossRef]

14. Aslantas, K.; Ucun, I.; Çicek, A. Tool life and wear mechanism of coated and uncoated Al_2O_3/TiCN mixed ceramic tools in turning hardened alloy steel. *Wear* **2012**, *274–275*, 442–451. [CrossRef]

15. Wang, C.Y.; Xie, Y.X.; Qin, Z.; Lin, H.S.; Yuan, Y.H.; Wang, Q.M. Wear and breakage of TiAlN- and TiSiN-coated carbide tools during high-speed milling of hardened steel. *Wear* **2015**, *336–337*, 29–42. [CrossRef]

16. Chen, J.; Yu, W.; Zuo, Z.; Li, Y.; Chen, D.; An, Q.; Chen, M. Tribological properties and tool wear in milling of in-situ TiB2/7075 Al composite under various cryogenic MQL conditions. *Tribol. Int.* **2021**, *160*, 107021. [CrossRef]

17. Venugopal, K.A.; Paul, S.; Chattopadhyay, A.B. Growth of tool wear in turning of Ti-6Al-4V alloy under cryogenic cooling. *Wear* **2007**, *262*, 1071–1078. [CrossRef]

18. Li, X.; Zhou, Y.; Liu, J.; Gao, J. Research on wear mechanism of carbide tool for high-speed cutting aluminum alloy. *Powder Metall. Technol.* **2018**, *36*, 256–260.

19. Rimpault, X.; Il, A.; Chatelain, J.F.; Lalonde, J.F.; Balazinski, M. Workpiece subsurface temperature study during aluminum skin milling in slotting and ramping. *Procedia CIRP* **2018**, *77*, 417–420. [CrossRef]

20. Coz, G.L.; Marinescu, M.; Devillez, A.; Dudzinski, D.; Velnom, L. Measuring temperature of rotating cutting tools: Application to MQL drilling and dry milling of aerospace alloys. *Appl. Therm. Eng.* **2012**, *36*, 434–441. [CrossRef]

21. Karaguzel, U.; Bakkal, M.; Budak, E. Modeling and Measurement of Cutting Temperatures in Milling. *Procedia CIRP* **2016**, *46*, 173–176. [CrossRef]

22. Davoodi, B.; Hosseinzadeh, H. A new method for heat measurement during high speed machining. *Measurement* **2012**, *45*, 2135–2140. [CrossRef]

23. Gekonde, H.O.; Subramanian, S.V. Tribology of tool-chip interface and tool wear mechanisms. *Surf. Coat. Technol.* **2002**, *149*, 151–160. [CrossRef]

24. Shaw, M.C.; Cookson, J.O. Metal Cutting Principles. *Tribol. Int.* **1985**, *18*, 55. [CrossRef]

25. Cui, X.; Guo, J.; Zhao, J.; Yan, Y. Chip temperature and its effects on chip morphology, cutting forces, and surface roughness in high-speed face milling of hardened steel. *Int. J. Adv. Manuf. Technol.* **2015**, *77*, 2209–2219. [CrossRef]

26. Cui, D.; Zhang, D.; Wu, B.; Luo, M. An investigation of tool temperature in end milling considering the flank wear effect. *Int. J. Mech. Sci.* **2017**, *131–132*, 613–624. [CrossRef]

27. Altintas, Y. *Manufacturing Automation: Metal Cutting Mechanics, Machine Tool Vibrations, and CNC Design*, 2nd ed.; Cambridge University Press: Cambridge, MA, USA, 2012.

28. Tsai, M.Y.; Chang, S.Y.; Hung, J.P.; Wang, C. Investigation of milling cutting forces and cutting coefficient for aluminum 6060-T6. *Comput. Electr. Eng.* **2016**, *51*, 320–330. [CrossRef]

29. Wan, M.; Ye, X.Y.; Yang, Y.; Zhang, W.H. Theoretical prediction of machining-induced residual stresses in three-dimensional oblique milling processes. *Int. J. Mech. Sci.* **2017**, *133*, 426–437. [CrossRef]

30. Zhou, Z. *Theory of Metal Cutting*; China Machine Press: Beijing, China, 1992.

31. Bi, J.; Cong, M.; Liu, D.; Xu, X.; Zhao, X. Simulation and experimental analysis of cutting of ADC12 Al-Si alloy. *Modul. Mach. Tool Autom. Manuf. Tech.* **2017**, *1*, 127–130.

32. Schulz, H.; Abele, E.; He, N. *The High Speed Machining-Fundamentals and Applications*; Science Press: Beijing, China, 2010.

33. Quan, Y.; Zhao, J.; Li, Y. Surface emissivity calibration for metal cutting tool and workpiece materials with infrared imager. *J. Mech. Eng.* **2009**, *12*, 188–192. [CrossRef]

 *lubricants*

Article

Simulation of Microscopic Fracture Behavior in Nanocomposite Ceramic Tool Materials

Tingting Zhou [1,2,*], **Lingpeng Meng** [1,2], **Mingdong Yi** [1,2] **and Chonghai Xu** [1,2]

1   School of Mechanical Engineering, Qilu University of Technology (Shandong Academy of Sciences), Jinan 250353, China
2   Shandong Institute of Mechanical Design and Research, Jinan 250031, China
*   Correspondence: zhoutingting506@163.com

Abstract: In this paper, the microstructures of nanocomposite ceramic tool materials are represented through Voronoi tessellation. A cohesive element model is established to perform the crack propagation simulation by introducing cohesive elements with fracture criteria into microstructure models. Both intergranular and transgranular cracking are considered in this work. The influences of nanoparticle size, microstructure type, nanoparticle volume content and interface fracture energy are analyzed, respectively. The simulation results show that the nanoparticles have changed the fracture pattern from intergranular mode in single-phase materials to intergranular–transgranular–mixed mode. It is mainly the nanoparticles along grain boundaries that have an impact on the fracture pattern change in nanocomposite ceramic tool materials. Microstructures with smaller nanoparticles, in which there are more nanoparticles dispersed along matrix grain boundaries, have higher fracture toughness. Microstructures with a nanoparticle volume content of 15% have the most obvious transgranular fracture phenomenon and the highest critical fracture energy release rate. A strong interface is useful for enhancing the fracture toughness of nanocomposite ceramic tool materials.

Keywords: nanocomposite ceramic tool material; microstructure; fracture behavior; Voronoi tessellation; cohesive element

1. Introduction

Nanocomposite ceramic tool materials possess excellent mechanical properties, along with nanoparticles that have great specific surface area, high chemical activity and large dispersivity [1,2]. Fracture is a primary failure mechanism for nanocomposite ceramic tool materials, just like ordinary ceramic materials. Grain boundaries (GBs), nanoparticles and thermal expansion mismatch between matrix grains and nanoparticles may lead to crack deflection, pinning and bridging, which tend to improve the fracture energy of materials [3]. Therefore, investigating the fracture behavior of nanocomposite ceramic tool materials in microscale contributes to the exploration of the relationship between microstructures and macroscopic properties, like fracture strength and toughness.

Due to the heterogeneous microstructures and the mechanical property anisotropies of constituent phases, the microscopic fracture behavior of ceramic tool materials cannot be reflected by means of continuum mechanics, which are suitable for predicting the average response of heterogeneous materials. Therefore, a proper approach is needed to perform the cracking simulation of polycrystalline materials; which does not only characterize microstructural heterogeneities, but also tracks the microscopic damage process, such as crack initiation, propagation and coalescence [4]. The cohesive element model is an approach that could achieve both objectives.

Dugdale [5] conducted the earliest studies on the cohesive element theory and found that cohesive tractions along the fracture process zone were constant for metallic materials under the condition of ideal elastic–plasticity. Xu and Needleman [6] carried out the dynamic crack growth simulation of isotropic elastic solids based on the cohesive element

Citation: Zhou, T.; Meng, L.; Yi, M.; Xu, C. Simulation of Microscopic Fracture Behavior in Nanocomposite Ceramic Tool Materials. *Lubricants* **2023**, *11*, 489. https://doi.org/10.3390/lubricants11110489

Received: 20 September 2023
Revised: 4 November 2023
Accepted: 7 November 2023
Published: 11 November 2023

model they proposed previously. Wang et al. [7] proposed a computational framework for the impact fracture of laminated glass based on the cohesive element model, which was applied to the impact fracture simulation of a laminated glass plate. The simulation results were found to be in agreement with the experimental ones. Geng et al. [8] conducted numerical simulations on two-dimensional standard concrete specimens under axial compression and tension based on the cohesive element theory. Their results indicated that the proposed method exhibited exceptional capabilities in predicting initial stiffness, uniaxial strength and corresponding strain. Xu et al. [9] studied the interface cracking mechanism of calcium-magnesium-alumino-silicates (CMAS)-corroded thermal barrier coatings based on the cohesive element model. Their theoretical model validated the spalling phenomena of thermal barrier coatings caused by CMAS corrosion at high temperatures. Zavattieri and Espinosa [10,11] developed a grain level micromechanical model of brittle materials in the context of cohesive elements to model crack initiation and propagation. Simulations of wave propagation experiments on alumina are used to illustrate the model capabilities in the framework of experimental measurements. Zhou, Tomar and Zhai [4,12] investigated the dynamic fracture process of Al_2O_3/TiB_2 microstructures and discussed the effects of heterogeneous phases on crack paths. Sun et al. [13] investigated the fracture behavior of graphene-toughened Al_2O_3 ceramic materials and characterized the Al_2O_3/graphene interface properties using the cohesive element method. The effects of phase interface bonding strength and graphene content on fracture behavior were analyzed and some toughening mechanisms induced by nano-graphene, such as crack bridging, crack branching, microcracks and pull-out of graphene, were discussed.

This paper mainly deals with microscopic cracking simulations based on nanocomposite ceramic tool materials. Although many studies have been carried out regarding the microscopic cracking behavior of both single phase ceramic tool materials and composite ceramic tool materials reinforced with microsized particles, little attention has been paid to the work on the microscopic fracture mechanisms of nanocomposite ceramic tool materials. It is meaningful for the optimization of mechanical properties to investigate the microscopic fracture behavior of nanocomposite ceramic tool materials, and to analyze the effects of microstructure heterogeneities such as nanoparticle distribution and the nanoparticle size on cracking paths.

In this paper, a micromechanical model of $Al_2O_3/SiCn$ nanocomposite is developed by embedding cohesive elements with fracture criteria into microstructures represented using the Voronoi tessellation. Then crack propagation modeling is conducted in order to explore the effects of microstructural morphologies, like nanoparticle size, microstructure type, nanoparticle volume content and interface fracture energy, on fracture patterns and mechanical responses.

2. The Cohesive Element Model for Microscopic Cracking Simulation

2.1. Traction–Separation Law

In the cohesive element model, it is assumed that crack surfaces carry cohesive tractions that resist normal separation (T_n) and tangential sliding (T_t) before cracking [14]. The relationship between cohesive traction (T) and cracking displacement (δ) can be described using the traction–separation law.

In this work, the linear traction–separation law for brittle materials (Figure 1) is utilized to characterize the properties of cohesive elements in ceramic tool materials. When the cohesive traction increases to a maximum value T_{max}, the damage process begins. As the traction gradually reduces to zero, the cohesive interface will fail completely and reach the maximum displacement (δ_{max}). The area under the traction–separation curve equals the interface fracture energy Γ. It is obvious that the traction–separation law can be determined by two parameters: the maximum interface traction T_{max} and the interface fracture energy Γ. This makes the cohesive element model very attractive for practical applications.

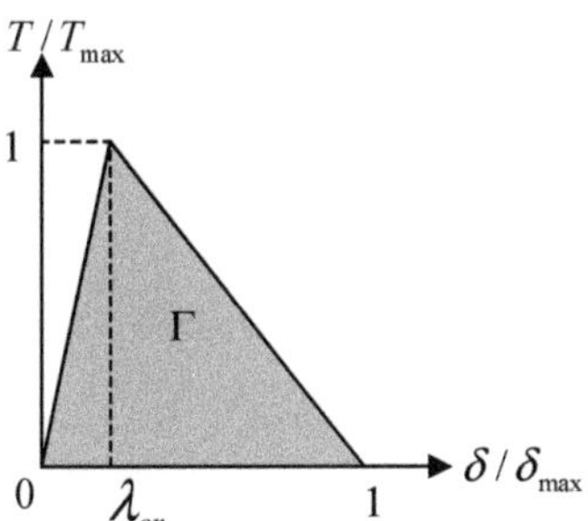

Figure 1. Traction–separation law for brittle fractures.

2.2. Development of the Cohesive Element Model

In this work, all the crack propagation simulation jobs are performed based on the microstructures of ceramic tool materials, which are represented by a Voronoi tessellation. Many studies have shown that the grain level structures of polycrystalline materials can be modeled very well by the Voronoi tessellation [15], and the influences of microstructure heterogeneities and neighboring elements on grain shapes can be reflected by the Voronoi tessellation. This feature gives the Voronoi tessellation the ability to quantitatively describe the characteristics of microstructures, such as the volume content, the size, the shape and the orientation of each composition phase. Due to the high computational efficiency and the simplicity of programming, the Voronoi tessellation has been widely used to characterize microstructures of polycrystalline materials [16–18].

In order to conduct the microscopic fracture simulation, a cohesive element model should be established by embedding cohesive elements with cracking law among adjacent volume elements using the Voronoi tessellation. Therefore, the Voronoi tessellation should be meshed with triangular volume elements first. In contrast with the simulation procedures for single phase ceramic tool materials described in [19], nanoparticles should be generated after the microstructure models are divided by triangular elements. In order to conduct the simulation of intergranular and transgranular cracking in nanocomposites, rectangular cohesive elements with cracking law should be embedded among all triangular volume elements to implement the arbitrary extension of the crack. The process of generating the Voronoi tessellation, the nanoparticles and the cohesive elements is carried out through programing with MATLAB 2020. The simulation procedures are illustrated in Figure 2.

Figure 3a shows the microstructure model represented with the Voronoi tessellation, in which each Voronoi polygon represents a matrix grain. And Figure 3b is the Voronoi diagram after embedding cohesive elements and triangular elements. It is observed from Figure 3b that white triangles indicate volume elements of the matrix and blue triangles denote volume elements of nanoparticles. Cohesive elements in matrix grains and those along matrix GBs are represented by yellow and red quadrilaterals, respectively. Cohesive elements in nanoparticles and along interfaces are characterized by green and black quadrilaterals, respectively. For presentation purposes, the thickness of cohesive elements is magnified in Figure 3b.

The addition of cohesive elements among volume elements has a negative effect on the stiffness of the FEM model. The density of cohesive elements increases as finite elements are refined, which will intensify the reduction of stiffness [12]. However, this issue can be avoided by carefully choosing a lower bound of finite element size and the constitutive parameters of elements. On the other hand, for the purpose of ensuring computational accuracy, there should be several cohesive elements distributed in the cohesive zone, which imposes an upper limit of finite element size. According to the reference [12], in order to ensure simulation accuracy and calculation convergence, the finite element size h takes the form:

$$\left(\frac{\lambda_{cr}\delta_0}{T_{max}}\right)\frac{E'\left(\sqrt{2}+1\right)}{(1-\nu)} \leq h \leq \frac{9\pi E\Gamma}{32(1-\nu^2)T_{max}^2} \tag{1}$$

where $E' = E$ for plane stress and $E' = E/(1-\nu^2)$ for plane strain. E and ν are the elasticity modulus and Poisson's ratio, respectively. $\lambda_{cr} = \delta_0/\delta_{max}$ is the critical effective displacement jump.

Substituting the constitutive parameters of cohesive elements into Equation (1), the range of finite element size is specialized to 0.0016 μm $\leq h \leq$ 2.54 μm. In our work, the finite element size of h = 0.073 μm is adopted, which is reasonable for achieving accurate modeling.

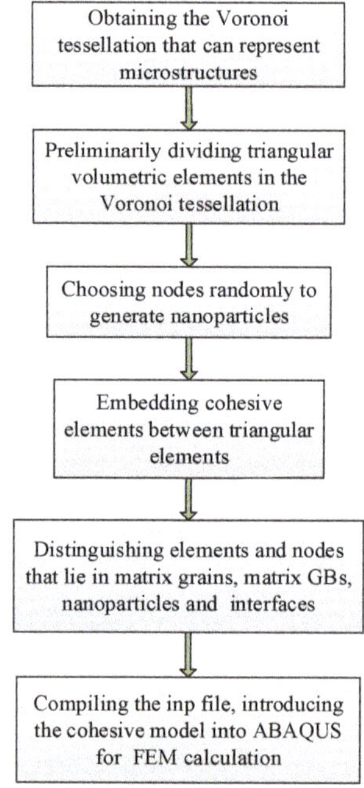

Figure 2. Flowchart of simulation procedures.

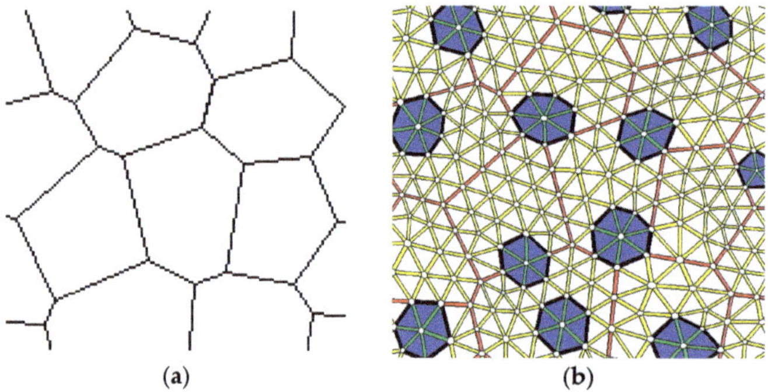

(a)   (b)

Figure 3. Cohesive element model. (**a**) Before the addition of cohesive elements. (**b**) After the addition of cohesive elements.

3. Implementation of FEM Simulation

The microscopic fracture problems depicted in Figure 4 can be solved with the cohesive element model. An initial crack added in the simulation model aims to form the stress concentration, which can form regular crack paths under loading and is convenient for analyzing the influences of microstructures on cracking behaviors. In this work, the initial crack is set along grain boundaries and its length is equivalent to grain size. The effect of the initial crack on simulation results can be ignored, since the length of the initial crack is smaller than 1/10 of the microstructure model's length. The constant velocity load is imposed symmetrically at the top and bottom edges of the model. The corresponding velocity load can be obtained using the expression $v = \dot{\varepsilon} \cdot h$, while the constant strain rate is $\dot{\varepsilon} = 5 \times 10^3$ and h is the microstructure height. The total simulation time is 0.5 μs.

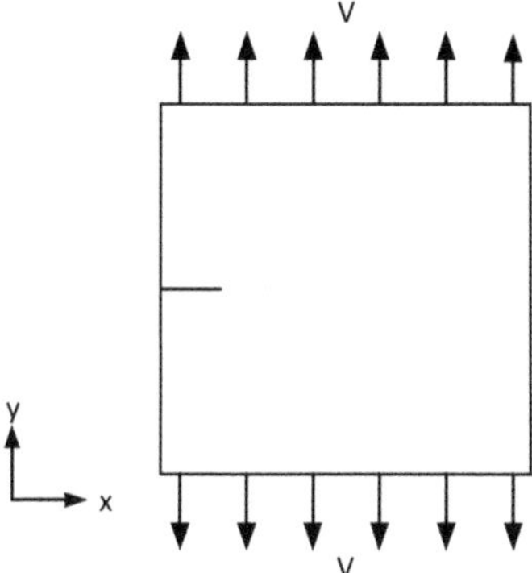

Figure 4. Boundary conditions.

The linear decaying cohesive law (Figure 1) is used to characterize the properties of cohesive elements. In this work, all modeling calculations are based on Al_2O_3/SiC_n nanocomposite ceramic tool materials. The material parameters for various elements are listed in Table 1.

Table 1. Material properties for the microstructure of Al_2O_3/SiC_n.

Element Type	E (GPa)	v	T_{max} (MPa)	Γ (J/m^2)
Triangular elements in Al_2O_3	390	0.23	—	—
Cohesive elements in Al_2O_3	390	0.23	644	2.3
Cohesive elements in GBs	78	0.25	186	1
Triangular elements in SiC	430	0.2	—	—
Cohesive elements in SiC	430	0.2	950	5.8
Cohesive elements in interfaces	250	0.25	712	2.5

4. Simulation Results and Discussion

4.1. Influence of Microstructure Type

In nanocomposite ceramic tool materials, nanoparticle size is under nanoscale and matrix grain size is under microscale or sub-microscale. Therefore, nanoparticles distribute in matrix grains as well as along GBs. According to the distribution of nanoparticles in matrix phases, nanocomposite ceramic tool materials can be divided into three microstructure types, which are intragranular, intergranular and intragranular/intergranular structures. In order to explore the influences of nanoparticle distribution on mechanical properties, models of the intergranular, intragranular and intragranular/intergranular microstructures for Al_2O_3/SiC_n have been established, as shown in Figure 5. The average size of Al_2O_3 matrix grains is 1 μm, while the average size of nano SiC is $d_{nano} = 70$ nm and the volume content of nano SiC is $V_{nano} = 10\%$.

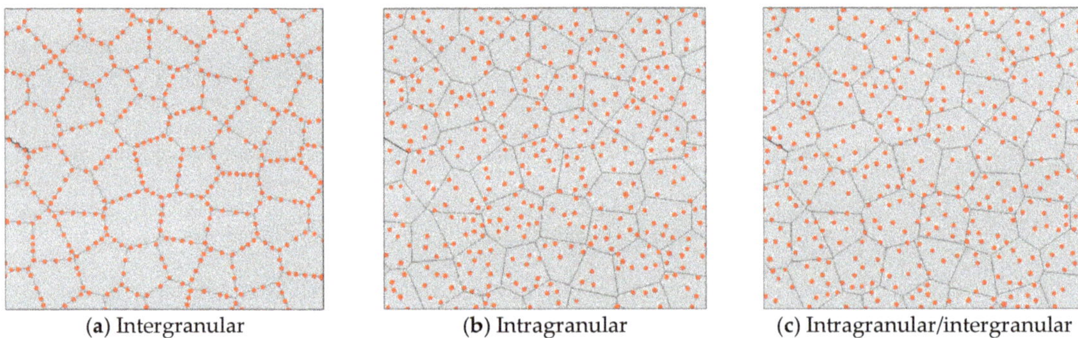

(**a**) Intergranular   (**b**) Intragranular   (**c**) Intragranular/intergranular

Figure 5. Microstructures of different types ($d_{nano} = 70$ nm).

The cracking paths of microstructure models are shown in Figure 6. It can be observed that transgranular cracking is predominant in the intergranular structure (Figure 6a), and the crack is deflected continuously into Al_2O_3 grains by SiC nanoparticles located along matrix GBs. For the intragranular structure (Figure 6b), only intergranular cracking happens. The transgranular fracture in the intragranular/intergranular structure is less remarkable than that in the intergranular structure (Figure 6c). As for Al_2O_3/SiC_n nanocomposite ceramic tool materials, SiC nanoparticles and Al_2O_3 matrix grains usually bond tightly together and no glass phases exist. Many researchers have observed this phenomenon by means of high resolution TEM. Figure 7 displays the interface between the Al_2O_3 matrix and the SiC nanoparticle in Al_2O_3/SiC_n [20]. In addition, S. Jiao [21] concluded that the fracture surface energy of interfaces between Al_2O_3 and SiC in Al_2O_3/SiC_n is higher than that of Al_2O_3 grains at home temperature, according to experimental results and empirical formulae. The high-strength interfaces between nanoparticles and matrix grains have a pinning effect on the intergranular crack. At matrix GBs without nano SiC particles, the intergranular cracking is predominant. When the intergranular crack encounters nano SiC located along GBs, it is liable to be deflected into Al_2O_3 matrix grains, and transgranular cracking occurs. As the crack travels along GBs without nanoparticle distribution, the intergranular cracking continues until it is deflected into matrix grains again by another nanoparticle located at GBs. Figure 8 shows the SEM morphologies of Al_2O_3-based nanocomposite ceramic tool materials, in which typical cleavage steps caused by transgranular fractures are clearly observed, and nanoparticles located at GBs are indicated by red arrows [22]. In the case of the intragranular structure (Figure 6b), the intergranular cracking will not be hindered by nanoparticles located in matrix grains. Therefore, the transgranular fracture hardly arises.

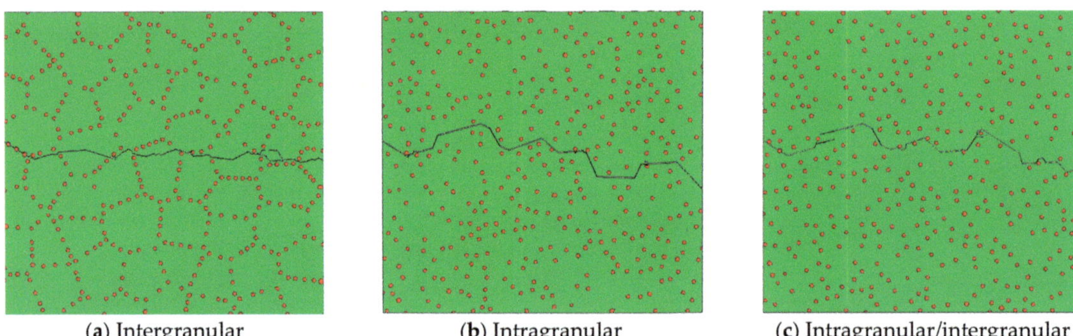

(**a**) Intergranular   (**b**) Intragranular   (**c**) Intragranular/intergranular

Figure 6. Crack growth paths of different microstructure models.

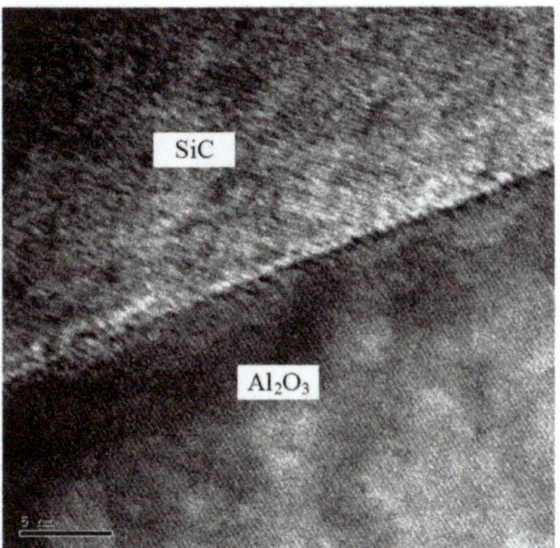

Figure 7. Interface between Al$_2$O$_3$ matrix and SiC nanoparticle [20].

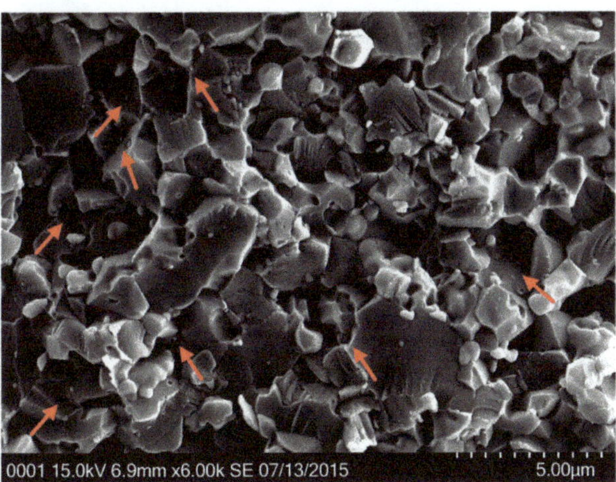

Figure 8. SEM morphologies of Al$_2$O$_3$-based nanocomposite ceramic tool materials [22].

Based on the energy equilibrium theory of cracking systems [23], as a crack propagates for the length of c, the critical fracture energy release rate G_C can be expressed as:

$$\frac{dU_S}{dc} = -\frac{d(U_E - W)}{dc} = G_c \, , \tag{2}$$

where U_S is the fracture energy dissipated during facture, which can be obtained through simulation.

The critical fracture energy release rate G_C is usually regarded as a quantitative description of the fracture resistance of materials. The relationship between fracture toughness K_{IC} and the critical fracture energy release rate G_c is given by:

$$K_{IC} = \begin{cases} \sqrt{G_c E} & Plane\ Stress \\ \sqrt{\frac{G_c E}{1-v^2}} & Plane\ Strain \end{cases}. \tag{3}$$

Figure 11 shows the calculation results of G_C during the fracture process. The G_C of the intergranular structure is higher than that of the intragranular/intergranular structure, and the intragranular structure exhibits the lowest level of G_C. Since the fracture surface energy of matrix grains is above that of matrix GBs (shown in Table 1), the transgranular fracture will consume more energy than the intergranular fracture during the cracking process. Therefore, if taking no consideration of residual stress and merely analyzing the effect of interface bonding strength, nanoparticles located along GBs dominate in the fracture mode transformation from intergranular to transgranular cracking, and in the improvement of fracture toughness for nanocomposite ceramic tool materials.

4.2. Influence of Nanoparticle Size

Figure 9 shows two intergranular microstructure models with different nanoparticle sizes ($d_{nano} = 150$ nm in Figure 9a and $d_{nano} = 70$ nm in Figure 9b). The two models have the same matrix morphologies, with the average matrix grain size $d_{matrix} = 1 \mu m$ and the volume content of nanoparticles $V_{nano} = 10\%$.

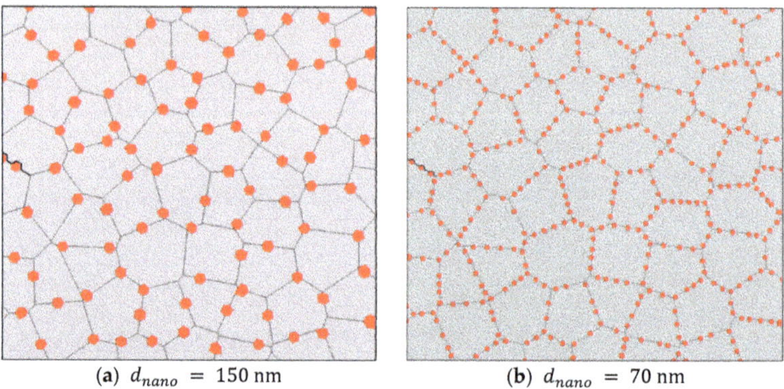

(a) $d_{nano} = 150$ nm   (b) $d_{nano} = 70$ nm

Figure 9. Microstructures with different nanoparticle sizes.

The cracking paths of the two microstructure models after FEM calculation are shown in Figure 10. For the single phase ceramic tool materials, the main fracture mode is basically intergranular cracking due to weak grain boundaries, while transgranular cracking is more liable to occur in metals with strong grain boundaries [24–26]. Compared with the intergranular cracking in single phase ceramic tool materials, the fracture mode of nanocomposite ceramic tool materials has changed to the mixed mode of intergranular and transgranular cracking. And the transgranular cracking is more obvious in the microstructure with smaller nanoparticles ($d_{nano} = 70$nm). Nanoparticles located along GBs represent strong obstacles to crack propagation due to their higher mechanical properties. The intergranular crack tends to be deflected into matrix phases when meeting nanoparticles at GBs.

The results of G_C for the intragranular, intergranular and intragranular/intergranular microstructures with two different sizes of nanoparticles (150 nm and 70 nm) are summarized in Figure 11. It can be found that microstructure models with smaller nanoparticles generally exhibit higher levels of G_C, and the characteristics of G_C resulting from different nanoparticle sizes are most remarkable in the intergranular microstructure. As for nanocomposite with the same content of nanoparticles, the toughening effect of microstructures with smaller nanoparticles is improved. That is because microstructures with smaller nanoparticles have nanoparticles that are more widely dispersed along GBs, which results in more significant effects on crack pinning and deflection into matrix grains. Crack deflection will absorb extra fracture energy, and the transgranular cracking consumes more energy compared with the intergranular cracking during the fracture process.

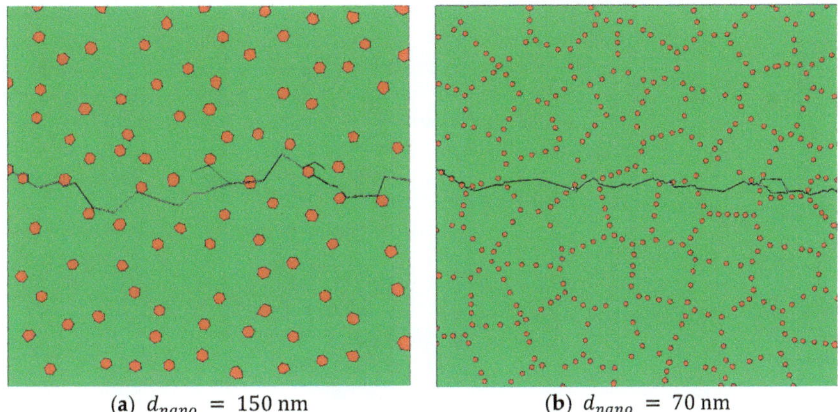

Figure 10. Crack growth paths of microstructures with different nanoparticle sizes.

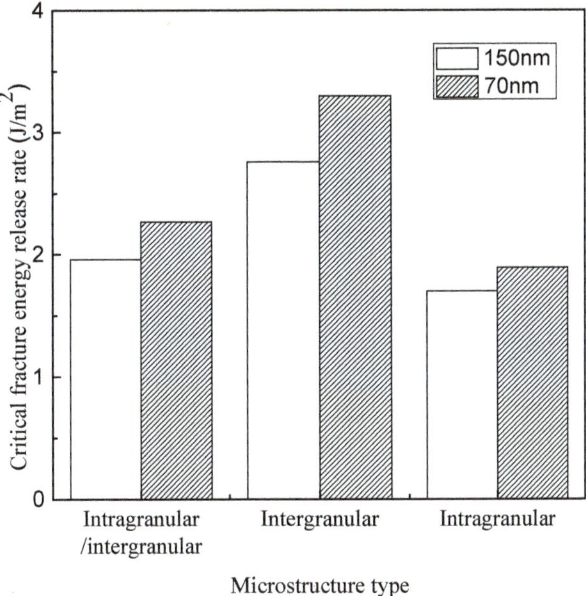

Figure 11. Comparison of G_C for microstructures with different nanoparticle sizes.

In addition, based on the grain growth theory, nanoparticles located along matrix GBs have a pinning influence on the migration of matrix GBs. The relationship between average matrix grain size and nanoparticle size is given as follows [27]:

$$D \propto R/V_f, \tag{4}$$

where D is the average matrix grain size, R is the nanoparticle size and V_f is the volume content of nanoparticles.

Smaller nanoparticles have a stronger pinning effect on matrix GB migration during the sintering procedure, which will lead to a refinement of matrix grains in microstructures. Cheng [28] observed this phenomenon by fabricating nanocomposite ceramic tool materials with different sizes of nanoparticles at the same volume content. It is known that grain refinement contributes to improving the fracture strength of materials, since the length of inherent cracks in ceramic materials is commonly the same as the maximum grain size, and materials containing shorter inherent cracks exhibit higher fracture strength based on the Griffith fracture theory [3,29]. From the above, conclusions can be drawn that smaller nanoparticles are useful for the enhancement of the fracture strength and toughness of nanocomposite ceramic tool materials.

4.3. Influence of Nanoparticle Volume Content

Figure 12 shows microstructure models for nanocomposite with V_{nano} = 3%, 5%, 10%, 15%, 20% and 30%, respectively. The microstructures have the same matrix morphology, with $d_{matrix} = 1$ μm and $d_{nano} = 70$ nm.

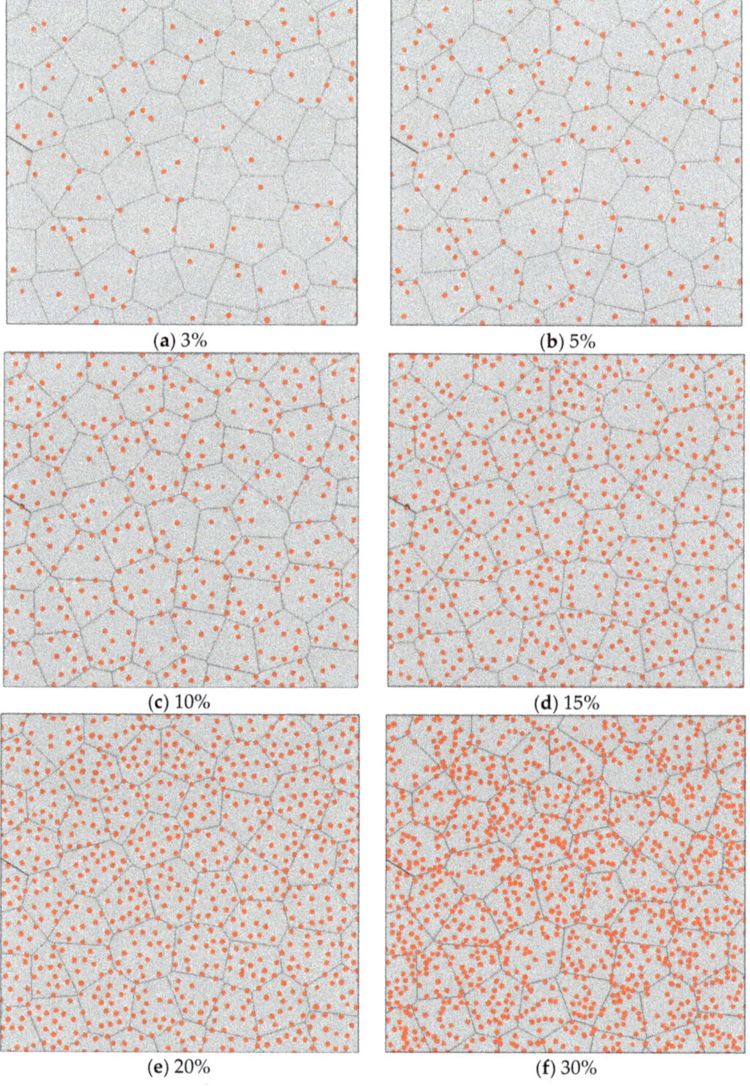

Figure 12. Microstructures with different nanoparticle content.

The cracking paths for the microstructures are shown in Figure 13. When the volume content of nanoparticles (V_{nano}) is lower (3%~10%), the intergranular cracking dominates. The microstructure with 15 vol% nanoparticles exhibits obvious transgranular cracking. As V_{nano} arrives at 20%, nanoparticle agglomeration does not arise and the transgranular cracking appears at some sites of the crack path, although V_{nano} is at a higher level and the distances among nanoparticles are relatively small. Some nanoparticles agglomerate together when V_{nano} reaches 30%. The main crack propagates from some region with weaker properties at the bottom of the microstructure model, rather than from the initial crack. The transgranular cracking still exists at some sites of the crack path, but is at a relatively lower level.

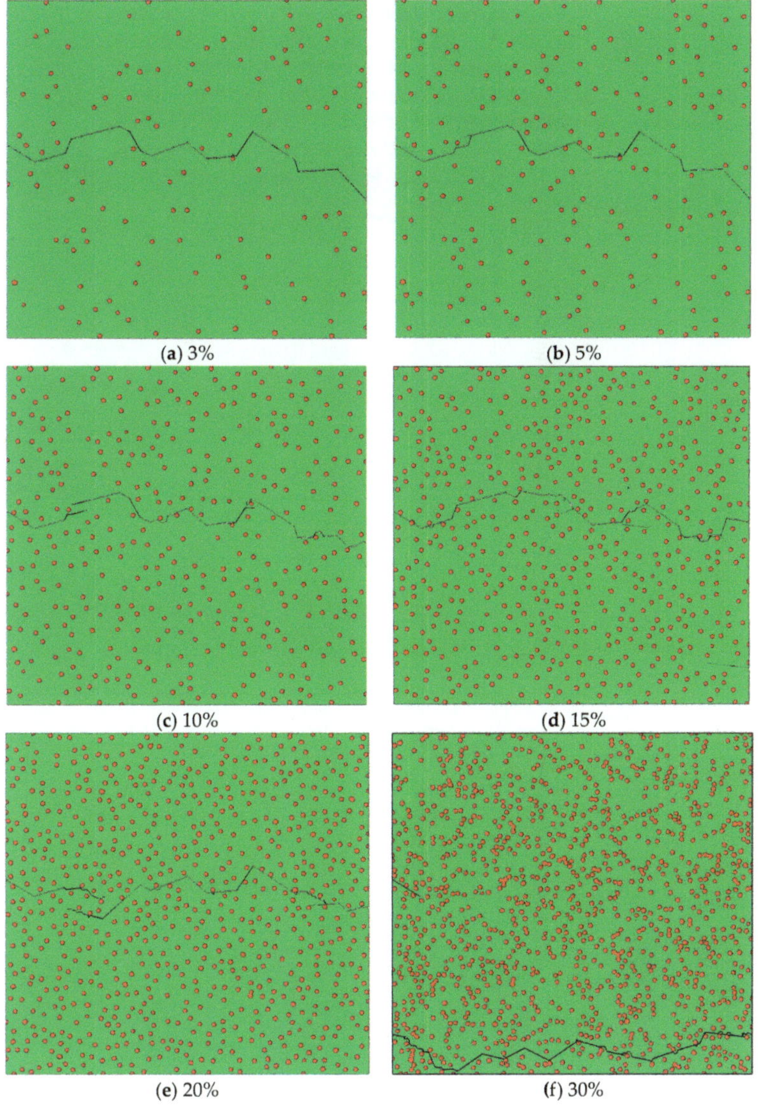

Figure 13. Crack growth paths of microstructures with different nanoparticle content.

The results of G_C are plotted in Figure 14. As depicted in Figure 14, G_C firstly shows an increase, and then a decrease trend as the nanoparticle content increases. G_C reaches the peak at $V_{nano} = 15\%$, which indicates that the fracture resistance is at its highest. Many researchers carried out theoretical and experimental investigations on the effect of the volume content of nano SiC on the fracture toughness. Niihara et al. [30] found that the fracture toughness of Al_2O_3/SiC_n reached the maximum with 5 vol% SiC through experiments, as depicted in Figure 15. Levin et al. [31] investigated the relationship between the fracture toughness of Al_2O_3/SiC_n and the SiC volume content, and obtained similar results to Niihara. Conclusions can be drawn from the modeling results in this paper that the microstructure with 15 vol% SiC has the maximum fracture toughness, which is obviously higher than the results published by other researchers. The primary reason for this difference is that the residual stress caused by the thermal expansion mismatch between nano SiC and matrix Al_2O_3 is not taken into account in our model. As the nanoparticle content is at a relatively higher level, there are more nanoparticles in the matrix grains, and the area of residual tensile stress (the thermal expansion coefficient of the Al_2O_3 matrix is higher than that of the SiC nanoparticle) increases as well. Cracking happens easily under the residual tensile stress. That is why the predicted nano SiC content with the maximum fracture toughness is higher in this work compared with other researchers' results.

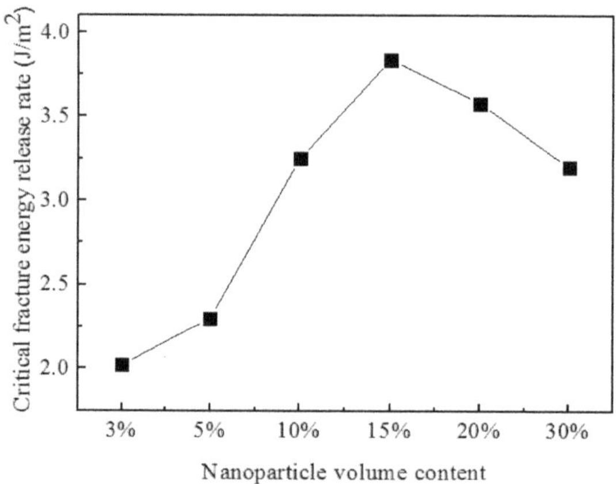

Figure 14. G_C versus nanoparticle volume content (simulation results in this work).

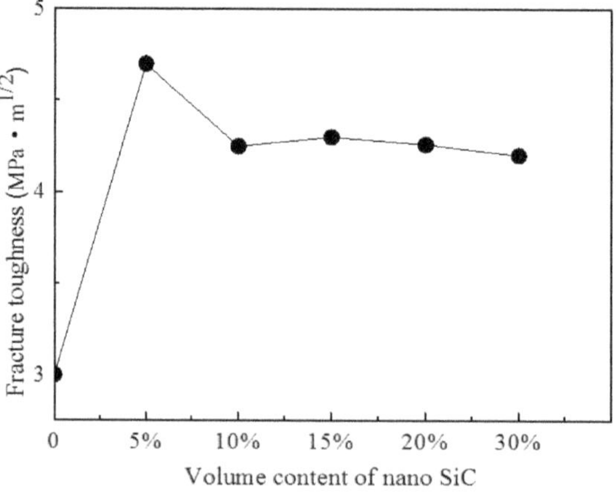

Figure 15. Relationship between the fracture toughness of Al_2O_3/SiC_n and the volume content of SiC (published by Niihara et al.) [24].

The dispersion of nanoparticles in this work is well controlled by programming. For instance, nanoparticles' dispersion remains excellent in the microstructure as nano SiC content is 20 vol% (Figure 13e), while the agglomeration usually occurs as the nano SiC content arrives at 10% and higher in sintering experiments. Compared with the sintering experiments, more excellent nanoparticle dispersion in our work may be another reason why the predicted nanoparticle volume content with optimum toughening effect is relatively higher. Under the condition of excellent nanoparticle dispersion, fracture toughness increases as the nanoparticle content rises. All of the above analysis leads up to the conclusion that excellent dispersion procedures contribute to improving the fracture toughness of materials.

4.4. Influence of Interface Fracture Energy

In nanocomposite ceramic tool materials, there exist not only GBs among matrix grains, but also interfaces between nanoparticles and matrix grains. Interface bonding strength, matrix GB strength and matrix grain strength will affect the macroscopic strength and fracture toughness of materials through their effects on cracking paths. Interface bonding strength between matrix grains and nanoparticles can vary from different processing methods and conditions, and many investigations have shown that it has a significant effect on the fracture mode. Due to the small grain size and complicated microstructure morphologies, it is very difficult to measure the interface bonding strength by means of experiments. In this section, the influence of the interface bonding strength between nanoparticles and matrix grains on cracking paths and fracture toughness will be discussed.

In actual ceramic tool materials, the strength of nanoparticles is usually higher than that of matrix grains in order to obtain a good toughening effect. Therefore, only the influence of the fracture energy proportions among the interface, matrix grain and matrix GB on the fracture mode is taken into account in this work. Three levels of the proportions of the fracture energy of the interface Γ_{inf}, the matrix grain Γ_{grain} and the matrix GB Γ_{gb} ($\Gamma_{inf} : \Gamma_{grain} : \Gamma_{gb} = 3 : 2 : 1$, $\Gamma_{inf} : \Gamma_{grain} : \Gamma_{gb} = 2 : 3 : 1$ and $\Gamma_{inf} : \Gamma_{grain} : \Gamma_{gb} = 1 : 3 : 2$) are considered. These values indicate strong, intermediate and weak interfaces, respectively. Figure 16 represents three different microstructure models, which are the intergranular, intragranular/intergranular and intragranular structures. The average size of the matrix grain and the average nanoparticle size are 1 µm and 150 nm, respectively. The volume content of nanoparticles is equal to 10%.

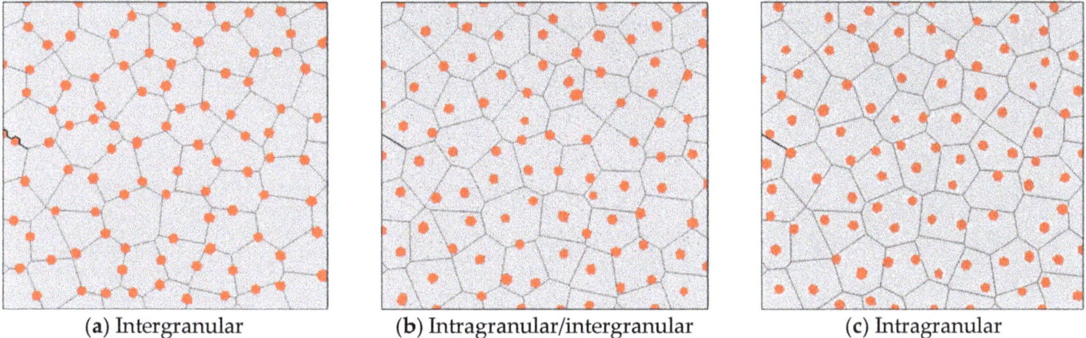

(**a**) Intergranular   (**b**) Intragranular/intergranular   (**c**) Intragranular

Figure 16. Different types of microstructure models ($d_{nano} = 150$ nm).

Figure 17 shows the cracking paths of all microstructures with different fracture energy proportions. It can be seen from Figure 17(a1) that transgranular cracking occurs in the intergranular microstructures with strong interfaces ($\Gamma_{inf} : \Gamma_{grain} : \Gamma_{gb} = 3 : 2 : 1$). Since the interface fracture energy is higher than that of the matrix grain, the intergranular crack is liable to extend into the matrix when encountering nanoparticles located at GBs. Comparing Figure 17(a1) with Figure 18 ($\Gamma_{inf} : \Gamma_{grain} : \Gamma_{gb} = 4 : 2 : 1$), it can be concluded that the higher the interface bonding strength, the stronger the pinning and deflecting effect of nanoparticles at GBs on the intergranular crack, and the more remarkable the transgranular fracture phenomenon. In Figure 17(b1), where the interfaces are stronger than GBs but weaker than matrix grains ($\Gamma_{inf} : \Gamma_{grain} : \Gamma_{gb} = 2 : 3 : 1$), the intergranular crack tends to surround nanoparticles as it encounters those located at GBs, and continue travelling along GBs when meeting matrix GBs. Similar fracture patterns appear in the intergranular microstructures with weaker interfaces (as shown in Figure 17(c1) with $\Gamma_{inf} : \Gamma_{grain} : \Gamma_{gb} = 1 : 3 : 2$).

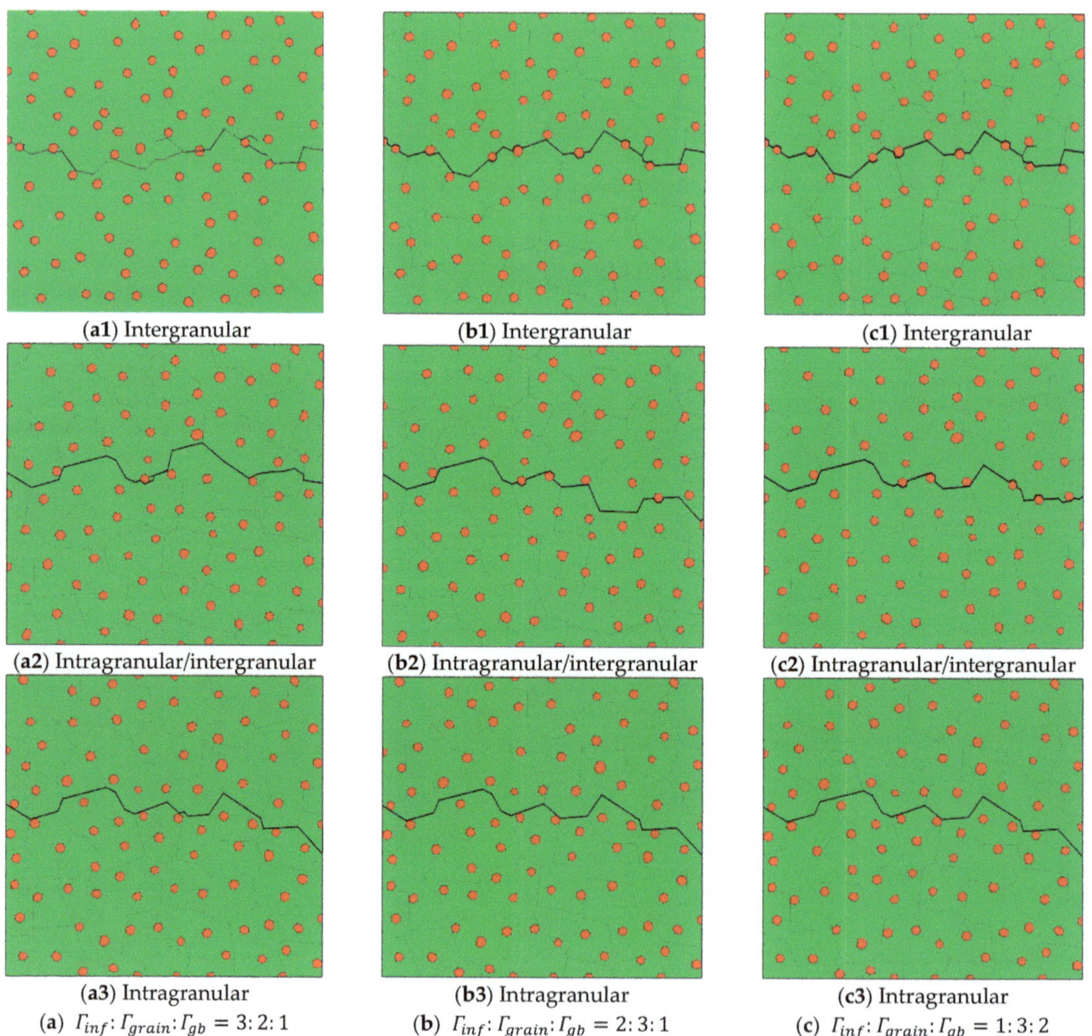

(**a1**) Intergranular   (**b1**) Intergranular   (**c1**) Intergranular

(**a2**) Intragranular/intergranular   (**b2**) Intragranular/intergranular   (**c2**) Intragranular/intergranular

(**a3**) Intragranular   (**b3**) Intragranular   (**c3**) Intragranular

(**a**) $\Gamma_{inf} : \Gamma_{grain} : \Gamma_{gb} = 3 : 2 : 1$ (**b**) $\Gamma_{inf} : \Gamma_{grain} : \Gamma_{gb} = 2 : 3 : 1$ (**c**) $\Gamma_{inf} : \Gamma_{grain} : \Gamma_{gb} = 1 : 3 : 2$

Figure 17. Crack paths in different types of microstructures with different fracture energy proportions.

Figure 18. Crack path in the intergranular microstructure with $\Gamma_{inf} : \Gamma_{grain} : \Gamma_{gb} = 4 : 2 : 1$.

As for the intragranular/intergranular microstructures, transgranular cracking still exists in the model with strong interfaces (Figure 17(a2)). Due to relatively fewer nanoparticles located at GBs in the intragranular/intergranular microstructure, the transgranular fracture phenomenon is less pronounced compared with that in the intergranular microstructure with the same material parameters. In microstructures with intermediate $\Gamma_{inf} : \Gamma_{grain} : \Gamma_{gb} = 2 : 3 : 1$ and $\Gamma_{inf} : \Gamma_{grain} : \Gamma_{gb} = 1 : 3 : 2$ (Figure 17(b2,c2)), where the interfaces are weaker than matrix grains, cracks are liable to propagate around nanoparticles lying at GBs if they encounter them. And the intergranular fracture continues once the crack meets the matrix GB.

In the case of the intragranular microstructures, the intergranular cracking is predominant for the three fracture energy proportions (Figure 17(a3–c3)), which indicates that nanoparticles distributed in matrix grains have little pinning and deflecting effect on the intergranular crack.

Figure 19 shows the results of G_C for microstructures with different interface strengths. It can be found that for the same microstructure type, materials with strong interfaces ($\Gamma_{inf} : \Gamma_{grain} : \Gamma_{gb} = 3 : 2 : 1$) exhibit the highest level of G_C, as well as the highest fracture toughness. This is in good agreement with the simulation results, that the most remarkable transgranular fracture can be observed in microstructures with $\Gamma_{inf} : \Gamma_{grain} : \Gamma_{gb} = 3 : 2 : 1$. In models with weak interfaces ($\Gamma_{inf} : \Gamma_{grain} : \Gamma_{gb} = 1 : 3 : 2$), although a few transgranular cracks arise, G_C is the lowest due to the low interface fracture energy. From the above, strong interfaces ($\Gamma_{inf} : \Gamma_{grain} : \Gamma_{gb} = 3 : 2 : 1$) seem to have a positive effect on the fracture toughness of materials, while weak interfaces ($\Gamma_{inf} : \Gamma_{grain} : \Gamma_{gb} = 1 : 3 : 2$) may be disadvantageous for material toughening. The results of G_C are in accordance with the results of cracking paths presented in Figure 17. The results of G_C in the intergranular models are relatively higher than other microstructures with the same interface bonding strength. Compared with other microstructure types, G_C in intragranular microstructures is at the lowest level no matter how strong the interface bonding effect is, since nanoparticles in this matrix have negligible impact on the transition of the intergranular cracking to the mixed mode of intergranular and transgranular cracking. Actually, nanocomposite ceramic materials prepared using the sintering method primarily possess the intragranular/intergranular microstructure. Thus, in actual nanocomposite ceramic tool materials, strong interfaces have a significant effect on the fracture mode transition to the mixed mode of intergranular and transgranular cracking, which is advantageous for the enhancement of fracture toughness of materials.

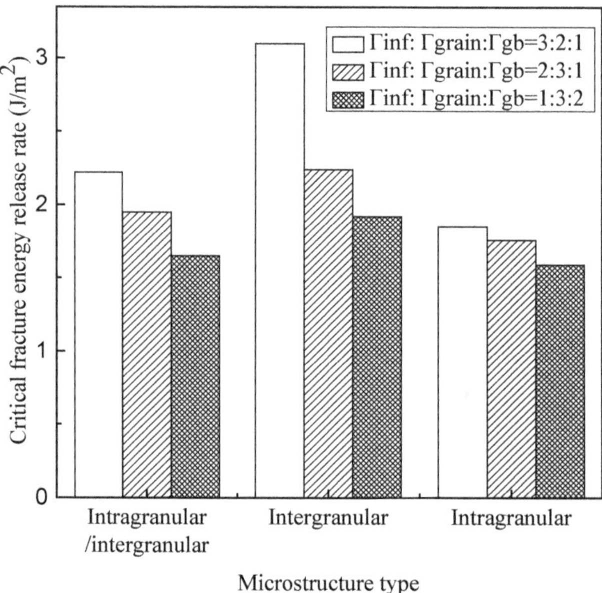

Figure 19. G_C in different types of microstructures with different fracture energy proportions.

Yu [32] modeled the cracking extension in Al_2O_3/TiB_2 composite ceramic tool materials and investigated the relation between Al_2O_3/TiB_2 interface bonding strength and cracking behavior. He found that high interface bonding strength (higher than that of matrix grains) would result in the transition of the fracture mode from purely intergranular cracking to a mix of intergranular and transgranular cracking, as well as the enhancement of the fracture toughness of materials. Zhai [4] studied the influences of interface bonding strength on the cracking behavior of Al_2O_3/TiB_2 composite ceramic materials and reached the conclusion that strong interfaces contribute to enhancing the critical energy release rate of microstructures. These results agreed well with our conclusion.

It should be noted that the conclusion drawn from the numerical results—that weak interfaces are disadvantageous for enhancing the fracture toughness of materials—is based on the premise of the simulation model, in which the interfaces between second phases and matrix phases are all weakly bonded and an intergranular main crack is easily formed under an applied load. However, things may be different in materials with weak-interface-bonding particles independently distributed through the strong interface network. On one hand, weak interfaces are liable to fracture to form microcrack zones under an applied load, which can facilitate microcrack toughening. On the other hand, strong interfaces are able to prevent the coalescence of the main crack and microcracks, and the material strength will not be reduced [33,34]. This is the toughening mechanism for the weak-interface-toughening materials, which is not dealt with in this paper.

5. Conclusions

The macroscopic mechanical properties of ceramic tool materials are determined by microstructures. In order to reduce the brittleness and enhance the fracture toughness of materials, it is necessary to carry out a simulation in microscale and investigate the strengthening and toughening mechanisms in essence. In this work, the cohesive element model of nanocomposite ceramic tool materials is built up by embedding cohesive elements with fracture criteria into microstructures represented with the Voronoi tessellation. Then, the microscopic fracture behavior simulation is carried out to explore the relationship between microstructural morphologies and material responses. This research provides some meaningful investigation results for the design and development of nanocomposite ceramic tool materials, which aims to optimize the tools' fracture toughness. The effect of microstructure type on the cracking patterns and critical energy release rate G_C is analyzed. It shows that the existence of nanoparticles changes the fracture mode from the intergranular cracking of single phase ceramic tool materials into mixed intergranular/transgranular cracking. Nanoparticles located at matrix GBs play a predominant role in the fracture mode transition of nanocomposite, while nanoparticles located in matrix grains have little effect on the intergranular cracking. The influence of nanoparticle size on microscopic fracture behavior is also discussed. Under the condition of the same nanoparticle content, the microstructure with smaller nanoparticles possesses more remarkable transgranular cracking and higher fracture resistance. The relationship between G_C and nanoparticle content is also examined. Finally, the influence of interface bonding strength between nanoparticles and matrix grains on cracking patterns and fracture toughness is investigated, resulting in the conclusion that strong interfaces contribute to improving the fracture toughness of nanocomposite ceramic tool materials.

Author Contributions: Conceptualization, methodology, software, writing—original draft preparation, T.Z.; investigation, L.M.; validation, M.Y.; writing—review and editing, C.X. All authors have read and agreed to the published version of the manuscript.

Funding: This research was funded by the National Natural Science Foundation of China grant number [U22A20201] and [51405253].

Conflicts of Interest: The authors declare no conflict of interest.

References

1. Karch, J.; Birringer, R.; Gleiter, H. Ceramics ductile at low temperature. *Nature* **1987**, *330*, 556–558. [CrossRef]
2. Nakahira, A.; Niihara, K.; Hirai, T. Microstructure and mechanical properties of Al_2O_3-SiC composites. *J. Ceram. Soc. Jpn.* **1986**, *94*, 767. [CrossRef]
3. Gong, J. *Fracture Mechanics of Ceramic Material*; Press of Tsinghua University: Beijing, China, 2001; pp. 234–235.
4. Zhai, J.; Tomar, V.; Zhou, M. Micromechanical simulation of dynamic fracture using the cohesive finite element method. *J. Eng. Mater. Technol.* **2004**, *126*, 179–191. [CrossRef]
5. Dugdale, D. Yielding of Steel Sheets Containing Slits. *J. Mech. Phys. Solids* **1960**, *8*, 100–104. [CrossRef]
6. Xu, X.P.; Needleman, A. Numerical simulations of fast crack growth in brittle solids. *J. Mech. Phys. Solids* **1994**, *42*, 1397–1434. [CrossRef]
7. Wang, D.; Chen, S.; Xu, W.; Zang, M.; Yoshimur, S. A computational framework for impact fracture analysis of laminated glass: An extrinsic cohesive shell approach. *Comput. Struct.* **2020**, *233*, 106238. [CrossRef]
8. Geng, J.; Liu, C. A methodology for parameter identification and calibration of the cohesive element based meso-scale concrete model. *Constr. Build. Mater.* **2023**, *393*, 132075. [CrossRef]
9. Xu, G.N.; Yang, L.; Zhou, Y.C. Investigating the interface cracking mechanism of CMAS-corroded thermal barrier coatings based on the cohesive zone model. *Corros. Sci.* **2022**, *204*, 110337. [CrossRef]
10. Zavattieri, P.D.; Espinosa, H.D. Grain level analysis of crack initiation and propagation in brittle materials. *Acta Mater.* **2001**, *49*, 4291–4311. [CrossRef]
11. Espinosa, H.D.; Zavattieri, P.D. A grain level model for the study of failure initiation and evolution in polycrystalline brittle materials. Part II: Numerical examples. *Mech. Mater.* **2003**, *35*, 365–394. [CrossRef]
12. Tomar, V.; Zhai, J.; Zhou, M. Bounds for element size in a variable stiffness cohesive finite element model. *Int. J. Numer. Methods Eng.* **2004**, *61*, 1894–1920. [CrossRef]
13. Sun, Z.; Zhao, J. Nano-graphene toughened Al_2O_3 ceramic materials: 3D simulation of the fracture behavior. *Ceram. Int.* **2020**, *46*, 28569–28577. [CrossRef]
14. Alfred, C.; Ingo, S.; Karl-Heinz, S. On the practical application of the cohesive model. *Eng. Fract. Mech.* **2003**, *70*, 1963–1987.
15. Espinosa, H.D.; Zavattieri, P.D. A grain level model for the study of failure initiation and evolution in polycrystalline brittle materials. Part I: Theory and numerical implementation. *Mech. Mater.* **2003**, *35*, 333–364. [CrossRef]
16. Chen, F.; Yan, K.; Zhu, Y.; Hong, J. Mechanical properties of hybrid multilayer graphene/whisker-reinforced ceramic composites: Simulation and experimental investigation. *Ceram. Int.* **2022**, *48*, 25673–25680. [CrossRef]
17. Wang, D.; Xue, C.; Cao, Y.; Zhao, J. Microstructure design and preparation of Al_2O_3/TiC/TiN micro-nanocomposite ceramic tool materials based on properties prediction with finite element method. *Ceram. Int.* **2018**, *44*, 5093–5101. [CrossRef]
18. Falco, S.; Jiang, J.; De Cola, F.; Petrinic, N. Generation of 3D polycrystalline microstructures with a conditioned Laguerre-Voronoi tessellation technique. *Comput. Mater. Sci.* **2017**, *136*, 20–28. [CrossRef]

19. Zhou, T.; Huang, C. Simulation of Crack Propagation in Single Phase Ceramic Tool Materials. *Comput. Mater. Sci.* **2015**, *104*, 177–184. [CrossRef]
20. Liu, H. Study on the Fabrication and Cutting Performance of Ceramic Tool Materials Based on Multi-Phase and Multi-Scale Nanocomposite. Ph.D. Thesis, Shandong University, Jinan, China, 2005.
21. Jiao, S.; Jenkins, M.L.; Davidge, R.W. Interfacial fracture energy-mechanical behaviour relationship in Al_2O_3/SiC and Al_2O_3/TiN nanocomposites. *Acta Mater.* **1997**, *45*, 149–156. [CrossRef]
22. Yin, Z.; Yuan, J.; Wang, Z.; Hu, H.; Cheng, Y.; Hu, X. Preparation and properties of an Al_2O_3/Ti(C,N) micro-nano-composite ceramic tool material by microwave sintering. *Ceram. Int.* **2016**, *42*, 4099–4106. [CrossRef]
23. Zhao, H.; Jin, Z. The analysis of residual stress and toughening mechanism for particle-reinforced composite ceramic material. *J. Chin. Ceram. Soc.* **1996**, *24*, 491–497.
24. Sokoluk, M.; Yuan, J.; Pan, S.; Li, X. Nanoparticles enabled mechanism for hot cracking elimination in aluminum alloys. *Metall. Mater. Trans. A* **2021**, *52*, 3083–3096. [CrossRef]
25. Quaresimin, M.; Salviato, M.; Zappalorto, M. A multi-scale and multi-mechanism approach for the fracture toughness assessment of polymer nanocomposites. *Compos. Sci. Technol.* **2014**, *91*, 16–21. [CrossRef]
26. Pan, S.; Yao, G.; Cui, Y.; Meng, F.; Luo, C.; Zheng, T.; Singh, G. Additive manufacturing of tungsten, tungsten-based alloys, and tungsten matrix composites. *Tungsten* **2023**, *5*, 1–31. [CrossRef]
27. Smith, C.S. Grains, phases and interphases: An interpretation of the microstructure. *Trans. Metall. Soc. AIME* **1948**, *175*, 15–51.
28. Cheng, H. Multi-Scale Simulation Study on Nanocomposite Ceramic Tool Materials. Ph.D. Thesis, Shandong University, Jinan, China, 2011.
29. Evans, A.G.; Davidge, R.W. The strength and fracture of fully dense polycrystalline magnesium oxide. *Philos. Mag.* **1969**, *20*, 373–388. [CrossRef]
30. Niihara, K.; Nakahira, A. Strengthening and toughening mechanisms in nanocomposite ceramics. *Ann. Chim. Fr.* **1991**, *16*, 479–486.
31. Levin, I.; Kaplan, W.D.; Brandon, D.G.; Layyous, A.A. Effect of SiC submicrometer particle size and content on fracture toughness of alumina–SiC "nanocomposites". *J. Am. Ceram. Soc.* **1995**, *78*, 254–256. [CrossRef]
32. Yu, H. Simulation Study on Crack Propagation in 3D Microstructure of Ceramic Tool Materials. Master Thesis, Shandong University, Jinan, China, 2013.
33. Liu, Z.; Qiu, S.; Liu, J. Development of weak-bonding interface ceramics and toughening mechanism. *Aerosp. Mater. Technol.* **2002**, *32*, 6–9.
34. Xu, C.; Huang, C.; Li, Z.; Ai, X. Toughening mechanisms of Al_2O_3/TiC ceramic composite with carbon additive. *J. Inorg. Mater.* **2001**, *16*, 256–262.

Article

Effect of Composite Bionic Micro-Texture on Cutting Performance of Tools

Tiantian Xu [1], Chunlu Ma [1], Hu Shi [1], Kai Xiao [1], Jinpeng Liu [2] and Qinghua Li [1,*]

[1] School of Mechanical and Vehicle Engineering, Changchun University, Changchun 130022, China
[2] Technology Centre, CNAD Changchun Control Technology Co., Changchun 130102, China
* Correspondence: liqh@ccu.edu.cn

Abstract: Dry cutting is an effective method to realize the concept of green cutting today. However, in the process of cutting bearing steel, the high temperatures and high pressures produced by the cutting tool and chip under dry friction seriously affect the machining performance of the tool. Therefore, a bionic microstructure tool based on bionics is proposed to improve the cutting performance and reduce friction by changing the size parameters of the microstructure. On the basis of finite element simulation and cutting tests, the cutting force, surface roughness, and chip shape are used to evaluate the cutting performance. It is found that composite bionic micro-textured tools have a significantly reduced cutting force compared with non-micro-textured tools; composite bionic micro-textured tools lead to a reduction in surface roughness of 10–25%; and composite bionic micro-textured tools are more prone to enhancing the curling and breaking of chips. In addition, with the increase in the microstructure area occupancy, the cutting performance of the tool was also significantly improved. Moreover, it was found that the cutting performance of the tool was improved when the area occupancy of the micro-texture on the front face of the tool was increased.

Keywords: bionic micro-textured; cutting force; cutting temperature; surface roughness; chip

1. Introduction

Cutting-edge technology, known for its accuracy, high efficiency, affordability, and reliability, finds extensive applications in sectors like aerospace, energy, automotive, and defense, establishing itself as a key component in the primary technology sector of advanced manufacturing [1,2]. CBN (Cubic Boron Nitride) tools are widely used in cutting difficult-to-machine materials due to characteristics such as ultra-high hardness and chemical stability at high temperatures [3,4]. However, with the rising speed of machining, the wear, bonding, and chipping of the tool at high temperatures and high pressures seriously affect the accuracy of the product. Therefore, the question of how to extend tool life and improve machining accuracy has become a hot issue in today's research [5].

The theory of friction reduction on non-smooth surfaces has gained a lot of attention in the manufacturing industry since the proposal of texturing the surface of the tool to extend tool life, reduce friction, and reduce cutting forces [6–8]. Khani, S, et al. studied the cutting of 7075 Aluminum Alloy with a micro-textured tool and found that the micro-textured tool was more efficient in comparison to the conventional tool, and the best cutting performance was achieved when the micro-texture width, depth, and edge pitch were 93.4 μm, 15 μm, and 50 μm, respectively [9]. Kümmel, J, et al. investigated the wear and tear of micro-textured carbide tools under dry cutting conditions and found that micro-textures with grooves on the tool's front surface reduced chip tumor production during cutting and improved the surface quality of processed workpieces [10]. Salem, A, et al. investigated ways to improve the surface quality of the machined workpiece under dry cutting conditions by using micro-textures. The study found that when cutting AISI 10 steel at different cutting speeds, micro-textures on the tool's surface reduced the length of contact

Citation: Xu, T.; Ma, C.; Shi, H.; Xiao, K.; Liu, J.; Li, Q. Effect of Composite Bionic Micro-Texture on Cutting Performance of Tools. *Lubricants* **2024**, *12*, 4. https://doi.org/10.3390/lubricants12010004

Received: 20 November 2023
Revised: 15 December 2023
Accepted: 19 December 2023
Published: 22 December 2023

between the tool and the chip, thus helping to reduce the cutting force and cutting heat [11]. Fouathiya et al. studied the micro-texturing of tools used to process titanium alloys and found that micro-texturing processes were more efficient during processing when microstructural types were crosslinked. It was found that the cutting forces, coefficient of friction, and tool wear generated by the tool were minimized when the micro-texture type was cross-micro-textured [12]. Therefore, it is particularly important to construct a suitable micro-texture on the tool surface for high-quality cutting. For the surface of the frictional contact sub-surface, most of the wear originates from abrasive and adhesive wear during the friction process. The micro-texture not only reduces friction but also accommodates abrasive particles and adhesion to improve the surface quality of the contact friction pair and reduce the wear rate. However, the surface structures of some organisms in nature show great potential for wear reduction and anti-wear effects; researchers have found that some animal body surfaces are not completely smooth but have certain geometric patterns of rows of tiny structural units, and these biological surfaces tend to have less friction. This discovery on the optimization of the cutting performance of the tool provides a new direction of research [13–15]. Yu, H, et al. found that a bionic tool alleviated the adhesion phenomenon and frictional wear of the tool in machining compared to a conventional tool [16]. Cui, X, et al. investigated the effect of micro-texturing of bionic multifunctional surfaces on the cutting performance of ceramic tools and found that micro-texturing could improve the impact resistance of the tool and allow it to store lubricating oil and promote lubrication spontaneously [17]. You, C, et al. designed a micro-texture based on the geometry of the head of the biological mantis and combined the micro-texture with the tool, finding that the bionic surface improved the cutting performance of the tool and decreased the cutting force under low-speed turning conditions [18]. Green manufacturing in the field of machining has become a research hotspot, but dry cutting conditions, due to the deterioration of the cutting conditions, will lead to increased tool wear, which seriously affects the machining accuracy, so the development of high-performance dry cutting tools is an effective way to improve cutting [19–21].

Previous studies have focused on optimizing the cutting performance of tools using the mechanism of action of micro-texturing to achieve a reduction in cutting forces and the extension of tool life, and despite the great potential of bionics, little mention has been made of bionics in the design and fabrication of micro-textures on tool surfaces. By observing the structure distribution of pearlite on the surfaces of shells, three kinds of micro-textured CBN tools were extracted and designed. The influence of the micro-textured structures on the cutting performance of the tool's front surface on bearing steel was studied. The proposed micro-textured biomimetic tool has the functions of impact resistance, cutting force generation, storage of solid lubricant, and self-lubrication of expelled chips. Firstly, the influence of three biomimetic micro-textured materials on stress, cutting force, and cutting temperature was compared and analyzed through finite element simulation experiments, and the most optimal biomimetic micro-textured tool was selected. On the basis of this, laser machining was carried out for different morphologies of microstructure tools, and several cutting experiments were carried out. The cutting performance of bionic micro-textured tools was compared by using the cutting force, surface roughness, and chip shape as evaluation criteria.

2. Theoretical Models

2.1. Metal Cutting Theory

Metal cutting is the process of interaction between the tool and the workpiece in which the tool cuts off the excess layer of metal from the workpiece, during which cutting deformation occurs, chips are formed, and cutting forces, cutting heat, cutting temperatures, etc., are generated [22,23].

The relationship between the real contact area of elastic–plastic materials and the load can be expressed as Equation (1):

$$F_N = A_r \times \sigma_S \tag{1}$$

where F_N is the load, A_r is the contact area, and σ_S is the compressive yield strength.

The formula for friction is shown in (2):

$$F_f = A_r \tau \prime_b \tag{2}$$

where F_f is the friction force; $\tau \prime_b$ is the shear strength of the contact at the bonding point; and A_r is the contact area of the bonding point.

From the aforementioned Equations, it is evident that altering the contact area during metal cutting machining directly impacts the load and the friction's intensity; thus, the micro-texture formation on the tool's front side can alter the friction condition between the tool and the chip.

Substituting Equations (1) and (2) into Equation (3) yields an expression for the coefficient of friction:

$$u = \frac{F_f}{F_N} = \tau \prime_b / \sigma_s \tag{3}$$

Consequently, based on the equation, friction correlates with the load in cases where the workpiece's material is definite. Consequently, altering the proximity zone between the tool and the workpiece can potentially reduce the creation of cutting forces. The process of the micro-textured tool's action is depicted in Figure 1.

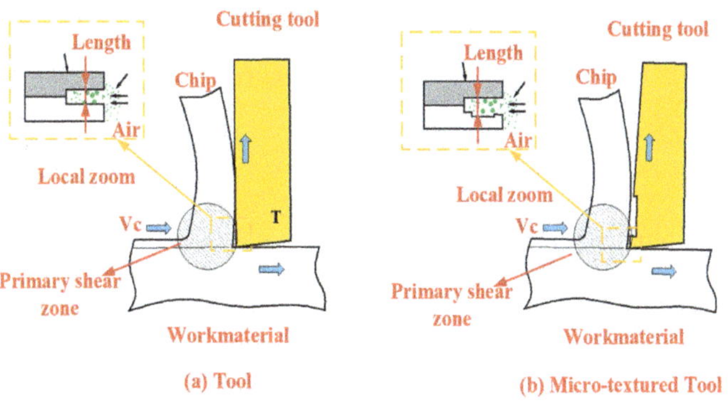

Figure 1. Mechanism of micro-textured tool.

The diagram illustrates that the actual length of contact between the tool and the workpiece is determined by an equation when the micro-textured design is assembled on the tool's front:

$$L = L_S - nM \tag{4}$$

where L is the micro-textured tool contact length; L_S is the actual cutting length; n is the number of micro-textures; and M is the micro-texture width.

So, the actual contact area of the micro-textured tool with the workpiece is represented by Equation (5):

$$A\prime_r = W \times L \tag{5}$$

where W is the cutting width; L is the cutting length.

Consequently, theoretical evaluations in crafting the micro-texture framework suggest that the dimensions of the micro-textured structure at the tool's front significantly impact its cutting efficacy.

2.2. Sources of Biomimetic Micro-Textures

The dimensions and form of the micro-texture formations in the tool's frontal working zone directly affect its wear, and especially when the micro-textured structures' size and form are impractical, attaining friction reduction proves challenging, potentially worsening the tool's wear and diminishing the workpiece's surface quality post-machining. Bionics analysis revealed that the perlite structure on the shell's surface is calcium-based, enabling it to evade marine predators from encroaching on it. The unique, consistent configuration of the surface perlite exhibits distinct mechanical characteristics, demonstrating resilience to external pressures, and the buffering impact on certain geological stresses is notably effective [24,25]. The surface configuration of the shell is depicted in Figure 2. Consequently, two varieties of micro-textures, namely, rectangular and curved, were derived from the shell's surface design and applied to the tool's front to enhance its cutting efficiency. The shapes of the micro-textures are shown in Figure 2a,b.

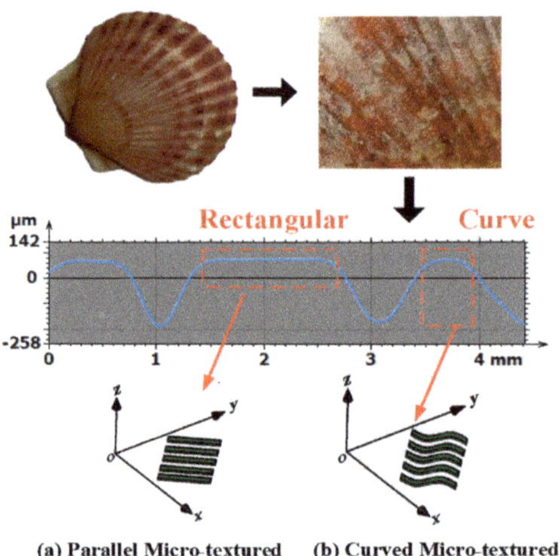

Figure 2. Sources of bionic micro-textures.

3. Finite Element Simulation Experiments of Bionic Micro-Textured Tools

Notable variations exist in the tool design and the materials used in diverse operational scenarios. Consequently, the bonding, deterioration, and frictional properties of tools used for slicing various materials vary [26–30]. Taking these elements into account, the surfaces of tools ought to possess attributes like enhanced impact resistance and encourage the movement of chips.

3.1. Three-Dimensional Modeling

At the forefront of the tool, two types of micro-texturing forms (parallel and arcuate grooves) were fashioned from the top shell surface, and the 3D modeling software developed three distinct micro-texturing instruments, as shown in Figure 3. The H tool (a curved micro-texturing tool), the P tool (a grooved micro-texturing tool), and the Z tool (a composite micro-texturing tool) symbolize a mix of curved and grooved shapes. The characteristics of the three micro-textured shapes are as follows: the dimensions of the micro-texture are 200 µm in spacing, 50 µm in width, 1200 µm in length, and 70 µm in depth.

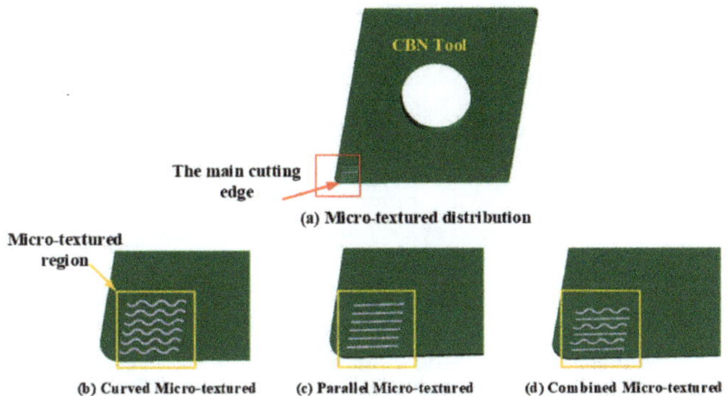

Figure 3. Establishment of three-dimensional model of micro-textured tool.

3.2. Bionic Micro-Texture Finite Element Simulation Experiment

The crafted three-dimensional bionic instrument was integrated into the finite element simulation software, where the tool's and workpiece's meshes were segmented. The tool's largest mesh dimension was set at 0.1 mm, and micro-textured components were segmented into finer mesh dimensions, with the smallest mesh size set at 0.01 mm; the largest mesh size was set at 3 mm for the workpiece, and the smallest was set at 0.01 mm. In the process of cutting, the tool's circular movement in relation to the workpiece leads to significant friction between the tool's tip and the material of the workpiece. During the cutting phase, the cutting instrument rotates in a circular path in relation to the workpiece. The tool's tip makes direct contact with the material of the workpiece, creating significant friction that leads to cutting forces and heat. To achieve more precise simulation outcomes, the space surrounding the tool's tip was characterized by an extremely accurate mesh. The material of the workpiece was identified as bearing steel; it was a round bar that measured 30 mm in diameter and 6 mm in thickness. The material used for the tool was CBN (Cubic Boron Nitride), the starting environmental temperature was set at 20 °C, and the simulation duration involved rotating the tool for one week in relation to the workpiece. During machining, the impact of the friction cutting force and cutting heat is substantial. Nonetheless, the way in which the cutting tool and workpiece interact, shaped by the process of cutting, results in a variety of phenomena. Through the analysis of relevant data, the friction coefficient was modified to 0.4, as depicted in the tool–workpiece finite element simulation mesh model in Figure 4.

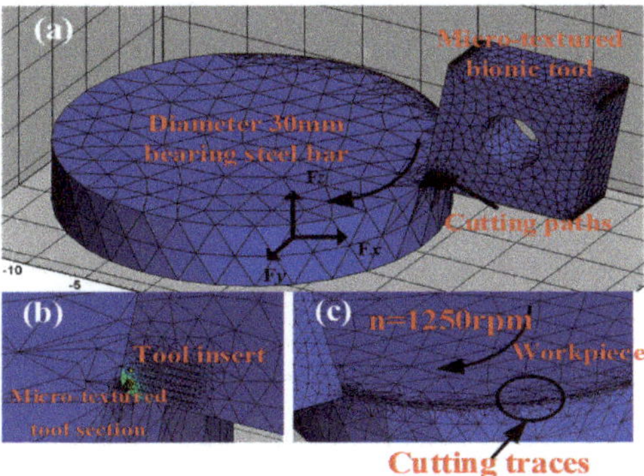

Figure 4. Workpiece–tool meshing. (**a**) Finite Element Simulation of Tool Cutting and Machining, (**b**) Partial enlargement of the front face, (**c**) Cutting track.

In the realm of metal cutting simulations, the material constitutive model is crucial, and the Power-Law and User-Defined Yield Surface constitutive models are primarily utilized. The document opts for the User-Defined Yield Surface model, which details the transition of materials from elastic to plastic properties in certain strain conditions. This study employed the selected material's Johnson–Cook model, which is adept at simulating strain hardening and thermal softening at various stages of cutting collisions and mirroring the actual phenomena, making it prevalent in the cutting field [31,32].

In the context of the Johnson–Cook material model, the yield stress is characterized as follows:

$$\sigma = \left[A + B(\varepsilon)^n\right] \times \left[1 + C\ln\left(\frac{\varepsilon_p}{\dot\varepsilon_0}\right)\right] \times \left\{1 - \left(\frac{t - t_0}{t_m - t_0}\right)^m\right\} \tag{6}$$

where σ is plastic stress; A is the yield strength; B is the strain hardening constant; C is the strain rate strengthening coefficient; m is the thermal softening parameter; n is the strain hardening parameter; ε is the plastic strain, $\dot\varepsilon_p$ is the plastic strain rate, $\dot\varepsilon_0$ is the reference strain rate; t is the material dynamic temperature; t_0 is the ambient reference temperature; and t_m is the material melting point.

Table 1 displays the material characteristics of bearing steel, while Table 2 presents the thermophysical properties and the modulus of elasticity parameters of the material.

Table 1. Material property parameters of the bearing steel Johnson–Cook model.

A (GPa)	B (GPa)	C	n	m	t_0 (°C)	t_m (°C)
1.204	1.208	0.036	0.12	0.89	20	1180

Table 2. Physical property parameters of bearing steel material.

Material Properties	Young's Modulus (Gpa)	Thermal Conductivity (W/m·K)	Poisson Ratio	Density (g/cm^3)	Specific Heat (J/kg k)
Value	210	43	0.3	7.85	458

To investigate the impact of cutting parameters on the forces generated in the process, finite element simulation experiments were performed on three preexisting bionic micro-textured tool models and a collection of micro-textured-free tools for external experimentation, with a cutting dosage of n = 1250 rpm, f = 0.15 mm/r, and a_p = 0.2 mm.

3.3. Experimental Results and Analyses

The results of the finite element simulation of data collection are illustrated in Figure 5, which shows the cutting force fluctuation during the microtome tool's cutting process, and the fluctuation map after noise reduction via processing filtering is also depicted in Figure 5's micro-textured tool cutting force fluctuation map. The illustration shows that when the tool first touches the workpiece, there is a rapid escalation in the cutting force. As the tool's processing progresses into the workpiece material, collisions with the cutting layer's tip and the cutting force escalate significantly. As the material of the workpiece undergoes shear failure due to equivalent plastic strain, causing it to break again, the metal's cutting layer and the substrate of the workpiece separate, leading to a reduction in the primary cutting force to a stable level, where the value of the cutting force generally stabilizes.

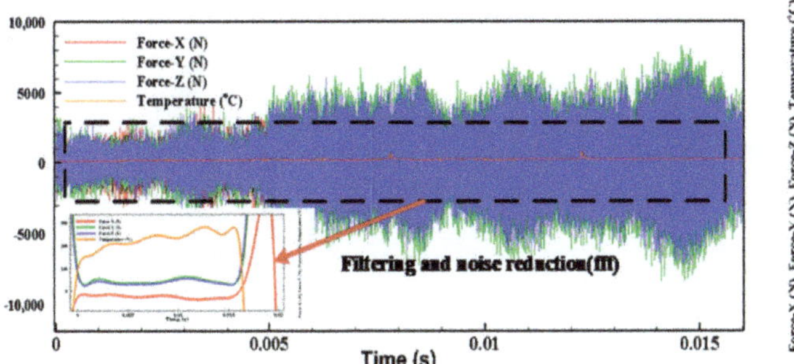

Figure 5. Cutting force fluctuation diagram.

3.3.1. Impact of Micro-Textured Material on Cutting Forces

In Figure 6a, the outcomes of the cutting force simulation for steel bearing cuts using four varieties of CBN tools are depicted. From the figure, it can be seen that when the spindle speed n = 1250 rpm, the cutting forces of the three kinds of micro-textured structure tools, compared to the traditional tool, were reduced by 10.4%, 12.8%, and 19.2%, of which the Z tool prompted the most significant reduction in the cutting force effect. The reason lies in the micro-textured nature of the tool's front surface, which diminishes the contact area between the tool and the workpiece and lowers the friction coefficient, thereby lessening the cutting force. To verify the simulation's accuracy, which expanded the experimental group, cutting experiments were performed at speeds of 800 and 1000 rpm, with the results showing the Z tool's ability to decrease the cutting force by 22% and 21.18% at 800 rpm and 1000 rpm, respectively. This demonstrates the potential of the composite bionic micro-weave device to improve the minimization of the cutting force.

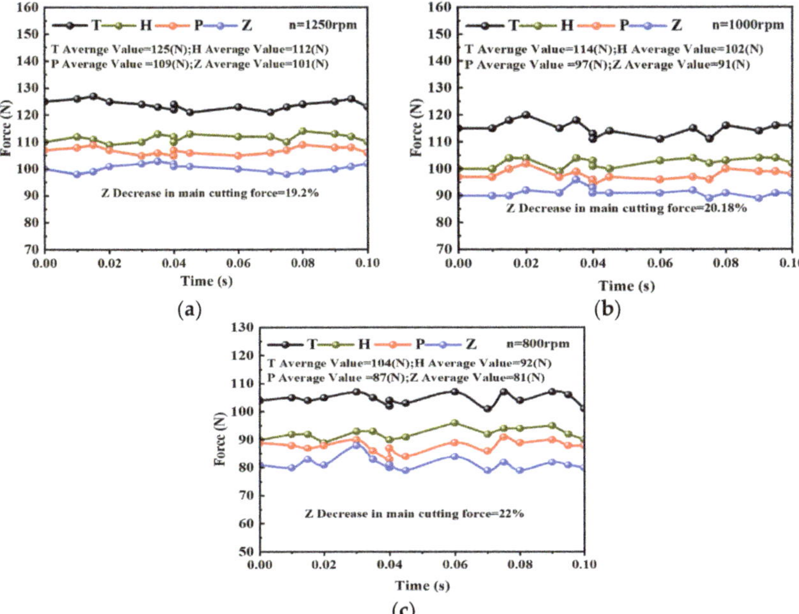

Figure 6. Comparison of cutting forces of 4 kinds of tools. (**a**) The impact of micro-textured instruments on slicing forces at n = 1250 rpm. (**b**) The impact of micro-textured instruments on slicing forces at n = 1000 rpm. (**c**) The impact of micro-textured instruments on slicing forces at n = 800 rpm.

3.3.2. Impact of Micro-Textured Surfaces on Stress Levels

Figure 7 illustrates the stress profiles of the four types of tools juxtaposed with non-micro-textured tools. H, P, and Z show stress diminution rates of 8.6%, 7.5%, and 15.22%, respectively, underscoring the critical role of composite biomimetic micro-textured tools in stress absorption. In the process of cutting with the four types of tools, especially those lacking micro-textures, a consistent advancement in the cutting method is observed, starting at the tip of the tool and gradually increasing to the internal cutting stress level; in the case of bionic micro-textured tools, stress slowly permeated the micro-textured sections near the tool's tip, accumulating predominantly in the micro-textured areas distant from the tool's location. Regarding the Z tool, the stress distribution within its interior was quicker compared to the other two tools.

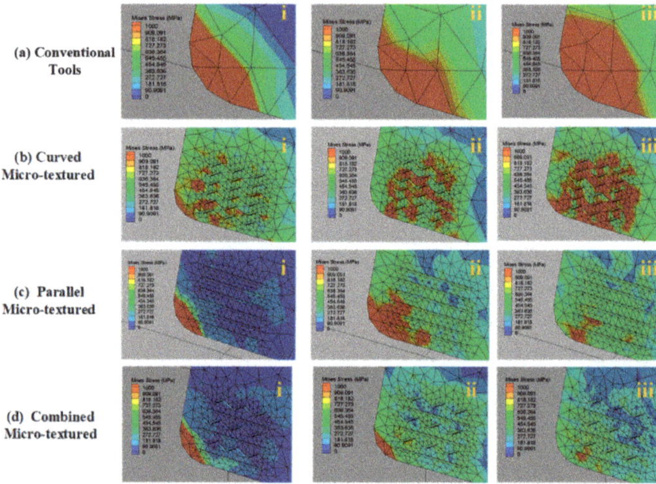

Figure 7. Stress distribution of 4 kinds of tools (**i–iii**) are the number of simulation steps in the cutting process).

3.3.3. Impact of Micro-Textures on the Temperature at Which Cutting Occurs

Information regarding the cutting temperature at the tip of the tool was collected, and Figure 8 visually depicts how the micro-textured structure tool affects this temperature. The illustration reveals that at an identical spindle velocity, the bionic micro-textured tool's heat dissipation efficiency markedly surpasses that of the non-micro-textured tool, where the Z tool's cutting temperature drops by 11.32%. The micro-textured nature of the tool's front surface enhances the frictional heat produced by the friction between the tool and the chip; conversely, the presence of a micro-textured material lessens the cutting force from the tool's direct interaction with the workpiece material, thereby decreasing the plastic deformation in the workpiece material's cutting layer, leading to a decrease in the temperature needed for cutting. The Z tool, having a lower overall cutting force than the other micro-textured tools, results in lower cutting heat generation.

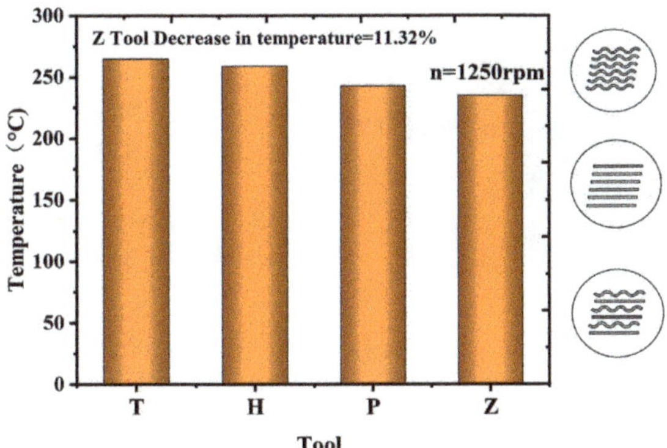

Figure 8. Comparison of cutting temperatures of 4 kinds of tools.

In Figure 9, it is observed that the tool's cutting heat primarily accumulates on the tool tip's front surface and the workpiece, facilitating the removal of material layers in close proximity to the tool's components, particularly for the bionic micro-textured tool. The micro-textured nature of the tool's surface results in an expanded heat dissipation area and a reduced cutting temperature.

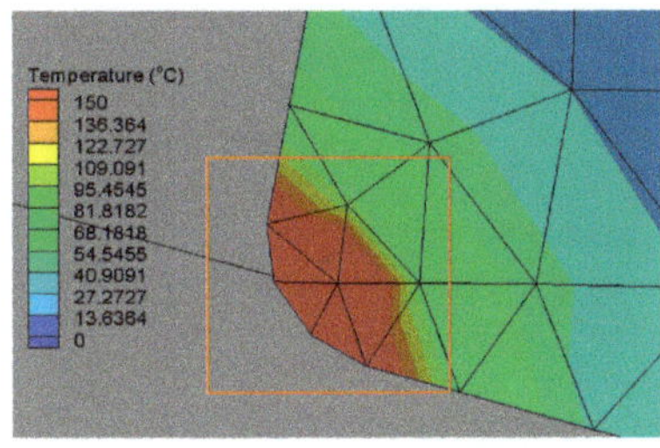

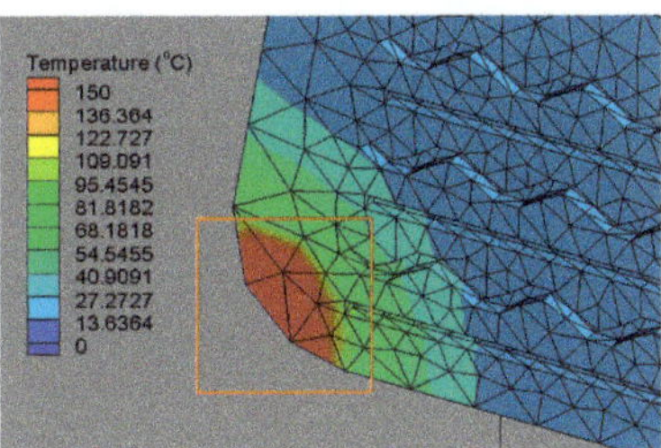

(**a**) Temperature of the front face of micro-textured tools

(**b**) Temperature of the front face of micro-textured tools

Figure 9. Cutting temperature distribution of the tools. (**a**) No micro-textured tools, (**b**) micro-textured tools.

Simulated tests on bionic micro-textured instruments demonstrate their superiority over the other three in terms of cutting force, stress distribution, and temperature, establishing a theoretical basis for the subsequent bionic micro-textured cutting experiments.

4. Bionic Micro-Textured Tool Cutting Experiments

4.1. Composite Biomimetic Micro-Textured Tool Preparation

The laser processing method has the advantages of high energy density, good processing controllability, fast processing speed, and easy-to-achieve precision processing. Therefore, in this study, the laser processing method was used to prepare the surface micro-texture of the tool. The default parameters were selected to process the tool, and the cutting performance of the Z tool was found to be superior through the above simulation experiments, so the composite biomimetic micro-texture of the Z tool was investigated in depth. Z tools (named Z1, Z2, and Z3, respectively) with spacings of 150 μm, 200 μm, and 250 μm were constructed using a laser machining method, as shown in Figure 10.

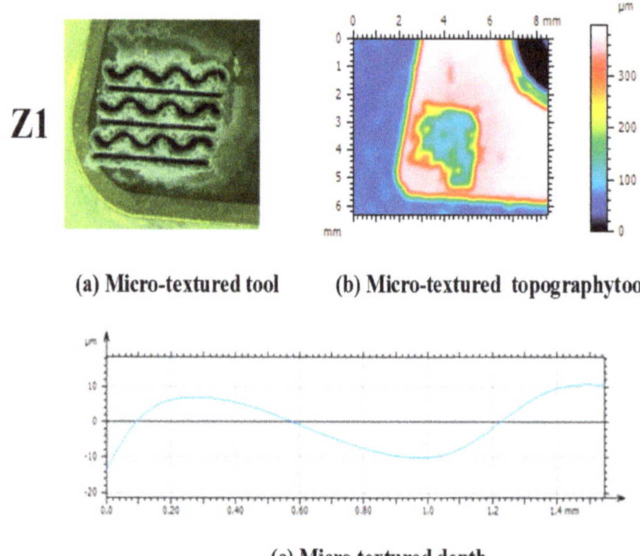

(a) Micro-textured tool   **(b) Micro-textured topographytool**

(c) Micro-textured depth

Figure 10. Composite bionic micro-textured tool preparation.

When utilizing laser machining on the tool's front, the debris produced is prone to melting on its micro-textured surfaces; thus, the laser-marked blade was placed in an ethanol ultrasonic bath and vibrated for thirty minutes to eliminate contaminants from the tool's exterior. Table 3 presents the characteristics of the micro-texture dimensions. On the basis of the micro-texture data in Table 3, the micro-textured samples were prepared, as illustrated in Figure 11.

Table 3. Parameters of micro-texture size.

Tool	Distance (μm)	Depth (μm)	Width (μm)	Length (μm)
Z1	150	70	50	1200
Z2	200	70	50	1200
Z3	300	70	50	1200

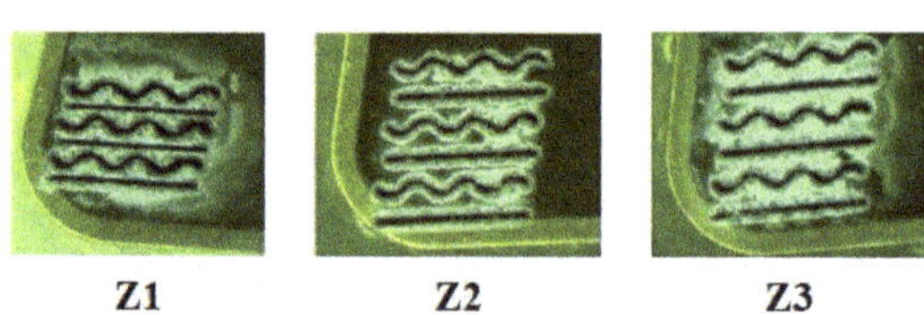

Z1   **Z2**   **Z3**

Micro-textured tool

Figure 11. Shows the microstructure morphology with different microstructure spacing.

4.2. Bionic Micro-Textured Tool Cutting Laboratory

Furthermore, to confirm the outcomes of the cutting simulation tests, CBN tool cutting experiments on bearing steel were conducted under identical cutting conditions. The object of study was a workpiece made of bearing steel in the shape of a round bar with a diameter of 50 mm and a length of 200 mm; both ends were machined to produce top holes, and the surface was pre-turned. The chemical composition of GCr15, a widely utilized type of bearing steel known for its excellent tool characteristics and mechanical stability, is detailed in Table 4 below. The cutting tool material was CBN. A CBN cutting tool is used in the cutting of ferrous metals when the chemical stability is very strong; its hardness is second only to that of a diamond tool, and it is considered a super-hard tool (brand CNGA120404-4N, Manufacturer Funek, Origin Henan, China, tool front angle $Y = -6°$). The CA6140 standard lathe was chosen, and the tool was fitted into a Kistler 9527B type tool holder to measure the cutting force in three directions utilizing the ZT-Y-50W laser marking machine (Manufacturer Shanghai Sanko, Place of Origin Shanghai, China) The Leica DVM250 super-depth-of-field microscope (Supplier Leica Microsystems, Shanghai, China) was employed to scrutinize and examine the wear patterns of the chips and tools. The experimental timetable is displayed in Table 5, while Figure 12 illustrates the necessary cutting tests for the experimental setup.

Table 4. The chemical makeup of steel.

C	Si	Mn	S	P	Cr	Mo
0.95–1.05%	0.15–0.35%	0.20–0.40%	≤0.020%	≤0.027%	1.40–1.65%	0.08–0.15%

Table 5. Parameters for processing in finite element analysis.

Group	Cutting Speed (rpm)	Feed f (mm/r)	Depth of Cut a_p (mm)
1	800	0.15	0.2
2	1000	0.15	0.2
3	1250	0.15	0.2

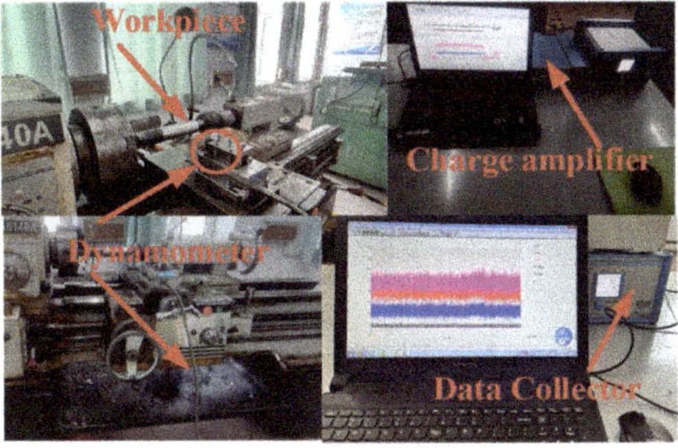

Figure 12. Experimental equipment.

4.3. Experimental Results and Analysis

4.3.1. Impact of Combined Bionic Micro-Texture Elements on Slicing Forces

Figure 13 depicts the cutting forces generated by the four tools at different cutting speeds. The depiction shows that at $n = 1250$ rpm, the slicing power of the composite bionic micro-textured instrument is significantly lower compared to the non-micro-textured one, resulting in enhanced uniformity in the simulation. Usually, under dry friction conditions, the main source of the cutting force stems from the distortion of the material's cutting layer and the resulting staining effect. In the case of the composite bionic micro-textured instrument, diminishing the tool's size and the chip's proximity results in decreased friction, thereby lowering the cutting force; conversely, this also affects the cutting procedure. The micro-textured segment of the chip held on the tool's front will be captured and preserved within the micro-textured interior, which acts as a storage unit for the chip, preventing staining from knives. This research revealed a 26.08% reduction in the cutting power of the Z3 tool relative to the non-micro-textured tool, markedly outperforming the other two composite bionic micro-textured tools. This is due to the Z3 tool's front having a more extensive micro-textured surface compared to the other two types of tools, leading to an expanded micro-textured area that minimizes the contact area between the tool and the workpiece material, thereby reducing friction in the Z3 tool and reducing the cutting force. Figure 14 shows the graph of the total area of the micro-texture.

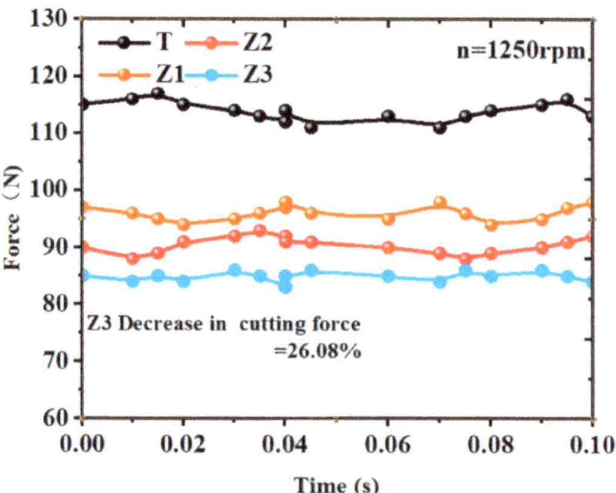

Figure 13. Effect of combined bionic micro-texture parameters on cutting force at *n* = 1250 rpm.

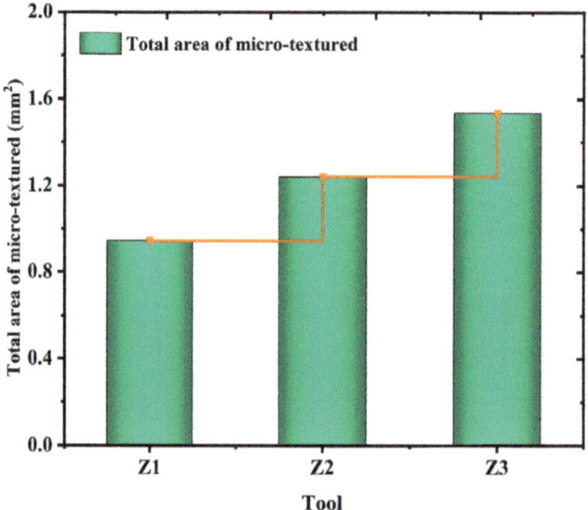

Figure 14. Area of three bionic micro-textured structures (the step line is a line connecting the areas of the three micro-textured configurations).

Figure 15 illustrates how the parameters of the composite bionic micro-textured structure affect the cutting force at different speeds, showing a steady decrease in the cutting force as the rotational speed lessens. When *n* = 800 rpm, the cutting force of the Z3 tool was reduced by 25.51% relative to the non-micro-textured structure tool; when *n* = 1000 rpm, the cutting force of the Z3 tool decreased by 28.03% in comparison to the non-micro-textured structure tool. With the reduction in speed, there is a decrease in the rate at which materials are removed per time unit, and as the deformation of the workpiece material lessens, the cutting force diminishes.

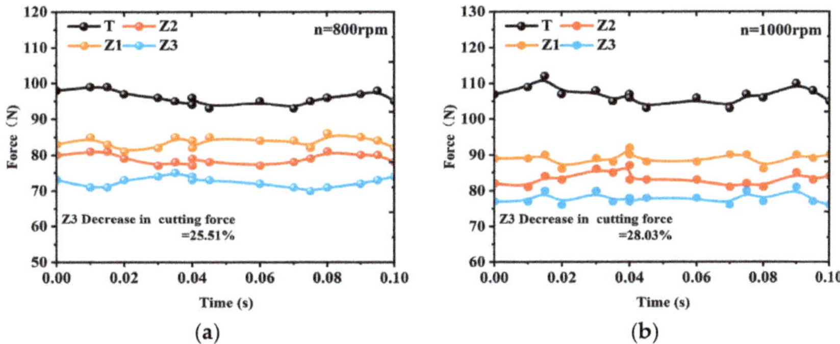

Figure 15. Impact of combined bionic micro-textured elements on slicing forces across varying spindle velocities. (**a**) n = 800 rpm, (**b**) n = 1000 rpm.

4.3.2. Impact of Combined Bionic Micro-Textured Factors on the Texture of Surface

Surface roughness is an important indicator for evaluating surface integrity and affects the service life of the workpiece. In this study, the surface quality of the workpiece after machining was observed with an optical microscope. Figure 16 shows the roughness data of the workpiece after machining with four types of tools at a spindle speed of 1250 rpm. From the figure, it can be seen that the average surface roughness of Sa decreases by 13.29%, 12.21%, and 23.26% when the Z1, Z2, and Z3 tools are used, compared to that obtained with the tool without micro-texture, and the root mean square height of the Z3 tool is 1.298 lower than that of the other three tools. This is because the micro-texture can store the tiny debris generated by friction, reducing the intermittent contact between the tool surface and debris; avoiding contact between the debris and the cutter surface in the machining process prevents the kind of extrusion caused by the tool surface wear, which affects the machining quality. With the increase in micro-texture spacing, the micro-textured area occupancy increases, and the ability to capture debris increases, contributing to the decline in the surface roughness of the workpiece.

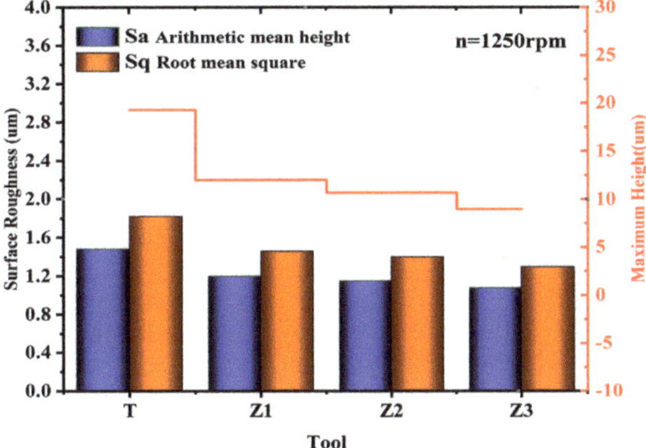

Figure 16. Effect of combined bionic micro-textured parameters on surface roughness for n = 1250 rpm (the step line (S_Z) is the average of the high and low points on the surface of the workpiece after machining).

A three-dimensional profilometer was employed to examine the workpiece's surface roughness post-machining, and Figure 17 illustrates the three-dimensional structure of the workpiece's surface following machining with four tools at an identical rotational velocity. The results reveal that the Z3 tool finalizes the machining process, and the tool's smooth surface is observed to bear clear tool marks. The reason lies in the fact that the cutting force produced by the tool's micro-textured structure is lower than that generated by a non-micro-textured structure, resulting in a micro-textured structure throughout the entire friction process, which in turn reduces friction and enhances its smoothness, and the stability of machine tool machining is enhanced.

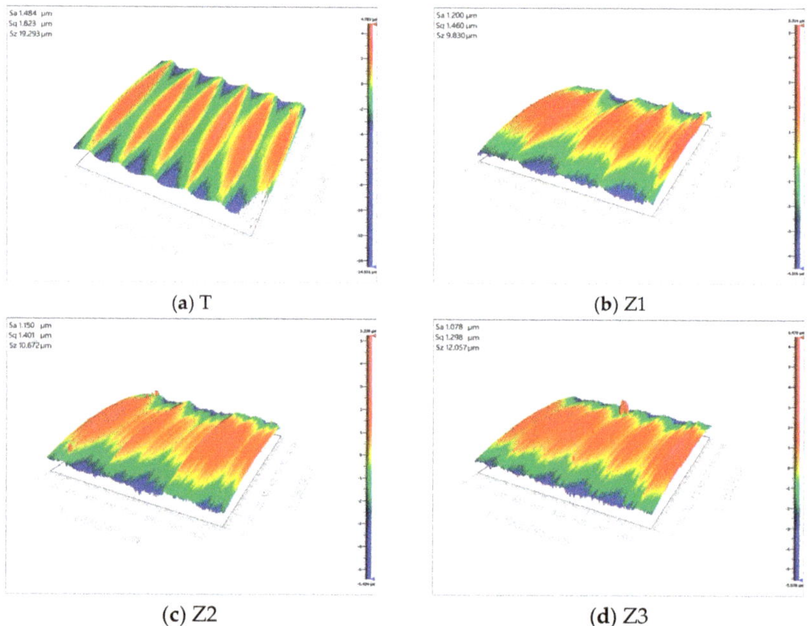

(a) T

(b) Z1

(c) Z2

(d) Z3

Figure 17. The three-dimensional structure of the workpiece's surface post-machining using a standard tool (T) and a variable-pitch micro-textured tool.

4.3.3. Impact of Combined Biomimetic Micro-Textured Variables on Microchips

In the process of cutting, a chip that does not break becomes uncontrollable, significantly hindering the machine tool's normal operation, causing damage to the tool, and changing the appearance of the processed product. This study revealed that using four different types of machining tools led to the development of four unique chip varieties, as depicted in Figure 18, and the surface of the workpiece following machining had a cutter mark topography. The diagram shows that when the T tool cuts into a stretched, ring-shaped spiral chip, the material of the workpiece continuously splits due to shear force, leading to the creation of chips along the outflow of the front face. When the chips are expelled during the collision of tool surface extrusions, they curl, resulting in the formation of curling strain inside the chip. The employment of a composite bionic micro-textured device leads to the formation of a short circular spiral chip characterized by a diminished circular radius. The resemblance of the tool's micro-textured part to the chip-breaking groove's design, along with the chip-breaking groove, leads to increased curling strain and a smaller curl radius, contributing to chip damage and leakage.

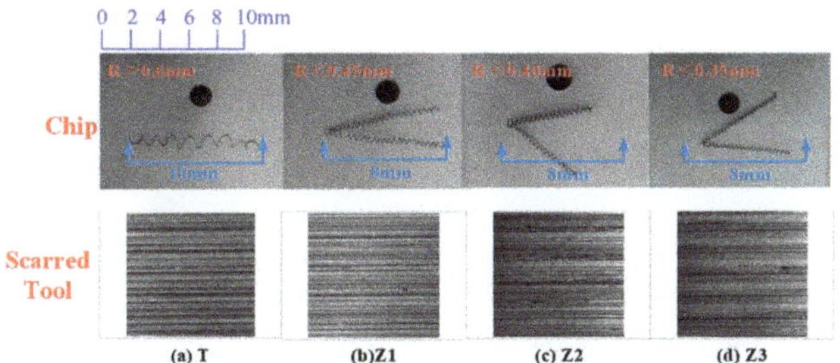

Figure 18. The intertwining of chips following machining with four instruments at $n = 1250$ rpm.

Differing from traditional tools, the micro-textured texture of the tool face facilitates the formation and storage of small chips within the micro-textured area while using the micro-textured cutting tool, thus averting severe wear due to the chip protruding between the tool face and the workpiece; this action results in the removal of the ultimate material. A micro-textured tool's front, known for its heat dissipation properties, leads to a reduced cutting force, thereby decreasing the wear of the tool.

The previously mentioned slicing experiments demonstrate that composite bionic micro-textured instruments surpass non-micro-textured ones in efficiency by lessening the force needed for cutting and diminishing the surface roughness of the machined workpiece, promoting the deterioration of chips and improving the slicing efficiency; these effects are optimal with the Z3 tool.

5. Conclusions

This research was carried out to explore how the micro-texture size factor (measured as micro-texture spacing) influences the tool's efficiency in slicing bearing steel with varying cutting settings. Drawing from bionic principles, this document introduces three varieties of bionic micro-textured structures: a curved bionic micro-textured tool, a parallel bionic micro-textured tool, and a composite bionic micro-textured tool. A multitude of finite element simulations and cutting tests were conducted to assess the micro-textured tool on the basis of the cutting force, temperature, chip bending radius, and surface texture, leading to the following findings:

1. Finite element simulation experiments reveal that the bionic micro-textured tool significantly outperforms the non-micro-textured tool in stress reduction. In the context of the cutting force and temperature, the composite bionic micro-textured tool enhances the cutting force to decrease by 19.2% and the cutting temperature to decrease by 11.32%.

2. The cutting tests revealed that the composite bionic micro-textured instrument successfully withstood impacts and lessened the tool–chip contact zone, and utilizing the micro-textured segment as the chip's storage unit lessened the friction between the chip and the tool, leading to a 26.08% decrease in the cutting force and a 23.26% reduction in surface roughness.

3. The experimental findings indicate the consistent enhancement of the tool's cutting efficiency as the proportion of its micro-textured area grows.

4. Demonstrations reveal that employing a composite bionic micro-textured instrument with a 300 μm micro-weave gap in bearing steel machining significantly enhances the tool's cutting efficiency, laying a theoretical groundwork for future superior bearing steel machining and highlighting the potential of micro-weave tools for machining.

Author Contributions: Conceptualization, T.X.; writing—original draft, C.M.; methodology, H.S.; software, K.X. and J.L.; review and editing, Q.L. All authors have read and agreed to the published version of the manuscript.

Funding: Jilin Provincial Department of Education "JJKH20240738KJ".

Data Availability Statement: Data will be made available on request.

Acknowledgments: The authors would like to thank the members of the project team for their dedication and effort and the teachers and schools for their help.

References

1. Fabián Pérez-Salinas, C.; Fernández-Lucio, P.; del Olmo, A.; Aldekoa-Gallarza, I.; Norberto López de Lacalle, L. The influence of cutting edge microgeometry on the broaching of Inconel 718 slots. *Eng. Sci. Technol. Int. J.* **2023**, *48*, 101563. [CrossRef]
2. Ghosh, S.; Rao, P.V. Application of sustainable techniques in metal cutting for enhanced machinability: A review. *J. Clean. Prod.* **2015**, *100*, 17–34.
3. Aouici, H.; Yallese, M.A.; Chaoui, K.; Mabrouki, T.; Rigal, J.F. Analysis of surface roughness and cutting force components in hard turning with CBN tool: Prediction model and cutting conditions optimization. *Measurement* **2012**, *45*, 344–353. [CrossRef]
4. Liu, K.; Li, X.P.; Rahman, M.; Liu, X.D. CBN tool wear in ductile cutting of tungsten carbide. *Wear* **2003**, *255*, 1344–1351. [CrossRef]
5. Stephenson, D.A.; Agapiou, J.S. *Metal Cutting Theory and Practice*; CRC Press: Boca Raton, FL, USA, 2018.
6. Li, B. A review of tool wear estimation using theoretical analysis and numerical simulation technologies. *Int. J. Refract. Met. Hard Mater.* **2012**, *35*, 143–151. [CrossRef]
7. Iliescu DM, F.D.; Gehin, D.; Gutierrez, M.E.; Girot, F. Modeling and tool wear in drilling of CFRP. *Int. J. Mach. Tools Manuf.* **2010**, *50*, 204–213. [CrossRef]
8. Patel, D.S.; Jain, V.K.; Shrivastava, A.; Ramkumar, J. Electrochemical micro texturing on flat and curved surfaces: Simulation and experiments. *Int. J. Adv. Manuf. Technol.* **2019**, *100*, 1269–1286. [CrossRef]
9. Khani, S.; Razfar, M.R.; Haghighi, S.S.; Farahnakian, M. Optimization of micro-textured tools parameters in thread turning process of aluminum 7075 aerospace alloy. *Mater. Manuf. Process.* **2020**, *35*, 1330–1338. [CrossRef]
10. Kümmel, J.; Braun, D.; Gibmeier, J.; Schneider, J.; Greiner, C.; Schulze, V.; Wanner, A. Study on micro texturing of uncoated cemented carbide cutting tools for wear improvement and built-up edge stabilisation. *J. Mater. Process. Technol.* **2015**, *215*, 62–70. [CrossRef]
11. Ellersiek, L.; Menze, C.; Sauer, F.; Denkena, B.; Möhring, H.C.; Schulze, V. Evaluation of methods for measuring tool-chip contact length in wet machining using different approaches (micro-textured tool, in-situ visualization and restricted contact tool). *Prod. Eng.* **2022**, *16*, 635–646. [CrossRef]
12. Fouathiya, A.; Meziani, S.; Sahli, M.; Barrière, T. Experimental investigation of micro-textured cutting tool performance in titanium alloy via turning. *J. Manuf. Process.* **2021**, *69*, 33–46. [CrossRef]
13. Du, H.Y.; He, L.; Du, H.S.; Ma, K.J.; Li, Y.P. Bionic tribological tool weaving design. *Comb. Mach. Tools Autom. Mach. Technol.* **2016**, *506*, 138–142. [CrossRef]
14. You, C.; Xie, C.; Chu, X.; Zhou, W.; Zhao, G.; Lian, Y. Cutting performance of bionic cutting tools based on surface microstructures of blood clam Tegillarca granosa in dry cutting of CFRP. *Int. J. Adv. Manuf. Technol.* **2022**, *119*, 2961–2969. [CrossRef]
15. Neugebauer, R.; Drossel, W.G.; Ihlenfeldt, S.; Harzbecker, C. Design method for machine tools with bionic inspired kinematics. *CIRP Ann.* **2009**, *58*, 371–374. [CrossRef]
16. Yu, H.; Han, Z.; Zhang, J.; Zhang, S. Bionic design of tools in cutting: Reducing adhesion, abrasion or friction. *Wear* **2021**, *482*, 203955. [CrossRef]
17. Cui, X.; Duan, S.; Guo, J.; Ming, P. Bionic multifunctional surface microstructure for efficient improvement of tool performance in green interrupted hard cutting. *J. Mater. Process. Technol.* **2022**, *305*, 117587. [CrossRef]
18. You, C.; Zhao, G.; Chu, X.; Zhou, W.; Long, Y.; Lian, Y. Design, preparation and cutting performance of bionic cutting tools based on head microstructures of dung beetle. *J. Manuf. Process.* **2020**, *58*, 129–135. [CrossRef]
19. Akhtar, S.S. A critical review on self-lubricating ceramic-composite cutting tools. *Ceram. Int.* **2021**, *47*, 20745–20767. [CrossRef]
20. Wu, Z.; Deng, J.; Chen, Y.; Xing, Y.; Zhao, J. Performance of the self-lubricating textured tools in dry cutting of Ti-6Al-4V. *Int. J. Adv. Manuf. Technol.* **2012**, *62*, 943–951.
21. Wu, Z.; Deng, J.; Su, C.; Luo, C.; Xia, D. Performance of the micro-texture self-lubricating and pulsating heat pipe self-cooling tools in dry cutting process. *Int. J. Refract. Met. Hard Mater.* **2014**, *45*, 238–248. [CrossRef]
22. Qin, S.; Shi, X.; Xue, Y.; Zhang, K.; Huang, Q.; Wu, C.; Shu, J. Coupling effects of bionic textures with composite solid lubricants to improve tribological properties of TC4 alloy. *Tribol. Int.* **2022**, *173*, 107691. [CrossRef]
23. Zheng, K.; Yang, F.; Pan, M.; Zhao, G.; Bian, D. Effect of surface line/regular hexagonal texture on tribological performance of cemented carbide tool for machining Ti-6Al-4V alloys. *Int. J. Adv. Manuf. Technol.* **2021**, *116*, 3149–3162. [CrossRef]
24. Zhao, C.; Long, R.; Zhang, Y.; Wang, Y.; Wang, Y. Influence of characteristic parameters on the tribological properties of vein-bionic textured cylindrical roller thrust bearings. *Tribol. Int.* **2022**, *175*, 107861. [CrossRef]
25. Liu, G.L.; Zheng, J.T.; Huang, C.Z.; Sun, S.F.; Liu, X.F.; Dai, L.J.; Wang, X.Y. Coupling effect of micro-textured tools and cooling conditions on the turning performance of aluminum alloy 6061. *Adv. Manuf.* **2023**, *11*, 663–681. [CrossRef]
26. Yu, Q.; Zhang, X.; Miao, X.; Liu, X.; Zhang, L. Performances of concave and convex microtexture tools in turning of Ti6Al4V with lubrication. *Int. J. Adv. Manuf. Technol.* **2020**, *109*, 1071–1092. [CrossRef]
27. Gajrani, K.K.; Sankar, M.R.; Dixit, U.S. Tribological performance of MoS2-filled microtextured cutting tools during dry sliding test. *J. Tribol.* **2018**, *140*, 021301. [CrossRef]

28. Gao, G.; Wang, Y.; Fu, Z.; Zhao, C.; Xiang, D.; Zhao, B. Review of multi-dimensional ultrasonic vibration machining for aeronautical hard-to-cut materials. *Int. J. Adv. Manuf. Technol.* **2023**, *124*, 681–707. [CrossRef]
29. Tapia, E.; Sastoque-Pinilla, L.; De Lacalle, N.L.; Lopez-Novoa, U. Towards real time monitoring of an aeronautical machining process using scalable technologies. In Proceedings of the 2022 7th International Conference on Smart and Sustainable Technologies (SpliTech), Split/Bol, Croatia, 5–8 July 2022; pp. 1–6.
30. Cui, X.; Li, C.; Ding, W.; Chen, Y.; Mao, C.; Xu, X.; Liu, B.; Wang, D.; Li, H.N.; Zhang, Y.; et al. Minimum quantity lubrication machining of aeronautical materials using carbon group nanolubricant: From mechanisms to application. *Chin. J. Aeronaut.* **2022**, *35*, 85–112. [CrossRef]
31. Pang, K.; Wang, D. Study on the performances of the drilling process of nickel-based superalloy Inconel 718 with differently micro-textured drilling tools. *Int. J. Mech. Sci.* **2020**, *180*, 105658. [CrossRef]
32. Fan, L.; Deng, Z.L.; He, Y.; Zhu, X.L.; Gao, X.J.; Jin, Z. The effects of micro-texture shape on serrated chip geometry in the hardened steel AISI D2 cutting process. *Surf. Topogr. Metrol. Prop.* **2022**, *10*, 015031. [CrossRef]

Article

Tool Wear Prediction Model Using Multi-Channel 1D Convolutional Neural Network and Temporal Convolutional Network

Min Huang [1,2], Xingang Xie [1,2,*], Weiwei Sun [2] and Yiming Li [2]

[1] School of Mechanical and Electrical Engineering, China University of Mining and Technology (Beijing), Beijing 100083, China; huangmin@bistu.edu.cn
[2] Mechanical Electrical Engineering School, Beijing Information Science and Technology University, Beijing 100192, China; sww@bistu.edu.cn (W.S.); liyimingxf@bistu.edu.cn (Y.L.)
* Correspondence: bqt2000401003@student.cumtb.edu.cn

Abstract: Tool wear prediction can ensure product quality and production efficiency during manufacturing. Although traditional methods have achieved some success, they often face accuracy and real-time performance limitations. The current study combines multi-channel 1D convolutional neural networks (1D-CNNs) with temporal convolutional networks (TCNs) to enhance the precision and efficiency of tool wear prediction. A multi-channel 1D-CNN architecture is constructed to extract features from multi-source data. Additionally, a TCN is utilized for time series analysis to establish long-term dependencies and achieve more accurate predictions. Moreover, considering the parallel computation of the designed architecture, the computational efficiency is significantly improved. The experimental results reveal the performance of the established model in forecasting tool wear and its superiority to the existing studies in all relevant evaluation indices.

Keywords: tool wear prediction; one-dimensional convolution; temporal convolutional network

1. Introduction

As the backbone of the modern economy, the manufacturing industry's production efficiency and product quality significantly depend on tool efficiency. Tool wear directly influences the machining precision, such as the surface roughness and dimensional stability, thus degrading the product quality. As tool wear increases, the processing efficiency decreases, and energy consumption increases, which can even lead to a costly production downtime and equipment damage. Therefore, accurate and timely tool wear prediction can ensure production continuity and quality control [1]. Nevertheless, tool wear prediction is a complex issue influenced by many factors, such as the tool material, cutting variables (like speed, feed rate, and cutting depth), and the kind of processing materials [2,3]. Traditional rule-based and periodic inspection approaches cannot handle these complex variables or adapt to rapidly changing manufacturing requirements.

Since traditional tool wear prediction approaches, which rely primarily on empirical rules and regular physical inspections [4], are simple and feasible, achieving an accurate real-time tool wear prediction is a challenge [5]. Subsequently, vibration [6], sound [7], and temperature sensors [8] were utilized to monitor tool conditions in real time [9]. Although these methods improve the monitoring accuracy, an effective data analysis method is still required to predict the wear state. Recently, data-driven-based tool wear prediction approaches, especially machine learning (ML) techniques, have been extensively utilized in the literature [10]. For example, support vector machines (SVMs) and neural networks were utilized to analyze sensor data to achieve more accurate wear predictions [11]. As we approach the Industry 4.0 era, manufacturing has accumulated huge amounts of data, such as machine performance, production process parameters, and tool wear rates. Feature

Citation: Huang, M.; Xie, X.; Sun, W.; Li, Y. Tool Wear Prediction Model Using Multi-Channel 1D Convolutional Neural Network and Temporal Convolutional Network. *Lubricants* **2024**, *12*, 36. https://doi.org/10.3390/lubricants12020036

Received: 28 December 2023
Revised: 18 January 2024
Accepted: 22 January 2024
Published: 26 January 2024

engineering is an indispensable upfront step to utilize ML techniques to effectively forecast and monitor tool wear. Li et al. presented a feature transfer learning-based tool wear prediction approach [12], in which multi-domain features were extracted from cutting force and vibration signals and were integrated into feature tensors for tool wear prediction [13]. However, since feature engineering is usually a manual, iterative, and time-consuming process that needs expertise, this leads to longer projects and increased costs [14]. Simultaneously, effective feature engineering needs a deep understanding of data and application domains, and a lack of relevant domain knowledge can lead to inappropriate feature selection or transformation.

Deep learning (DL) has recently become the core of study innovation in this field [15]. Due to their unique benefits in processing time series data, recurrent neural networks (RNNs) show significant potential in tool wear prediction. Due to their ability to retain historical information, RNNs can effectively deal with evolving manufacturing process data. RNNs [16] can efficiently utilize tool usage patterns, operating parameters, and wear states to predict future wear trends. For example, Liu et al. adopted the RNN model to analyze tool wear data under various operating situations [17]. The RNN model successfully predicts the future wear state by learning the relation between tool wear and the operating parameters, thus significantly improving the accuracy compared with the traditional prediction method. An RNN was utilized to analyze complex machining process data, including cutting forces, vibration, an electric current, and temperature. Wang et al. extracted wear-related signal features from various machining signals and employed Long Short-Term Memory (LSTM) for tool wear prediction [18]. Chan et al. introduced a tool wear prediction approach utilizing a global-local LSTM network [19]. Kolář et al. utilized the spindle drive current for tool condition monitoring [20]. Duan et al. established a parallel DL model with mixed attention for tool wear prediction [21]. The outcomes demonstrated the ability of RNNs to accurately identify specific situations, accelerating tool wear prediction and recognizing complex data patterns and relationships.

Despite its significant progress, the existing research still faces data diversity and complexity. For example, it is challenging to perform feature extraction and feature fusion in massive amounts of data [22]. At the same time, although RNNs perform well in tool wear prediction, RNN models may encounter gradient disappearing or explosion when processing long sequence data and cannot perform parallel computation, thus limiting their application in complex scenarios and in the effective handling of long-term dependencies.

To address the challenges identified, a comprehensive model, including a multi-channel one-dimensional convolutional neural network (1D-CNN) and a temporal convolutional network (TCN), is established to enhance the precision and efficiency of tool wear state prediction. This combination overcomes the limitations of traditional machine learning (ML) and deep learning approaches, particularly in handling large-scale data and long-term dependencies. This combined approach can effectively capture both the spatial and temporal features of tool wear data. The 1D-CNN component can extract spatial features from multi-channel sensor data, while the TCN component can capture temporal dependencies over long sequences. By employing this synergy, the model can more accurately predict tool wear states in various machining contexts. Furthermore, the model's architecture is designed to facilitate parallel computation, significantly enhancing the computational efficiency.

The rest of the current paper is structured as follows. Section 2 introduces the 1D-CNN and TCN corresponding theories, and Section 3 introduces the established model. Section 4 describes the experimental dataset, process, and results. The relevant summary is given in Section 5.

2. Relative Theories

2.1. One-Dimensional CNN Feature Extraction

A one-dimensional CNN [23] can automatically identify and extract key features from time series data, which is crucial for understanding the tool wear pattern and accurately predicting its future state. A one-dimensional CNN comprises the convolutional, pooling, and fully connected layers.

(1)   Convolutional layer

The core part of the 1D-CNN is the convolutional layer, which can perform the feature extraction of time series data or 1D spatial data [24]. Although the structure of the convolutional layer is similar to traditional 2D convolution, it is specifically designed to work with 1D data. As shown in Figure 1, these layers can capture local features and patterns in the time series by applying different convolution kernels to the input data, thereby extracting informative features. Accordingly, a 1D-CNN can extract complex and informative features from the raw data.

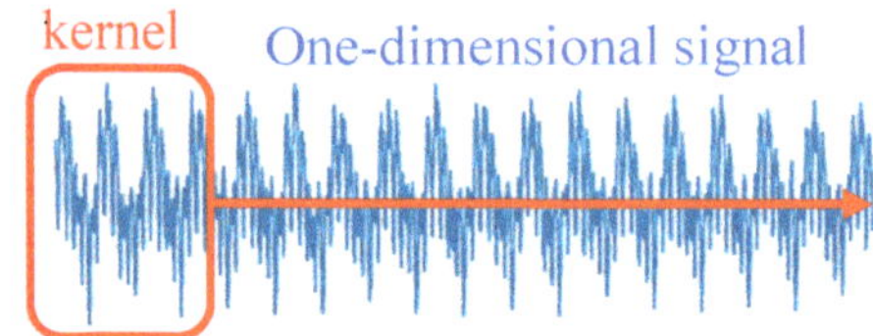

Figure 1. One-dimensional convolution extraction feature diagram.

The 1D convolution operation for a 1D input signal can be described as follows:

$$y[i] = x[i] \times k[i] = \sum_{j=-M}^{M} x[i-j] \times k[j] \tag{1}$$

where $y[i]$ describes the convolution operation output, M indicates the half-size of the convolution kernel, $x[i]$ describes a 1D signal (i is the time step), and $k[j]$ is the convolution kernel (j is a position in the convolution kernel).

A convolution operation is usually followed by applying an activation function ReLu to introduce nonlinearity:

$$a[i] = \max(0, y[i]) \tag{2}$$

where $a[i]$ indicates the output after applying the activation function.

(2)   Pooling layer

Pooling layers are often employed in 1D-CNNs to alleviate the features' dimension and the computational cost while retaining the main feature information. The selected maximum pooling operation reduces the data dimension by selecting the maximum value within a specific window size. Maximum pooling can be described as follows:

$$p[j] = \max(a[i]), \text{ for } i \in [j \times s, \ j \times s + f] \tag{3}$$

where $p[j]$ indicates the pooling layer output; s describes the step size, indicating the distance that the pooling window moves in the sequence; and f indicates the size of the pooling window. Maximum pooling selects the maximum value in the window as its representation.

(3)   Fully connected layer

The fully connected layer is placed at the end of the 1D-CNN, which helps the network synthesize multi-dimensional features from different sensors (such as vibration, sound, current, and cutting forces). Its operation can be described as follows:

$$y = Wp[j] + b \tag{4}$$

where y indicates the fully connected layer output, W describes the weight matrix, and b is a bias vector. Compared to traditional feature engineering methods, a 1D-CNN reduces the reliance on expert knowledge and the need for manual data pre-processing. Additionally, the pooling layer in a 1D-CNN reduces the dimension of features, thus reducing the computational cost while preserving critical information. This makes 1D-CNNs efficient and effective when processing tool wear data with complex time dependence.

2.2. Time Convolutional Network

Since RNN often faces gradient disappearing or explosion when dealing with long sequence data, it is difficult for the network to learn and retain long-term dependent information. Although RNN variants, such as LSTM [25] and GRU [26], can better handle long-term dependencies, they are still confined by inherently handling serialized data. Therefore, TCN is adopted for time series long-term dependency modeling because it can deal with long-term data dependency problems. TCN [27] employs causal convolution to make predictions dependent on past and current information, but not future data. This is particularly important in time series forecasting, where future data are unavailable in real-world scenarios. By stacking multiple convolutional layers and using dilated convolution, TCN effectively increases its receptive field, i.e., the range of historical data that the model can "see". This allows the TCN to capture long-term data dependencies without significantly increasing the computational costs. Compared with RNNs, such as LSTM or GRU, the convolutional operations of TCN can be more efficiently computed in parallel because the forward propagation of the convolutional network does not depend on the previous point in time in the sequence. This improves the efficiency of the TCN when dealing with large-scale data. The TCN mainly comprises the causal, dilated, and residual convolutions.

(1)   Causal convolution

The main idea of causal convolution is to make the prediction dependent only on the previous and current information in the time series analysis. This is achieved by controlling the interaction of the convolution kernel with the input data. For a 1D input sequence $x[t]$, where t is the time step, causal convolution is described as follows:

$$y[t] = \sum_{s=0}^{s-1} x[t - s] \cdot k[s] \tag{5}$$

where $y[t]$ describes the convolution output at time t, $k[s]$ is the convolution kernel, and s is the convolution kernel index.

As presented in Figure 2, considering that the input layer's last two nodes are x_{t-1} and x_t, the calculated output at time t is y_t, which is obtained as follows:

$$y_t = k_0 \times x_t + k_1 \times x_{t-1} \tag{6}$$

(2)   Dilated convolution

Simple causal convolution faces the conventional CNN issue; that is, the modeling length of time is confined by the convolution kernel size [28]. In order to grasp a longer dependency, it is necessary to stack many layers linearly. Therefore, the TCN usually employs dilatation convolution to handle long-term dependencies in the time series data while maintaining the model's causality, as shown in Figure 3.

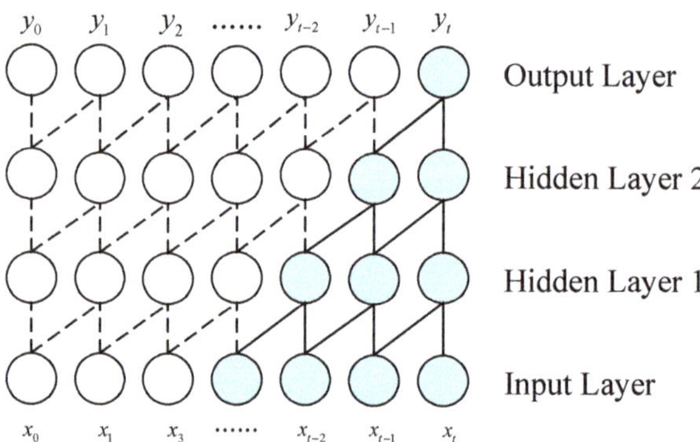

Figure 2. The causal convolution structure's schematic diagram.

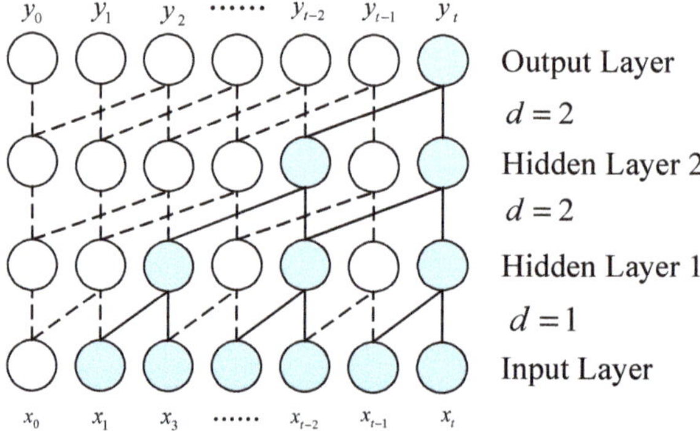

Figure 3. Dilated causal convolution structure diagram.

Dilated convolution introduces intervals inside the convolution kernel to cover longer input sequences. This allows the network to "see" and process further historical information, capturing long-term dependencies without increasing the convolution kernel size or the number of network layers [29]. The following relation describes the dilated convolution:

$$y[t] = \sum_{s=0}^{s-1} x[t - s \cdot d] \cdot k[s] \tag{7}$$

where d is the dilated factor, determining how the convolution kernel covers the input data. The more significant the d value, the more input data points the convolution kernel "skips", allowing the network to cover a longer input sequence.

(3)   Residual connections

Residual connections in the TCN improve the network's learning ability, especially when dealing with deep networks. Residual connections help solve gradient disappearing in deep networks and enhance information transmission by introducing feeder lines from the previous layer to the next.

As described in Figure 4, assuming that the input of a convolution block in a TCN is x, and the output after applying the convolution is $F(x)$, the residual connections' output is described as follows:

$$O(x) = Activation(x + F(x)) \tag{8}$$

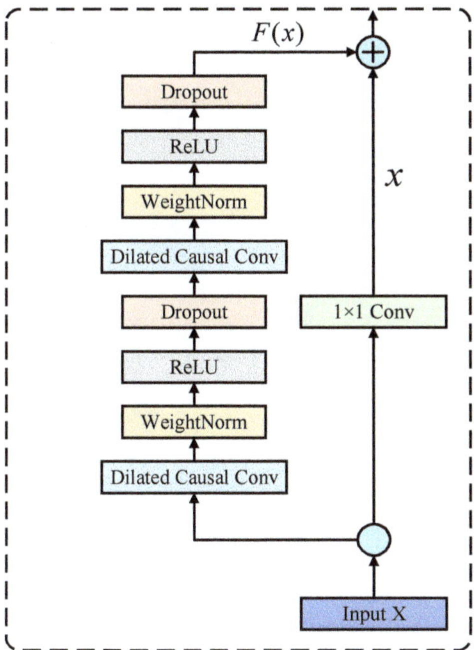

Figure 4. The TCN residual structure's schematic diagram.

Applying the residual connection to the TCN improves the model's capability to process long-term dependency and complex time series data, especially the residual structure, which is particularly important in deep network construction. This improves the efficiency and reliability of the TCN for prediction tasks for many complex time series.

3. Proposed Model

Tool wear is a complex process that considerably affects the production efficiency and quality during manufacturing. Tool wear monitoring and management are crucial for efficient, high-quality production. However, since tool wear cannot be predicted immediately in the machining procedure, sensors are utilized to monitor the cutting force, vibration, acoustic emission, and other signals in the machining procedure to forecast tool wear indirectly. In order to meet the tool wear prediction's precision and real-time requirements, a CTCN model is established using the 1D-CNN and TCN. Figure 5 shows the structure and prediction flow based on the CTCN model.

(1) A multi-channel 1D-CNN extracts features from the time series signals gathered using the sensor. Subsequently, the multi-feature integration is achieved by combining the feature maps derived from various convolution branches.

(2) The extracted features in step (1) are fed into the TCN network for long-term dependency modeling to learn time series features corresponding to tool wear features.

(3) In the end, a mapping relation between high-dimensional features and tool wear values is established through the fully connected layer to realize tool wear prediction.

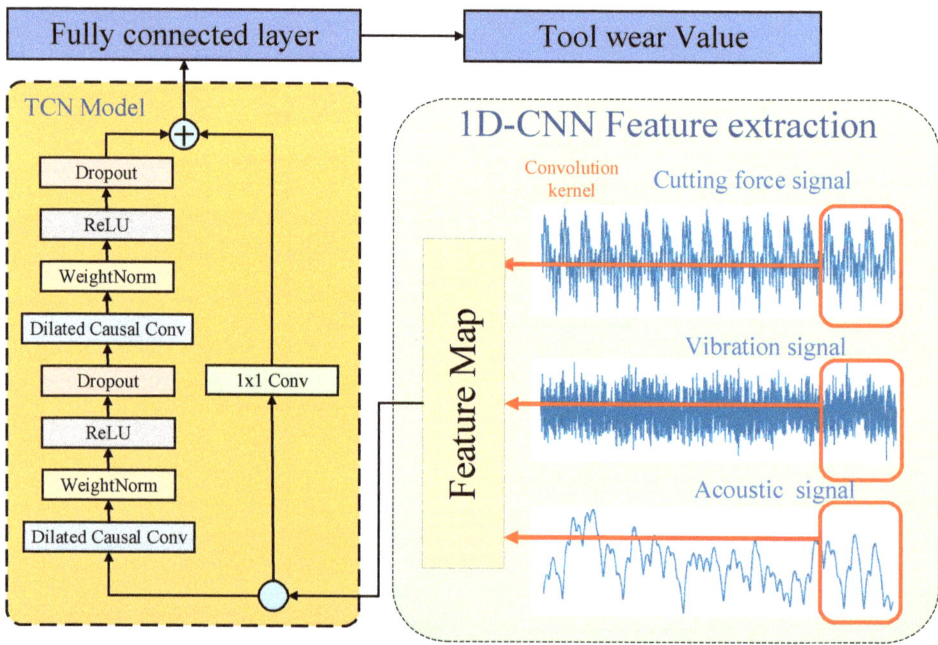

Figure 5. CTCN model network structure diagram.

4. Experimental Study

4.1. Dataset

In order to evaluate the established CTCN's effectiveness, the PHM2010 tool wear dataset was adopted for verification [30]. The PHM2010 tool wear dataset can be downloaded from https://phmsociety.org/phm_competition/2010-phm-society-conference-data-challenge/ (accessed on 1 December 2023). Figure 6 presents the PHM2010 data experimental setup. Under dry milling conditions, a high-speed CNC machine tool was utilized to test a three-edge ball-end mill with stainless steel (HRC-52) as the workpiece material. The spindle speed was 10,400 RPM, the sampling frequency was 50 KHz, the feed speed was 1555 mm/min, and the Z-axis and Y-axis cutting depths were 0.2 mm and 0.125 mm, respectively.

Table 1 shows the test equipment. The tool was milled along the X-axis direction during the test and removed after each milling. After each milling operation, the tool wear behind the three cutters was measured using a LEICA MZ12 microscope, recorded as flute1–3, and the wear measurement accuracy was 10^{-3} mm. Finally, a total of three cutters were tested under the same experimental conditions, designated C1, C4, and C6, and 315 cutting tests were performed on each cutter. Finally, 315 datasets and the corresponding wear values of three blades were collected for each cutter. The C1, C4, and C6 datasets recorded the full life cycle of the cutter wear. Figure 7 shows the trend of the three-blade wear for each cutter. For safety reasons, the maximum value of the three blades is chosen as the final wear label.

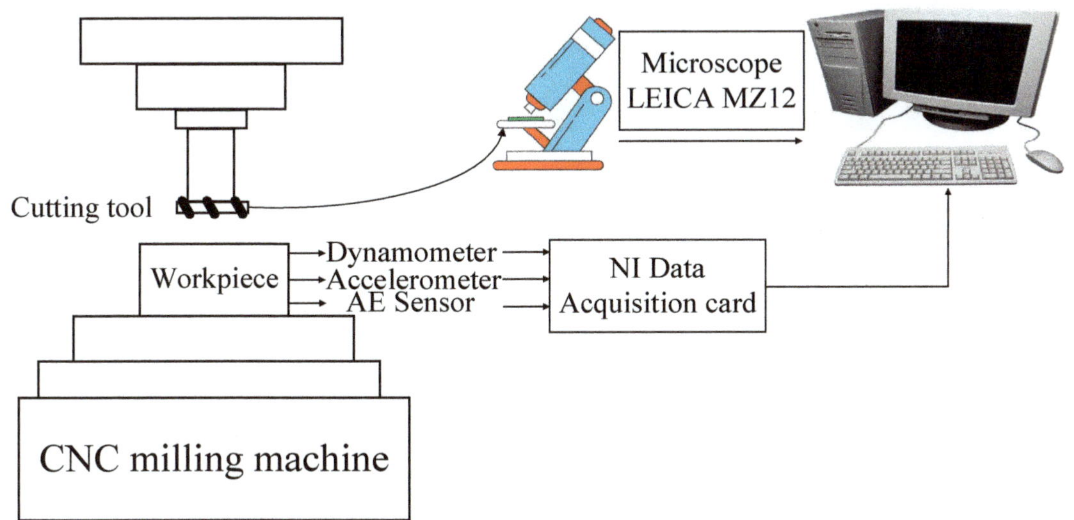

Figure 6. The experimental apparatus's PHM schematic diagram.

Table 1. Model of the experimental equipment.

Equipment	Type
CNC milling machine	Roders Tech RFM760
Dynamometer	Kistler 9265B
Charge amplifier	Kistler 5019A
Acoustic emission sensor	Kistler AE sensor
Cutters	3-flute ball carbide milling cutters
Data acquisition card	DAQ NI PCI 1200
Abrasion measuring apparatus	LEICA MZ12 microscope

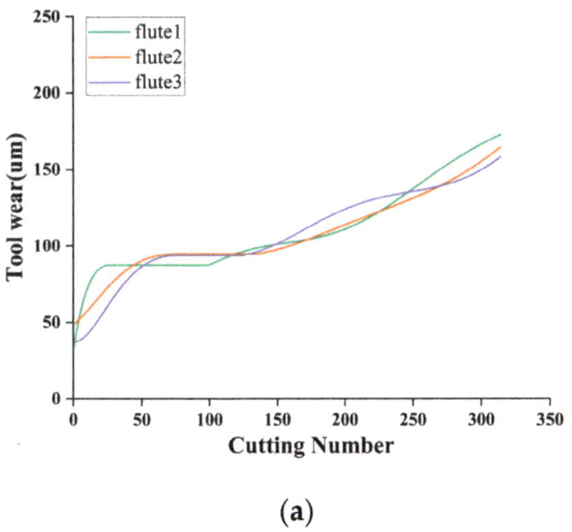

(**a**)

Figure 7. *Cont.*

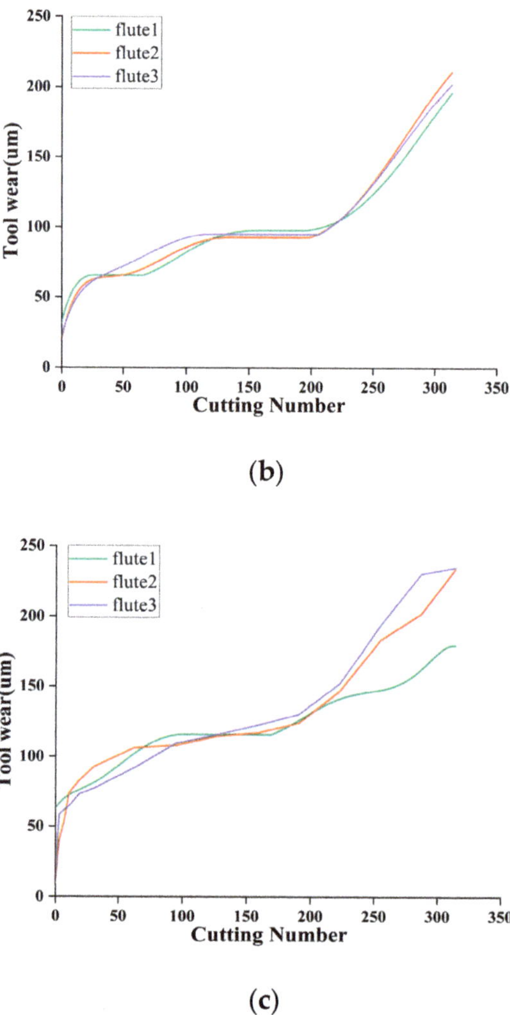

(b)

(c)

Figure 7. The PHM2010 dataset's tool wear value curves: (**a**) C1 tool wear value curve, (**b**) C4 tool wear value curve, and (**c**) C6 tool wear value curve.

4.2. Data Study and Preprocessing

During the tool wear experiment, each experiment covers three stages: tool feed, stable cutting, and tool retraction. Since the signal is relatively stable in the stable cutting stage, the signal collected from the tool can effectively reflect the degree of tool wear. As presented in Figure 8, the generated signals often need clarification during the tool feed and retraction stages. They cannot accurately represent the relation between the tool wear and the signal. Therefore, the unstable signal points of 3% before and after the start and end of each cut were excluded to enhance the data quality and the accuracy of the analysis. Considering the difference in the data scale between the original signals, this study adopted the Z-Score normalization to standardize the signal data to alleviate the effect of various signal scales. This step creates data with a uniform scale and distribution before entering the network model, thus improving the consistency and accuracy of the model when processing different signals.

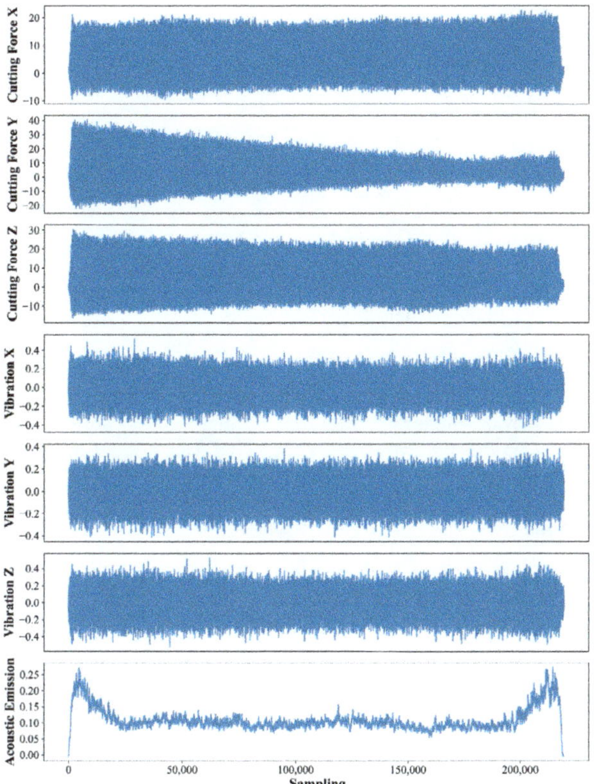

Figure 8. PHM dataset experimental process signal diagram.

4.3. Experimental Setup

Under the PyTorch framework, the CTCN model was programmed, and the NVIDIA GeForce RTX 3090 GPU was utilized to perform the learning process. The CTCN model was trained for 100 rounds, and the learning rate was chosen as 0.0001. In order to enhance the model's generalization ability, L2 regularization terms were incorporated into the training process. Additionally, the Adam optimizer, which adjusts and optimizes the parameters through a backpropagation mechanism, was employed to minimize the loss effectively.

4.4. Evaluation Index

The evaluation index can assess the model performance. In order to assess the CTCN model's correlation efficiency, the mean absolute error (MAE), root mean square error (RMSE), determination coefficient (R^2), and mean absolute percentage error (MAPE) were selected as assessment indices, as shown in Equations (9)–(12). The MAE is the average of the absolute error between the predicted and actual values. The RMSE is the square root of the MSE, providing an error measure on the same scale as the actual value. R^2 measures the ability of the model to describe the variables, i.e., how well the model fits the data. A value of R^2 closer to 1 reveals the better fitting of the model. The MAPE is the magnitude of the prediction error relative to the actual value. It calculates the mean ratio of the absolute value of the difference between the predicted and actual values to the actual one.

$$\text{MAE} = \frac{1}{n}\sum_{i=1}^{n}|y_i - \hat{y}_i| \tag{9}$$

$$\text{RMSE} = \sqrt{\frac{1}{n}\sum_{i=1}^{n}(y_i - \hat{y}_i)^2} \tag{10}$$

$$R^2 = 1 - \frac{\sum\limits_{i=1}^{n}(y_i - \hat{y}_i)^2}{\sum\limits_{i=1}^{n}(y_i - \bar{y})^2} \tag{11}$$

$$\text{MAPE} = \frac{100\%}{n}\sum_{i=1}^{n}\left|\frac{y_i - \hat{y}_i}{y_i}\right| \tag{12}$$

where n indicates the number of samples, and y_i and $\hat{y}_i$ describe the actual and predicted values of the ith sample, respectively.

4.5. Experimental Setup and Model Parameter Adjustment

The current test was performed using the PHM tool wear dataset, and the cross-validation strategy outlined in Table 2 was employed for model training and evaluation. The experiment process was divided into three groups, each adopting training and testing stages.

Table 2. Training and testing set settings.

Training Set		Testing Set
c4	c6	c1
c1	c6	c4
c1	c4	c6

This experiment established the CTCN model using the model parameters presented in Table 3. The model's parameters can be calculated as shown below.

Table 3. CTCN model parameters.

Layer	Kernel Shape	Output Shape
Layer1.CNN.Conv1d	[7, 64, 9]	[32, 64, 1024]
Layer2.CNNConv1d	[64, 64, 5]	[32, 64, 512]
Layer3.CNN.Conv1d	[64, 128, 3]	[32, 128, 258]
Layer4.TCN.Conv1d	/	[32, 7, 130]
Layer5.TCN.Chomp1d	/	[32, 7, 129]
Layer6.TCN.Conv1d	/	[32, 1024, 129]
Layer7.TCN.Chomp1d	/	[32, 1024, 129]
Layer8.Linear	[1024, 1]	[32, 129, 1]

4.6. Experimental Results and Discussion

4.6.1. Experimental Results

Figure 9 shows the experimental results, where the predicted and actual tool wear curves are indicated in blue and red, respectively, and the red histogram represents their difference. The experimental results indicate the compatibility of the predicted curve with the actual curve. Table 4 shows the evaluation indices corresponding to the three groups of experiments.

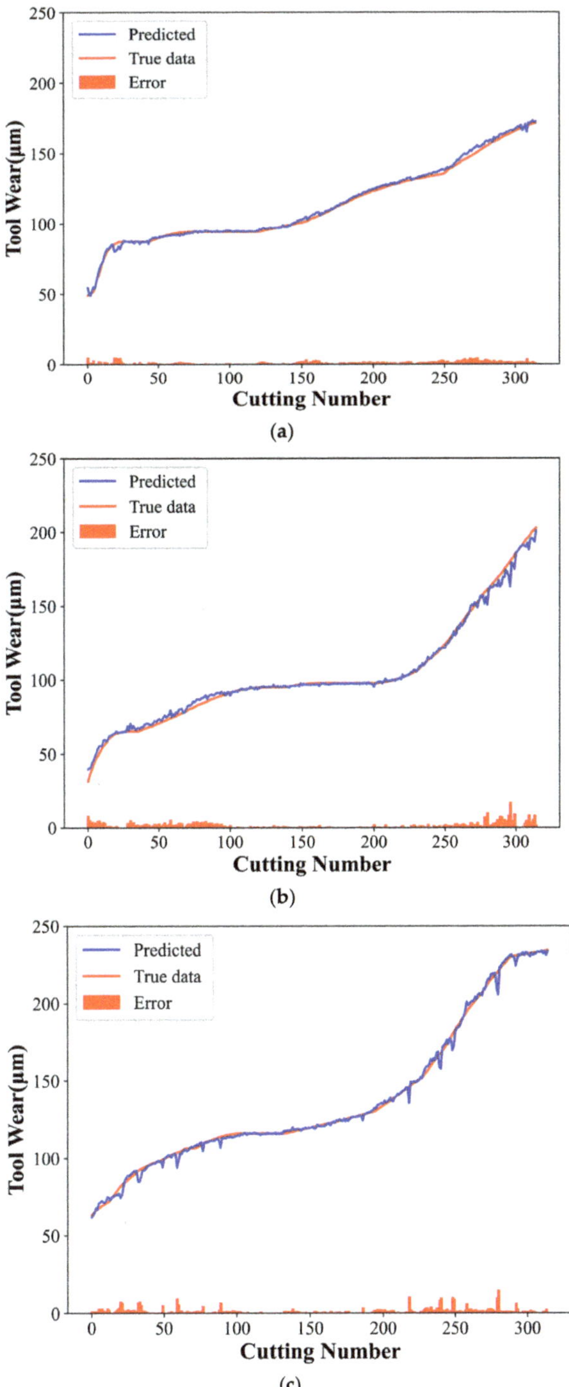

Figure 9. CTCN model experimental results: (**a**) C1 tool predicted wear, (**b**) C4 tool predicted wear, and (**c**) C6 tool predicted wear.

Table 4. Ablation results.

Models	Datasets											
	C1				C4				C6			
	MAE	RMSE	R^2	MAPE	MAE	RMSE	R^2	MAPE	MAE	RMSE	R^2	MAPE
CNN	3.609	6.476	0.946	0.035	5.201	8.131	0.955	0.052	4.423	6.388	0.981	0.033
TCN	2.551	5.198	0.965	0.024	5.273	8.729	0.942	0.048	6.834	9.271	0.961	0.047
CTCN	2.031	3.553	0.983	0.019	2.744	6.151	0.974	0.027	2.415	4.274	0.991	0.018

4.6.2. Ablation Experiment

The ablation experiments were designed to demonstrate the model's effectiveness and superiority. Experiments 1 and 2 employed the CNN and TCN models, respectively, with the same parameters as the CTCN model. Figures 10 and 11 present the experimental results of the CNN and TCN models, respectively. Table 4 shows the evaluation indices for different models.

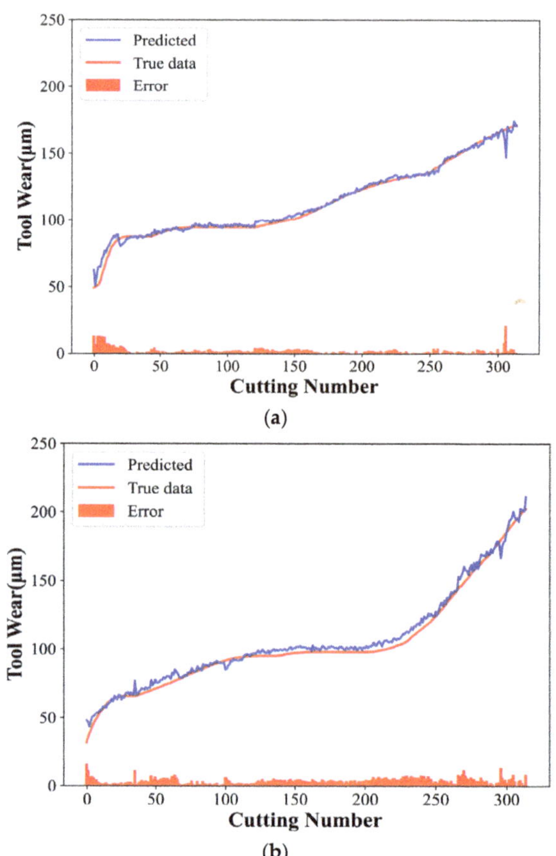

Figure 10. *Cont.*

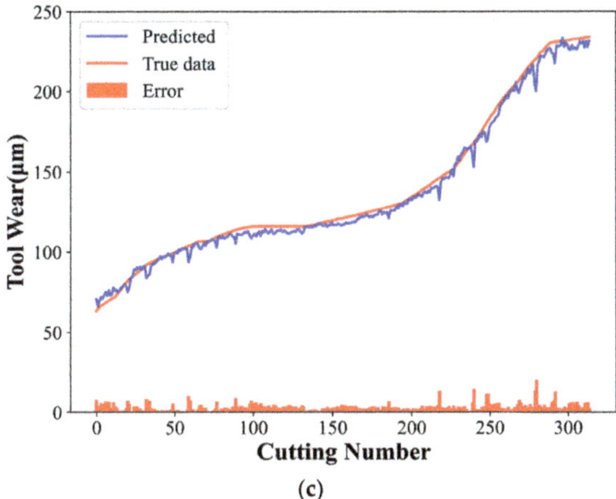

(c)

Figure 10. CNN model experimental results: (**a**) C1 tool predicted wear, (**b**) C4 tool predicted wear, and (**c**) C6 tool predicted wear.

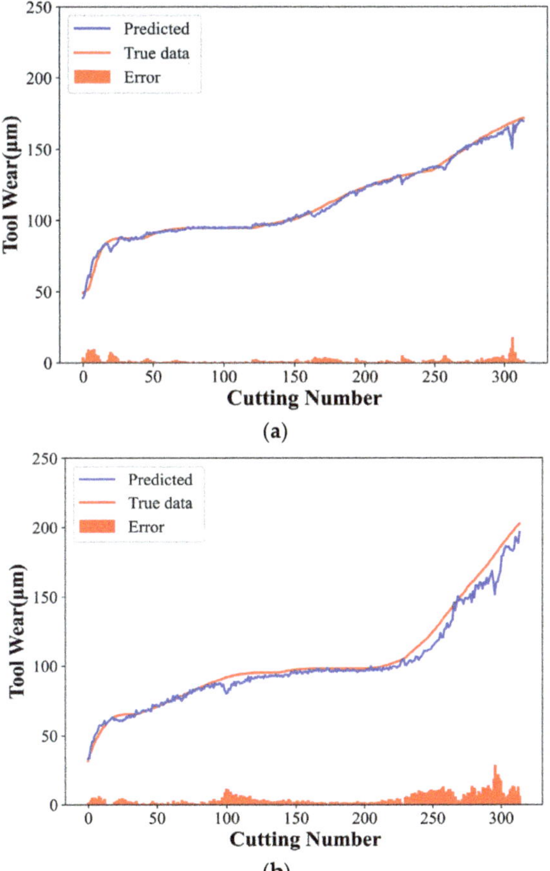

Figure 11. *Cont.*

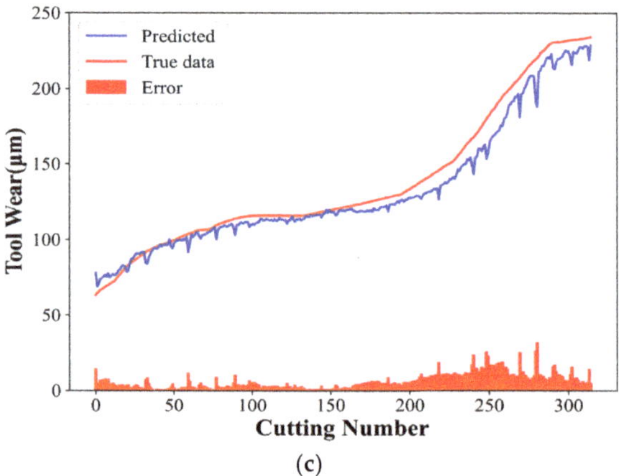

Figure 11. TCN model experimental results: (**a**) C1 tool predicted wear, (**b**) C4 tool predicted wear, and (**c**) C6 tool predicted wear.

The MAE, RMSE, and MAPE values of the CTCN model were lower than those of the other comparison models. In contrast, the R^2 values were higher than those of the comparison models in a similar group, reaching 98.38%, 97.48%, and 99.17%, respectively. The CNN mainly extracts local features from multi-channel sensor data. Additionally, its powerful spatial feature extraction capability helps the model identify and capture critical patterns during tool wear. These characteristics are crucial to understand and predict the tool wear process and are essential for the subsequent time series analysis.

The CNN-extracted features are then input into the TCN. The TCN can deal with the time dependence of these features, considering the tool wear process's time-varying behavior. The dilated causal convolution structure of the TCN helps the model capture long-term time series dependencies without losing causality. Accordingly, the model can make accurate wear predictions based on current and historical data and reveal the wear development trend.

The established model can effectively take advantage of the sensor data's spatial properties (captured using a CNN) and temporal properties (processed using a TCN). The ablation experiments show that combining a TCN with a CNN significantly improves the model's overall performance, especially the prediction accuracy and the ability to understand time-dependent relationships. This demonstrates the unique and complementary roles of CNNs and TCNs in the established model and their synergies in processing complex time series data.

4.6.3. Model Comparison

Since the PHM2010 tool wear dataset has become the benchmark dataset for many relevant scholars to perform experiments, several models using the PHM2010 dataset were selected for comparison to demonstrate the CTCN model's superiority. Table 5 presents the relevant assessment indices of various models under the PHM2010 dataset.

After a comprehensive performance comparison using the same dataset, the results indicate the superiority of the established model to other models in each evaluation index. This highlights the advantage of the established model in understanding and analyzing tool wear data, especially when dealing with complex data structures and variable data characteristics. Due to its unique algorithm design and efficient data processing strategy, this model stands out in competitive comparisons, demonstrating its potential in practical applications.

Table 5. Comparison results of different models.

Models	Datasets											
	C1				C4				C6			
	MAE	RMSE	R^2	MAPE	MAE	RMSE	R^2	MAPE	MAE	RMSE	R^2	MAPE
SSAE-BP [31]	12.00	13.53	0.91	/	12.44	14.62	0.93	/	7.29	10.08	0.94	/
Parallel-CNN [32]	3.89	5.54	/	/	4.53	5.27	/	/	3.52	4.90	/	/
CBLSTM [33]	7.5	10.8	/	/	6.1	7.1	/	/	8.1	9.8	/	/
PGGM [34]	4.3	5.0	/	/	8	9.6	/	/	5.9	13.9	/	/
MFPBM [35]	4.2	4.9	/	/	5.8	7.1	/	/	4.4	5.8	/	/
PRes–SBiLSTM [36]	4.7	8.5	/	/	5.6	7.9	/	/	4.7	5.9	/	/
CGRU [37]	3.32	5.33	0.962	/	4.95	7.43	0.962	/	4.45	6.47	0.974	/
CTCN	2.031	3.553	0.983	0.019	2.744	6.151	0.974	0.027	2.415	4.274	0.991	0.018

5. Conclusions

The current work combines multi-channel 1D-CNN with TCN models to establish a tool wear prediction model and verifies the tool wear prediction capability of the CTCN model. The model performs excellently in various evaluation indices, such as RMSE and MAPE. Compared with existing forecasting models, the established model provides higher precision and stability when dealing with complex datasets. The model outperforms the single CNN or TCN model in many experiments, and the following conclusions can be obtained:

(1) A 1D-CNN can effectively extract the 1D signal features, which can efficiently describe the tool wear state.

(2) Due to the dilated causal convolution structure of the TCN, long-term time series dependencies can be effectively captured without losing causality. Accordingly, the model can make accurate wear predictions based on current and historical data and reveal the development trend.

(3) This study demonstrates that combining a CNN and TCN can significantly improve tool wear prediction. A CNN's powerful feature extraction ability, combined with the efficient time series analysis of a TCN, helps the model effectively capture complex tool wear patterns.

The training and testing datasets for the CTCN model were specifically tailored to the conditions and environment of high-speed milling machining obtained from high-speed CNC machining centers. Therefore, the applicability of the CTCN model in tool wear prediction is mainly limited to similar high-speed milling scenarios such as those in the training data environment. Due to different mechanical properties, cutting parameters, and operating conditions in turning and grinding machining processes, these variations may degrade the tool wear prediction by the CTCN model in these scenarios. Since each machining method exhibits unique tool wear patterns and influencing factors, training and adapting the model using specific datasets from each machining process are necessary to enhance its prediction accuracy and generalization ability. Consequently, although the CTCN model may provide an excellent tool wear prediction for high-speed CNC milling machining, its predictive capability might be constrained when applied to other scenarios, such as turning and grinding. Further customization and optimization are necessary to adapt to these scenarios' specific requirements and conditions.

In future work, we will apply our model to larger and more diverse datasets, further validating its generalization and robustness, especially in different manufacturing environments and with more varied tool wear. Additionally, the model interpretation should be improved. As a future work, the interpretability of the model should be enhanced so that users can better understand the internal logic of the model's predictions.

Author Contributions: M.H. led the design and conceptualization of the entire study, conducted an in-depth analysis of the experimental data, and was responsible for managing the project funding and resources. X.X. is responsible for developing and debugging the computer models used in the experiments, analyzing the experimental results, and providing technical writing for the relevant chapters of the paper. W.S. led the implementation of the experiment, including data acquisition and preprocessing, and participated in the preliminary analysis of the data. Y.L. provided an important literature review during the paper writing process and participated in the writing and proofreading of the paper. All authors have read and agreed to the published version of the manuscript.

Funding: This research was funded by the Ministry of Industry and Information Technology's High-end Numerical Control Systems and Servo Motors Project (Grant No. ZTZB-22-009-001).

Data Availability Statement: All data are contained within the article.

Conflicts of Interest: The authors declare no conflicts of interest.

References

1. Xia, W.; Zhou, J.; Jia, W.; Guo, M. Milling Tool Wear Prediction Based on 1DCNN-LSTM. In Proceedings of the 8th International Conference on Mechanical, Automotive and Materials Engineering, Hanoi, Vietnam, 16–18 December 2022; Springer: Berlin/Heidelberg, Germany, 2022; pp. 77–91.
2. Knittel, D.; Nouari, M. Milling diagnosis using machine learning approaches. In Proceedings of the Surveillance, Vishno and AVE Conferences 2019, Lyon, France, 8–10 July 2019.
3. Zhou, L. Performance of Cellular-Based Positioning with Machine Learning (Dissertation). 2022. Available online: https://urn.kb.se/resolve?urn=urn:nbn:se:kth:diva-320524 (accessed on 1 December 2023).
4. Liang, X.; Liu, Z.; Wang, B. State-of-the-art of surface integrity induced by tool wear effects in machining process of titanium and nickel alloys: A review. *Measurement* **2019**, *132*, 150–181. [CrossRef]
5. Dantone, M.; Gall, J.; Fanelli, G.; Van Gool, L. Real-time facial feature detection using conditional regression forests. In Proceedings of the 2012 IEEE Conference on Computer Vision and Pattern Recognition, Providence, RI, USA, 16–21 June 2012; pp. 2578–2585.
6. Alonso, F.J.; Salgado, D.R. Analysis of the structure of vibration signals for tool wear detection. *Mech. Syst. Signal Process.* **2008**, *22*, 735–748. [CrossRef]
7. Zhou, Y.; Sun, B.; Sun, W.; Lei, Z. Tool wear condition monitoring based on a two-layer angle kernel extreme learning machine using sound sensor for milling process. *J. Intell. Manuf.* **2022**, *33*, 247–258. [CrossRef]
8. He, Z.; Shi, T.; Xuan, J.; Li, T. Research on tool wear prediction based on temperature signals and deep learning. *Wear* **2021**, *478*, 203902. [CrossRef]
9. Ma, W.; Liu, X.; Yue, C.; Wang, L.; Liang, S.Y. Multi-scale one-dimensional convolution tool wear monitoring based on multi-model fusion learning skills. *J. Manuf. Syst.* **2023**, *70*, 69–98. [CrossRef]
10. Li, C.; Zhao, X.; Cao, H.; Li, L.; Chen, X. A data and knowledge-driven cutting parameter adaptive optimization method considering dynamic tool wear. *Robot. Comput.-Integr. Manuf.* **2023**, *81*, 102491. [CrossRef]
11. Gomes, M.C.; Brito, L.C.; Da Silva, M.B.; Duarte, M.A.V. Tool wear monitoring in micromilling using support vector machine with vibration and sound sensors. *Precis. Eng.* **2021**, *67*, 137–151. [CrossRef]
12. Li, J.; Lu, J.; Chen, C.; Ma, J.; Liao, X. Tool wear state prediction based on feature-based transfer learning. *Int. J. Adv. Manuf. Technol.* **2021**, *113*, 3283–3301. [CrossRef]
13. Li, G.; Wang, Y.; Wang, J.; He, J.; Huo, Y. Tool wear prediction based on multidomain feature fusion by attention-based depth-wise separable convolutional neural network in manufacturing. *Int. J. Adv. Manuf. Technol.* **2023**, *124*, 3857–3874. [CrossRef]
14. Zhang, W.; Luktarhan, N.; Ding, C.; Lu, B. Android Malware Detection Using TCN with Bytecode Image. *Symmetry* **2021**, *13*, 1107. [CrossRef]
15. Hu, Y.; Feng, W.; Li, W.; Yi, X.; Liu, K.; Ye, L.; Zhao, J.; Lu, X.; Zhang, R. Morphological classification method and data-driven estimation of the joint roughness coefficient by consideration of two-order asperity. *Rev. Adv. Mater. Sci.* **2023**, *62*, 20220336. [CrossRef]
16. Elman, J.L. Learning and development in neural networks: The importance of starting small. *Cognition* **1993**, *48*, 71–99. [CrossRef]
17. Liu, C.; Li, Y.; Li, J.; Hua, J. A meta-invariant feature space method for accurate tool wear prediction under cross conditions. *IEEE Trans. Ind. Inf.* **2021**, *18*, 922–931. [CrossRef]
18. Wang, M.; Zhou, J.; Gao, J.; Li, Z.; Li, E. Milling Tool Wear Prediction Method Based on Deep Learning under Variable Working Conditions. *IEEE Access* **2020**, *8*, 140726–140735. [CrossRef]
19. Chan, Y.; Kang, T.; Yang, C.; Chang, C.; Huang, S.; Tsai, Y. Tool wear prediction using convolutional bidirectional LSTM networks. *J. Supercomput.* **2022**, *78*, 810–832. [CrossRef]
20. Kolář, P.; Burian, D.; Fojtů, P.; Mašek, P.; Fiala, Š.; Chládek, Š.; Petráček, P.; Švéda, J.; Rytíř, M. Indirect drill condition monitoring based on machine tool control system data. *MM Sci. J.* **2022**, *10*, 5905–5912. [CrossRef]
21. Duan, J.; Zhang, X.; Shi, T. A Hybrid Attention-Based Paralleled Deep Learning model for tool wear prediction. *Expert Syst. Appl.* **2023**, *211*, 118548. [CrossRef]

22. Tan, X.; Triggs, B. Enhanced local texture feature sets for face recognition under difficult lighting conditions. *IEEE Trans. Image Process.* **2010**, *19*, 1635–1650. [PubMed]

23. Kiranyaz, S.; Avci, O.; Abdeljaber, O.; Ince, T.; Gabbouj, M.; Inman, D.J. 1D convolutional neural networks and applications: A survey. *Mech. Syst. Signal Process.* **2021**, *151*, 107398. [CrossRef]

24. Xie, X.; Huang, M.; Liu, Y.; An, Q. Intelligent Tool-Wear Prediction Based on Informer Encoder and Bi-Directional Long Short-Term Memory. *Machines* **2023**, *11*, 94. [CrossRef]

25. Hochreiter, S.; Schmidhuber, J. Long short-term memory. *Neural Comput.* **1997**, *9*, 1735–1780. [CrossRef]

26. Cho, K.; Van Merriënboer, B.; Bahdanau, D.; Bengio, Y. On the properties of neural machine translation: Encoder-decoder approaches. *arXiv* **2014**, arXiv:1409.1259.

27. Bai, S.; Kolter, J.Z.; Koltun, V. An empirical evaluation of generic convolutional and recurrent networks for sequence modeling. *arXiv* **2018**, arXiv:1803.01271.

28. HKechang, H.; Jingjiao, L. Short-term Load. Forecasting Considering Demand Response. In Proceedings of the 2021 IEEE Sustainable Power and Energy Conference (iSPEC), Nanjing, China, 23–25 December 2021; pp. 2472–2476.

29. Zhang, H.; Chen, Y.; Chu, X.; Zhang, Z.; Hao, T.; Wu, Z.; Yang, Y. Neural Computing for Advanced Applications. In Proceedings of the Third International Conference, NCAA 2022, Jinan, China, 8–10 July 2022; Proceedings, Part I. Springer Nature: Berlin/Heidelberg, Germany, 2022.

30. Li, X.; Lim, B.S.; Zhou, J.H.; Huang, S.; Phua, S.J.; Shaw, K.C.; Er, M.J. Fuzzy neural network modelling for tool wear estimation in dry milling operation. In Proceedings of the Annual Conference of the PHM Society 2009, San Diego, CA, USA, 27 September–1 October 2009.

31. Qin, Y.; Liu, X.; Yue, C.; Zhao, M.; Wei, X.; Wang, L. Tool wear identification and prediction method based on stack sparse self-coding network. *J. Manuf. Syst.* **2023**, *68*, 72–84. [CrossRef]

32. Xu, X.; Wang, J.; Zhong, B.; Ming, W.; Chen, M. Deep learning-based tool wear prediction and its application for machining process using multi-scale feature fusion and channel attention mechanism. *Measurement* **2021**, *177*, 109254. [CrossRef]

33. Zhao, R.; Yan, R.; Wang, J.; Mao, K. Learning to monitor machine health with convolutional bi-directional LSTM networks. *Sensors* **2017**, *17*, 273. [CrossRef]

34. Wang, J.; Li, Y.; Zhao, R.; Gao, R.X. Physics guided neural network for machining tool wear prediction. *J. Manuf. Syst.* **2020**, *57*, 298–310. [CrossRef]

35. Zhang, K.; Zhou, D.; Zhou, C.A.; Hu, B.; Li, G.; Liu, X.; Guo, K. Tool wear monitoring using a novel parallel BiLSTM model with multi-domain features for robotic milling Al7050-T7451 workpiece. *Int. J. Adv. Manuf. Technol.* **2023**, *129*, 1883–1899. [CrossRef]

36. Liu, X.; Liu, S.; Li, X.; Zhang, B.; Yue, C.; Liang, S.Y. Intelligent tool wear monitoring based on parallel residual and stacked bidirectional long short-term memory network. *J. Manuf. Syst.* **2021**, *60*, 608–619. [CrossRef]

37. Yang, J.; Wu, J.; Li, X.; Qin, X. Tool wear prediction based on parallel dual-channel adaptive feature fusion. *Int. J. Adv. Manuf. Technol.* **2023**, *128*, 145–165. [CrossRef]

MDPI AG
Grosspeteranlage 5
4052 Basel
Switzerland
Tel.: +41 61 683 77 34
www.mdpi.com

Lubricants Editorial Office
E-mail: lubricants@mdpi.com
www.mdpi.com/journal/lubricants